Advanced Sciences and Technologies for Security Applications

The series Advanced Sciences and Technologies for Security Applications focuses on research monographs in the areas of

- Recognition and identification (including optical imaging, biometrics, authentication, verification, and smart surveillance systems)
- Biological and chemical threat detection (including biosensors, aerosols,materials detection and forensics),

and

- Secure information systems (including encryption, and optical and photonic systems).

The series is intended to give an overview at the highest research level at the frontier of research in the physical sciences.

The editors encourage prospective authors to correspond with them in advance of submitting a manuscript. Submission of manuscripts should be made to the Editor-in-Chief or one of the Editors.

More information about this series at http://www.springer.com/series/5540

Babak Akhgar · Ben Brewster
Editors

Combatting Cybercrime and Cyberterrorism

Challenges, Trends and Priorities

Springer

Editors
Babak Akhgar
CENTRIC (Centre of Excellence in Terrorism, Resilience, Intelligence and Organised Crime Research)
Sheffield Hallam University
Sheffield
UK

Ben Brewster
CENTRIC (Centre of Excellence in Terrorism, Resilience, Intelligence and Organised Crime Research)
Sheffield Hallam University
Sheffield
UK

ISSN 1613-5113 ISSN 2363-9466 (electronic)
Advanced Sciences and Technologies for Security Applications
ISBN 978-3-319-38929-5 ISBN 978-3-319-38930-1 (eBook)
DOI 10.1007/978-3-319-38930-1

Library of Congress Control Number: 2016941287

Printed on acid-free paper

This Springer imprint is published by Springer Nature
The registered company is Springer International Publishing AG Switzerland

Preface

It is with great privilege that we welcome you to the volume *Combatting Cybercrime and Cyberterrorism: Challenges, Trends and Priorities*. In this collection we provide an authoritative and accessible guide highlighting a broad range of challenges and complexities faced by modern society in relation to cybercrime and cyberterrorism.

At this point, we would like to take the opportunity to recognize the work of the contributors for allowing us to draw upon their expertise in order to shape the content of this book, a process that has enabled us to highlight many of the pressing cyber-related needs and requirements of society within its chapters. This interdisciplinary approach has helped us to bring together a wide range of organizations from large and small-to-medium enterprise, law enforcement and academia to present the reader with an analysis of current and relevant issues pertinent to cybercrime and cyberterrorism.

The growth in significance of cyberspace across society has opened up vectors for, and extended the scope of, many existing forms of criminality. As well as acting as an enabler for the globalization of business, cyberspace has created a truly global landscape for crime as individuals from across the globe are now able to utilize this environment to attack critical national infrastructure, governments and private business by stealing, compromising the integrity of, and destroying data. It has created new market places for the sale and exchange of illegal weapons and drugs, other illicit materials and even the trafficking and exploitation of human beings and provides a platform for the creation and exchange of materials associated with the solicitation and sexual exploitation of children.

However, cyberspace is not only a tool for business and criminal enterprise; citizens increasingly depend on it as a social mechanism, publicly exposing large amounts of information about themselves and those they interact with. For these reasons, it has become vitally important that we address and overcome these new challenges as a society, restoring the confidence we have in the networks and infrastructure that form the backbone of not just European, but global society. Ensuring the future of our economic welfare, privacy and collective security is a primary concern not limited to the idea of cybercrime. These threats extend beyond extending the reach and scope of traditional criminal motivations through to the emerging threats of cyberterrorism and cyberwarfare. In this context, the very nature of terrorism is evolving because of cyberspace, providing a mechanism for the propagation of ideology and extremist rhetoric, the recruitment, coercion and training of individuals, and a platform to plan and execute attacks against governments, business and critical infrastructure. It is particularly attractive to criminals and terrorists alike due to the potential for anonymity, making the job of investigators and prosecutors to prevent and respond to these activities increasingly difficult.

In response to the growing role cyberspace has across society, both in its ability facilitate new opportunities as well as opening up new threats, this volume covers a wide spectrum of challenges, from analyzing the legal and ethical issues associated with conducting research, to details regarding specific challenge areas such as public/private cooperation, attack attribution and standardization. These subject areas are enriched with contextual information and findings from the research projects contributing to it, providing the theoretical and practical frame for future research, practice and policy aimed at enhancing societal resilience to cyber-threats and contributing towards the overriding objective of supporting initiatives at both national and EU levels. Authored and edited by a multi-disciplinary team of practitioners, researchers and experts from academia, law enforcement and private industry, this new volume provides a welcome introduction to contemporary challenges we face in respect of cybercrime and cyberterrorism, providing a welcome point of reference to aid researchers, practitioners and policy makers in the development of their respective cyber security strategies.

Babak Akhgar
Ben Brewster

Acknowledgement

The editors would like to take this opportunity to thank the multidisciplinary team of contributors who dedicated their time, knowledge and experiences in preparing the chapters contained in this edited volume. In particular, we would like to recognise the dedication of Dr. Raluca-Elena Lefticaru, Constantinos Orphanides, Alison Lyle and the wider team at CENTRIC (Centre of Excellence in Terrorism, Resilience, Intelligence and Organised Crime Research, Sheffield Hallam University) without whom this edited volume would not have been possible.

We also extend our thanks to the consortium partners of the COURAGE (cybercrime and Terrorism European Research Agenda), CAMINO (Comprehensive Approach to Cyber Roadmap Coordination and Development) and CyberROAD (Development of the Cybercrime and Cyberterrorism Research Roadmap) FP7 Projects for their support of this book:

COURAGE

- Engineering ingegneria informatica
- CENTRIC (Centre of Excellence in Terrorism, Resilience, Intelligence and Organised Crime Research), Sheffield Hallam University
- European Organisation for Security
- UNICRI (United Nations Interregional Crime and Justice Research Institute)
- Cybercrime Research Institute
- TNO, Netherlands Organisation for applied Scientific Research
- FOI, Swedish Defence Research Agency
- Office of the Police and Crime Commissioner for West Yorkshire
- Aconite Internet Solutions
- EstEnter Polska
- Conceptivity SARL
- Institut Jožef Stefan
- Selex Sistemi Integrati

- Tilburg University
- fraunhofer Gesellschaft
- International Cyber Investigation Training Academy

CAMINO

- ITTI Sp. Z. o. o.
- CBRNE Ltd
- Consiglio Nazionale delle Ricerche
- DFRC AG
- Epsion Ltd
- Everis Aeroespacial y Defensa S.L.
- Universite Montpellier I
- Wyższa Szkoła Policji w Szczytnie
- S21sec Information Security Labs S.L.
- Sec-Control Finland Ltd

CyberROAD

- University of Cagliari, PRA Lab
- Technical University of Darmstadt
- INDRA
- Poste Italiane
- SecurityMatters
- Vitrociset
- FORTH, Foundation for Research and Technology
- INOV – Insec Inovação
- Demokritos National Center for Scientific Research
- SBA Research Austria
- Proprs Ltd.
- NASK, Research and Academic Computer Network
- Polícia Judiciária Portugal
- CEFRIEL Center of Excellence for Research, Innovation, Education and industrial Labs Partnerships
- SUPSI University of Applied Sciences and Arts
- CyberDefcon
- Royal Holloway, University of London
- Greek Ministry of National Defence
- McAfee UK
- MELANI, Reporting and Analysis Unit for Information Assurance

These projects received funding from the European Union's Seventh Framework Programme for research, technological development and demonstration (FP7-SEC-2013) under grant agreement no's 607406 (CAMINO), 607642 (CyberROAD) and 607949 (COURAGE).

Contents

Part I: Approaching Cybercrime and Cyberterrorism Research

Megatrends and Grand Challenges of Cybercrime and Cyberterrorism Policy and Research 3
Bert-Jaap Koops

Towards a Systematic View on Cybersecurity Ecology 17
Wojciech Mazurczyk, Szymon Drobniak and Sean Moore

Challenges Priorities and Policies: Mapping the Research Requirements of Cybercrime and Cyberterrorism Stakeholders 39
Douglas Wells, Ben Brewster and Babak Akhgar

A (Cyber)ROAD to the Future: A Methodology for Building Cybersecurity Research Roadmaps 53
Davide Ariu, Luca Didaci, Giorgio Fumera, Giorgio Giacinto, Fabio Roli, Enrico Frumento and Federica Freschi

Part II: Legal, Ethical and Privacy Considerations

Data Protection Law Compliance for Cybercrime and Cyberterrorism Research 81
Arnold Roosendaal, Mari Kert, Alison Lyle and Ulrich Gasper

Non-discrimination and Protection of Fundamental Rights in Cybercrime and Cyberterrorism Research 97
Francesca Bosco, Elise Vermeersch, Vittoria Luda, Giuseppe Vaciago, Ulrich Gasper and Alison Lyle

Risks Related to Illegal Content in Cybercrime and Cyberterrorism Research 117
Alison Lyle, Benn Kemp, Albena Spasova and Ulrich Gasper

Part III: Technologies, Scenarios and Best Practices

Cybercrime Economic Costs: No Measure No Solution 135
Jart Armin, Bryn Thompson and Piotr Kijewski

Towards the Development of a Research Agenda for Cybercrime and Cyberterrorism – Identifying the Technical Challenges and Missing Solutions 157
Borka Jerman-Blažič and Tomaž Klobučar

The Never-Ending Game of Cyberattack Attribution: Exploring the Threats, Defenses and Research Gaps 175
Piotr Kijewski, Przemyslaw Jaroszewski, Janusz A. Urbanowicz and Jart Armin

Emerging Cyber Security: Bio-inspired Techniques and MITM Detection in IoT 193
Michał Choraś, Rafał Kozik and Iwona Maciejewska

Cyber Situational Awareness Testing 209
Joel Brynielsson, Ulrik Franke and Stefan Varga

Part IV: Policy Development and Roadmaps for Cybercrime and Cyberterrorism Research

How the Evolution of Workforces Influences Cybercrime Strategies: The Example of Healthcare 237
Enrico Frumento and Federica Freschi

European Public-Private Partnerships on Cybersecurity - An Instrument to Support the Fight Against Cybercrime and Cyberterrorism 259
Nina Olesen

Are We Doing All the Right Things to Counter Cybercrime? 279
Michal Choraś, Rafal Kozik, Andrew Churchill and Artsiom Yautsiukhin

Consolidated Taxonomy and Research Roadmap for Cybercrime and Cyberterrorism 295
Babak Akhgar, Michał Choraś, Ben Brewster, Francesca Bosco, Elise Vermeersch, Vittoria Luda, Damian Puchalski and Douglas Wells

Author Index 323

Part I: Approaching Cybercrime and Cyberterrorism Research

Megatrends and Grand Challenges of Cybercrime and Cyberterrorism Policy and Research

Bert-Jaap Koops(✉)

TILT Tilburg Institute for Law, Technology, and Society, Tilburg University, Tilburg, The Netherlands
e.j.koops@tilburguniversity.edu

Abstract. What are grand challenges of cybercrime and cyberterrorism policy and research for the coming one or two decades? To answer this question, we first need to grasp some major trends that influence the future of cybercrime and cyberterrorism, and the combatting thereof, in fundamental ways. This chapter therefore starts with sketching seven megatrends in technology and society: Internet as the infrastructure of everything, autonomic technologies, datafication, the onlife world, the transformation of crime, the fourth generation of cybercrime as attacks on the Internet of Things and People, and the gradual erosion of privacy. Against this background, seven grand challenges for keeping societies secure and inclusive against the threats of CC/CT are presented: underground marketplaces, hiding technologies, ubiquitous data, smart regulation, smart organisation, designing technology, and preserving the human rights framework in a volatile context.

1 Introduction

Cybercrime and cyberterrorism (CC/CT) pose significant challenges to society challenges that are unlikely to decrease in the coming decades. Although much is being done, in policy and practice, to address these challenges, adequate measures remain difficult to conceive and implement, as the field is dynamic, complex, and global. Research is needed to help determine which measures are more likely to be adequate, i.e., both effective and legitimate, not only in the short term but also in the longer run. Policy-makers can, in turn, assist researchers by pointing out which research topics are most urgent and valuable for policy and practice. Thus, policy and research can both benefit from a research agenda that includes those issues that are most pressing to be addressed in public policy to combat CC and CT and that would most profit from high-level research.

Developing policies and research programmes to address the challenges of CC and CT requires insight into the major issues that need to be investigated and addressed. To fulfill this need, the chapters in this book offer an overview

B. Akhgar and B. Brewster (eds.), *Combatting Cybercrime and Cyberterrorism*,
Advanced Sciences and Technologies for Security Applications,
DOI 10.1007/978-3-319-38930-1_1

of significant, topical, and concrete issues and challenges that policy-makers and researchers can or should address. An overview of key issues and topical challenges is not enough, however. In order to be able to prioritise policy measures and research topics, and, perhaps more importantly, to be able to see the larger picture and develop policy and research programmes that are capable of addressing CC/CT challenges also in the longer run. A broader, high-level overview is needed, that shows how the many topics relate to each other and fit the broader landscape of CC/CT policy and research. In that light, this chapter aims to sketch the broader picture that puts the various topics discussed in this book in a wider and longer-term perspective. It is based on experience in CC research and many discussions with researchers, practitioners, and policy-makers over the past two decades, and it is therefore essayistic in character.

What, then, are the grand challenges of CC and CT policy and research for the coming one or two decades? To grasp what the grand challenges are, we can build on known and current challenges in the combating of CC and CT, which are likely to persist in the near future. But we also need some-thing more: a vision of the main trends that are affecting the landscape at large, and which bring along new, or shifted, challenges for policy and research. Major trends that have the potential to change society in fundamental ways are called megatrends, and it is an important exercise to have a timely vision on what today's megatrends are, in order to prepare for the future [13]. Therefore, before listing what is perceived as grand challenges of keeping societies secure and inclusive against the threats of CC/CT, megatrends are mapped that influence the future of CC and CT, and the combating thereof, in fundamental ways.

2 Megatrends

Identified are seven megatrends that have the potential to change the ways in which CC and CT can occur and can be combated. These are perhaps not radically moving away from the current situation, since megatrends take place over a longer period and we are already seeing the first effects of these trends, but they strengthen certain current developments and may require novel responses to the way CC/CT is currently approached in policy, practice and research.

The megatrends can roughly be clustered in two groups. First, we have megatrends taking place at the different layers of the Internet: its infrastructure, its applications, and its content. Second, as a consequence of the trends in the first cluster, we have megatrends associated with changes in society at large, with changes in how crime and terrorism occur in society, and with changes in CC, CT, and how these are combated in particular.

2.1 Megatrend 1: Internet as the Infrastructure of Everything

The Internet is rapidly becoming, or perhaps has in some countries already become, the backbone of everything in society. Not only do communications, media, and entertainment classic functions of the Internet in its earlier days rely on the Internet, also education, labour, healthcare, transport, public service-delivery, law and order, and virtually every other sector in society have come to crucially rely on the Internet to facilitate their everyday processes. While usually various critical or vital infrastructures are distinguished, for example water, energy, or banking, nowadays none of these could really function without the Internet. The Internet is thus not only becoming the backbone of all kinds of societal processes, it is also, more importantly, becoming the backbone of backbones. Colloquially often called the mother of all networks, the Internet might now be better described as the lifeblood of all networks. This makes society extremely vulnerable to attacks on or failures of the Internet infrastructure.

2.2 Megatrend 2: Autonomic Technologies

Where ICT, biotechnologies, and nanotechnologies can be seen as the primary enabling technologies of the past decades (which of course continue to develop and play a major role in future society), the current major technological fields in development are neurotechnologies and robotics. Although not necessarily converging, both of these rely on complex, self-learning processes. In the coming decades, we will see many applications being introduced in society that function highly or completely autonomically, i.e., responding to and interacting with their environment in flexible, resilient, and self-learning ways. Self-driving cars and (increasingly automated) Unmanned Aerial Vehicles are on the verge of breaking through, and service robots will start appearing in different contexts, including in domestic aid, healthcare and elderly care. The move from dumb tools that use software but are largely mechanical in operation (and thus relatively predictable) to smart tools that use sensors and software to respond to environmental stimuli (and thus become more unpredictable) will have many implications for the ways in which people act and are acted upon. Autonomic technologies create new opportunities but also new threats, not only for malicious attacks, but also in relation to malfunctioning and natural disasters, as it is unknown how autonomic devices such as self-driving cars will respond in extreme situations.

An important aspect of this megatrend is that, while the different enabling technologies are primarily associated with different applications, they are also converging, e.g., in hybrid applications such as bio-chips at an almost nano-scale, or bionic limbs connected to the nervous system. And just as the Internet is the backbone of all backbones, ICT is pervading all technological applications, including bio-, nano-, neuro- and robo-applications, as a primary enabling technology. This means that vulnerabilities in or caused by ICT and there are many of those also become intrinsically intertwined with almost all applications that individuals and organisations use.

2.3 Megatrend 3: The Datafication of Everything

At the content level, the current trend of datafication will continue and expand. Big Data is the new oil that is (or at least is said to be) driving the economy, showing the importance and value of data and information. But also everything is increasingly translated into, and reduced to data, today's Internet service providers business models (thriving on data-driven targeted advertising) is an example of this as well as wearables and health apps, with the quantified self movement that measures everything as an extreme but telling illustration of this trend. Datafication implies that huge amounts of information is available about individuals and organisations, which can not only be mined for profit, but also abused for criminal or terrorist purposes. Moreover, in Big Data Analytics, correlation is coming to replace causation as the primary driver of knowledge-based decisions and interventions. This has interesting applications in combating CC/CT, but also creates new risks of statistical or algorithm-driven decisions that no human, including those responsible for administering justice, can really understand the rationale of.

2.4 Megatrend 4: The Onlife World

Associated with the megatrends of the Internet as the lifeblood of all networks and the datafication of everything, life in society is transforming in important ways. In the past decades, policy and practice typically relied on a distinction between online and offline situations, with different expertise being required for online or offline issues, a situation that is still on-going in many countries. Today, we see an increasing merging of online and offline situations, to the extent that the distinction no longer makes sense. With the Internet of Things, physical space is being riddled by online connections. More importantly, people move around in physical space and cyberspace at the same time: within seconds, they switch seamlessly back and forth between navigating in their physical environment communicating with people around them, and looking up information on the web and communicating online. Smartphones are becoming not only vital instruments in daily life they are becoming an integral interface between people and their environment, to the extent that many people would feel, if you take away their smartphone, cut off from life. The seamless merger of online and offline life is best described by the term onlife, and the fact that society in the coming decades will be an onlife society has important implications for how people behave and interact and also for the ways in which they are vulnerable to crime and terrorism.

2.5 Megatrend 5: The Transformation of Crime and Terrorism

That crime is being transformed by the opportunities afforded by the Internet is a longer-known megatrend, related to developments in globalisation, datafication, and automation [9,16]. It means that profit-seeking criminals are shifting from classic forms of crime, such as drugs trafficking, to CC, because the profits are equally high (or possibly higher) and the risks of getting caught are lower. Although this trend has already been occurring for a decade or so, it will not only continue but is likely to become even stronger, given the increasingly important role that the Internet has as the backbone of everything. Thus, while CC and CT are already prominent forms of crime that are high on policy agendas, they have serious potential of becoming the primary ways in which crime and terrorism will occur. Indeed, just as life is becoming onlife, it may no longer make much sense to distinguish between offline and online forms of crime or terrorism, simply because the two spaces can no longer be really separated. As crime and terrorism transform into CC and CT, so CC/CT simply comes down to crime and terrorism. This has important implications not only for the ways in which (cyber)crime and (cyber)terrorism occur, but also for the ways in which these phenomena are to be combated. Policies and measures dedicated exclusively to off-line or physical crime and terrorism will risk underestimating the role that digital technologies play; but also, and more importantly, policies and measures dedicated exclusively to CC and CT will underestimate the physical component of attacks and threats, if they do not take into account the onlife character of today's world.

2.6 Megatrend 6: The Fourth Generation of Cybercrime: Attacks on IoT and IoP

In a generational approach to CC, David Wall has distinguished three generations to date. The first generation, of low-end CC, concerned traditional crimes in which computers were used as a mere tool, e.g., in computer-related fraud or forgery. The second generation, of hybrid CC, still consisted of classic crimes, but facilitated by computer networks to the extent that the scale and scope started to make important differences. The third generation, of high-end CC, concerns crimes targeted at computers or computer networks themselves, such as hacking or denial-of-service attacks [16]. While Wall himself speculated back in 2007 what the fourth generation might be, vacillating between completely virtual crime (taking place in virtual worlds) and ambient crime (targeted at Ambient Intelligence, or what is now usually referred to as the Internet of Things) [16], it is increasingly becoming apparent that it will be the latter. Given the first three megatrends, society is becoming extremely vulnerable in its move towards connecting everything and introducing autonomic devices; these vulnerabilities are bound to be exploited by criminals and terrorists. We have already seen cars being hacked and being remotely controlled by hackers [6], and that will happen to everything in the Internet of Things.

Moreover, it is not only things that will be cyber-attacked, it is also us, humans. Although somewhat further down the future than the Internet of Things, an Internet of People is looming on the horizon, driven by an increasing use of implants on or inside the body. From current pacemakers and cochlear implants via RFID implants to neural implants, bionic prostheses and neural prostheses, people will also become physically (and mentally) vulnerable to cyber-attacks [5,6]. Although the forms of these attacks on the Internet of Things and People will roughly be the same as known cyber-attacks (namely hacking, data interference, system interference, intercepting communications), the impact will be different in character, and particularly the fear that may be induced by cyber-terrorist attacks on cars, parcel-delivering drones, pacemakers or bionic limbs can hardly be overestimated.

2.7 Megatrend 7: The Gradual Erosion of Privacy

Somewhat different in character than the previous megatrends, there is another that merits mentioning in a high-level overview of CC and CT policy and research. Partly as a result of governments taking up the challenge of CC/CT combating and benefiting from the affordances of the datafication of every-thing (but by no means only because of this), privacy is gradually being eroded. A broad trend visible over the past decades (well before 9/11), and continuing into the coming decade, is that both governments and industry are increasingly gathering massive amounts of data that reveal much of individuals personal lives [11]. The possibilities that technological developments allow for collecting and analysing data often seem to outweigh the possibilities that technology also creates for securing and hiding data, at least for the large majority of citizens [8,11]. Thus, the gradual erosion of privacy is not only caused by governments introducing (ever) more intrusive investigation and intelligence powers in order to combat the threats of crime and terrorism; it is also caused by a seemingly natural mechanism at play in technological development. The mechanism is that, as the level of privacy protection that people have is associated with what people can reasonably expect to remain private, technology makes people's expectations of what can be kept private ever smaller (or less reasonable). In a datafied, onlife world, home walls and curtains no longer help to keep private life private, and the digital equivalents of walls and curtains keep private life, if used at all, translucent rather than opaque.

The relevance of this megatrend for CC/CT policy and research is that this mechanism must be recognised and taken into account: it is all too easy to argue in individual cases and for single policy measures that privacy should give way to other interests, but the cumulative effect of such argumentation will be that privacy continues to erode until there's nothing left of it. By then, it will be too late to recognise that we need privacy as an essential component of a livable society: Privacy is like oxygen. We really appreciate it only when it is gone [15].

3 Grand Challenges

With the above-described megatrends in mind, what can be said about the grand challenges for CC/CT policy and research? Distinguishing six grand challenges, again in two clusters (which do not mirror the megatrends themselves the challenges are driven by different combinations of megatrends). These are not the only challenges for research, but they are urgent and large hence grand and they can serve to illustrate different aspects of the complexities of combating CC/CT. The first cluster concerns different aspects of the ways in which CC and CT occur, each presenting particular challenges for policy and practice: the easy availability of cyberattack tools (the infrastructural level of an underground marketplace), the many possibilities of criminals and terrorists to remain under the radar (the application level of hiding tools), and the role of information in criminal practices (the content level of ubiquitous data). The second cluster concerns different aspects of response strategies, which can be distinguished given that responses need to combine legal, organisational, and technical measures in challenges for smart regulation, smart organisation, and smart technologies to address the threats of CC/CT. An overarching challenge in all these responses is to maintain respect for human rights, which are fundamental for a livable society, in a volatile context.

3.1 Grand Challenge 1: The Underground Marketplace

Committing a CC or cyber-terrorist attack requires cyber-tools. Although to be effective, such tools need to be sophisticated, at least to overcome basic levels of security measures, they do not necessarily need to be highly sophisticated: with a global network of potential targets, attackers can easily look for the weakest link and benefit from poor security in one place or another. But more importantly, tools are available also to would-be criminals or terrorists who have no technological expertise or skills, through the existence of a large underground marketplace where hacker tools are traded, in much the same way as legal online marketplaces function (along with vendor rating systems and helpdesks). A particularly challenging manifestation of this underground economy is the availability of botnets, which can be rented to commit distributed denial-of-service attacks and spread ransomware. The existence of such black markets was a primary reason to criminalise the misuse of devices in the Cybercrime Convention and Directive 2013/40/EU[1], but the effect of this penalisation on the factual easy availability of hacker tools remains to be seen. Combating the underground CC economy may be even more challenging than combating the narcotics economy, given the global and non-material character of the networks along which the CC market is functioning.

3.2 Grand Challenge 2: Technologies to Hide

A well-known but unchangeably relevant challenge for CC/CT combating is the many ways in which perpetrators can hide their operations, traces, and identities. The dark web, TOR, encryption, and bullet-proof hosting are key terms that keep turning up in this respect. The effectiveness of strategies to hide should not be overestimated; for instance, the Internet is not as anonymous as is often alleged, and cryptographic algorithms may be strong but their implementation or use can be weak. Nevertheless, the difficulty of tracing perpetrators remains one of the key challenges for cyber-investigators. Research can, and should, continue to identify particular challenges of specific hiding technologies. However, there are overarching questions associated with criminals using technologies to hide, because the trade-offs involved in diminishing the possibility for bad guys to hide while preserving the possibility for good guys (e.g., human-rights defenders) to use the same tools for legitimate purposes.

[1] Article 6, Convention of Cybercrime, CETS 185, Budapest 2001; Article 7 Directive 2013/40/EU on attacks against information systems, Official Journal 14 August 2013, L218/8.

Moreover, as is visible from the recent debates surrounding the United States FBI's attempts to force Apple to undo the security of iPhones through court orders[2], enlisting the aid of providers to break security of their products for government investigation or intelligence purposes can backfire, if technology providers decide in response to build in stronger security, possibly in such a way that they themselves can-not undo it. At the same time, cybersecurity researchers and civil liberties defendants opposing government attempts to break security technologies should be aware that this can also backfire on civil liberties. The more hiding technologies are used that are hard to uncover for investigative agencies, the stronger the call from state agencies, resonating in media and politics, will be for new or reinforced government powers. In political climates thriving on incident-driven law-making, risk aversion, and a culture of fear, such calls might easily lead to legislation that introduces highly intrusive policing powers, such as covert remote access to computers to install malware that intercepts passwords.

Therefore, addressing the overarching questions at issue in hiding technologies re-quire nuanced and extremely complex balancing acts.

3.3 Grand Challenge 3: Ubiquitous Data

At the content level, the ever-increasing role of information in society poses questions that have not been well researched. CC research has tended to focus on various types of content-related offenses, but has not addressed yet the challenges of ubiquitous data. The trend of datafication facilitates new or more sophisticated forms of CC, enabling in particular increasing personalisation of attacks (e.g., spear-phishing, ransomware), which can be the precursor not only to financial crimes but also to hacktivism or cyber-terrorism. At the same time, datafication also provides new opportunities for responses, as the same personalised attacks can be used to remotely infect perpetrators computers. In a similar vein, Big Data Analytics will enable new forms of profiling both potential victims and potential offenders. Thus, the datafication of society requires an overarching vision on the role of information in CC/CT that goes beyond content-related offenses and the rudimentary forms of intelligence-led policing [14] that have been developed until now. Similarly to the role of information in the new economy, the role of information in crime and terrorism needs to be far better understood at both theoretical and practical levels.

3.4 Grand Challenge 4: Smart Regulation

CC/CT combating is in urgent need of smarter forms of regulation, in at least two main senses. First, traditional approaches of command-and-control regulation en-forced by the government fall short. Inspiration can be drawn from notions of smart regulation, responsive regulation or regulatory innovation that have

[2] See https://en.wikipedia.org/wiki/FBI-Apple_encryption_dispute (accessed 15 April 2016).

been developed in regulation & governance studies [1,2]. These emphasise that regulation needs to become more responsive, reflexive, and flexible. Besides hard law (primarily legal, command-and-control regulation), also soft law is needed, such as standard-setting, codes of conduct, and sectoral guidelines (which draw on market and social forms of regulation). Although this is well recognised in CC and cybersecurity policy documents, in practice, regulators in the field of CC and CT often still follow the classic reflex of focusing on more law: expanding criminalisation of behavior, expanding government investigation powers citeref9. This may improve CC/CT combating in theory, and possibly in a few high-profile cases in practice, but it leaves the wide area of more mundane and existing CC/CT threats vulnerable to huge problems of enforcement, which not only have to do with jurisdiction issues (see challenge 5) but also with expertise and resources. This is where a broader, reflexive and self-learning, approach to regulation can come in but the insights from regulatory studies have so far hardly been applied to the CC/CT field.

Second, a particular challenge within (hard) law is regulatory connection [4]. It is a well-known challenge to keep the law up-to-date, particularly in a field with high technological turbulence. Significant efforts are being made to reconnect the law to the current state-of-the-art, for example with the smart set of minimum legal provisions that the harmonisation efforts of the Cybercrime Convention offers; however, in many countries, these still leave gaps, particularly in the regulation of digital investigation powers. Moreover, keeping the law connected to socio-technological developments requires a constant effort in reassessing and revising the law every few years. Especially challenging for regulators is to regulate with care and foresight, avoiding incident-driven law-making that tends to miss the larger picture and thus risks introducing new gaps or undesirable side-effects. Regulators should build on technology assessment and formulate laws with the right level of technological neutrality, that is, sufficiently abstract so as to cover also the technologies and applications of the short- or middle-term future, but not so abstract that it becomes unclear which technologies are covered by the law [7]. In addition, regulatory connection also involves a more fundamental level of reflection, since legal frameworks often are based on implicit assumptions dating from the time they were created, for example that most private things are stored in people's homes; as times change, such assumptions may lose validity (e.g., people nowadays carry most private things along, on smartphones or in the cloud), which requires a more thorough rethinking of the framework than simply adding or changing a few legal provisions.

3.5 Grand Challenge 5: Smart Organisation

CC/CT combating is also in urgent need of smart organisation. This is a threefold challenge. First, and perhaps relatively most feasible in the short term, is the internal organisation within government to combat CC/CT. Most countries have specialist units dedicated to CC, cybersecurity, and CT; these, however, face challenges of resources (they are often understaffed in light of the extent of CC/CT threats), of remit (limitations to what they are allowed to investigate or do), and of coordination of responses with other branches (both classical police or intelligence units, and other branches of government, cf. [3]). Moreover, also non-specialised units (who need to do much of the basic work in addressing high-volume or low-profile CC) need to be trained and periodically stimulated to update their practices in light of technological changes. The internal organisation of CC/CT responses can be optimised, but a challenge remains that there are few known best practices, and no metrics to determine how many resources have to be allocated in a risk-averse society, CC/CT counter-efforts will always need more resources.

Second, the collaboration between government and the private sector needs to be well-organised. Although Public-Private Partnerships and multi-stakeholder approaches feature prominently in CC/CT policy documents, and there are some good practices in, e.g., botnet mitigation, much remains to be done before coordinated multi-stakeholder approaches become really effective in addressing CC/CT threats while remaining within the boundaries of legitimacy at the same time. A general challenge, besides the need to overcome institutional and cultural barriers between public and private sectors, is that public bodies often have legitimacy to act but lack expertise in complex technological cases, while private bodies often have required expertise but lack authority to intervene; combining these is a complex puzzle.

Third, collaboration across borders needs to be organised. Mutual assistance procedures, despite on-going efforts to streamline them and despite many good contacts among states and practitioners, is still often slow and therewith ineffective. Unilateral actions to investigate across borders, although occurring in practice because of the inadequacy of Multi-Lateral Assistance Treaties, are highly contested in light of international law, and it will require a long-term effort at the highest political levels to come to agreements on conditions under which governments can unilaterally use cross-border investigations [10]. This makes the need to come up with more effective and efficient arrangements of cooperation between countries in CC/CT combating all the more pressing.

3.6 Grand Challenge 6: Designing Technology

CC/CT combating also requires doing something about technology itself. This challenge has at least two major aspects. The first is well-known but nevertheless still highly challenging: making technology less vulnerable to attacks. The notions of security by design and privacy by design which often, although not always, go hand in hand to address vulnerabilities in technology have been proposed and developed over the past decade, but yet remain to be made operational in significant ways. Computer technologies are notoriously difficult to make secure, with billions of lines of code, legacy problems, and high market pressure for short development cycles and high-frequent innovations; and they are almost intrinsically difficult to make privacy-friendly, as privacy is highly contextual and involves open norms, which are difficult to embed in technology design [12]. Nevertheless, progress is being made, and the field of privacy and security by design merits much research and innovation also in the coming decade.

The second aspect of doing something about technology is less recognised and all the more challenging, as it involves rowing against the tide: making society less dependent on technology. Given the megatrends of relying on the Internet as a backbone of all infrastructures and of datafication, society is rapidly making itself extremely vulnerable to cyber-attacks. The vulnerability does not lie in the threat of cyber-attacks per se, but rather in both the scale and the cascade effects that such attacks can have on many sectors, people, and activities. Although it is an unwelcome message in a risk-averse society, people must realise that it is impossible to live in a risk-free world and that we therefore have to learn to cope with adversities. This requires building in resilience in the face of the major cyber-attacks that are bound to take place sooner or later. Resilience implies not only early warning and quick response systems and procedures, but also mitigating the effects of attacks on critical infrastructures. An important part of the latter is to have adequate fall-back options, in particular to have functioning and (periodically) tested fall-back infrastructures in case Internet-based infrastructures are temporarily out of order. Besides measures focusing on preventing and mitigating cyber-attacks, the question must also be faced to what extent we want society to become totally dependent on the Internet as a backbone of all societal activities. If we want to preserve some possibilities to continue life if the Internet collapses, how are we going to achieve that?

3.7 Grand Challenge 7: Preserving Human Rights in a Volatile Context

One overarching challenge in all the responses to CC/CT threats is to preserve the human rights framework while taking measures that are aimed to be effective. This challenge relates to the security versus privacy frame that is often employed in debates, but it is broader and more profound than that. In fact, the security versus privacy frame is too simple, if not simplistic: it mistakenly assumes that security and privacy clash (while in fact, they often go together to a considerable extent) and that they are comparable units of measurement (while in fact, they are incommensurable). Moreover, privacy is not the only human right at issue: freedom of expression, non-discrimination, and the right to an effective remedy are equally relevant in anti-CC/CT policy and practice. A better way to look at questions of the compatibility of anti-CC/CT measures with the human rights framework is the three-prong test for privacy- and free speech-invasive measures that is embedded in the European Convention on Human Rights: (1) does the measure have a legal basis, (2) does it serve a legitimate purpose (such as crime-fighting or national security), and (3) is it a necessary measure, i.e., one that meets the requirements of proportionality and subsidiarity in light of the measure's foreseen benefits and effects on human rights? Thus, rather than weighing or balancing the goal of security against the impact on human rights as such, the human rights impact is to be assessed in light of the expected results to improve security against the requirements of proportionality (is the measure proportionate in relation to the goal?) and subsidiarity (is the measure the least intrusive one that can achieve the goal?). Yet, even in this more nuanced frame, this is easier said than done.

The challenge of preserving human rights is a continuing, and therewith grand, challenge, because the field is forever moving and the context of policy measures can shift very quickly. The political situation changes with every election, but also with every incident that captures media headlines and opinion polls, inviting incident-driven measures to show that something is being done, rather than evidence-based policy-making. Also the technological context is volatile, with relatively rapid innovation cycles. One of the consequences of this is that, when legal frameworks have gaps or lag behind, and thus do not provide clear answers on what is or is not allowed for investigative agencies, practitioners will experiment and push the boundaries of their investigation powers (which may in turn lead to incidents, media attention, and hasty legislation to fill presumed gaps, in a self-perpetuating cycle).

Privacy, data protection, freedom of expression, and non-discrimination are fundamental rights - fundamental in the sense that, although they can be infringed, they cannot be taken away. They have fundamental value not only for individuals (to preserve a space in which they can develop who they want to be), but also for society as the democratic, pluralistic societies we live in require cohesion (an inclusive society) and contestation (a pluralistic public debate), both of which require an adequate level of human rights protection. The challenge is how to effect that in a volatile context, which is compounded by the significant challenges that CC/CT pose to society, such as underground marketplaces and technologies to hide. The context is likely to necessitate intrusive government powers; such powers can meet the third prong of the European Convention on Human Right's test, as being necessary in a democratic society, since often no less intrusive measures may be available but only if the measures impact on human rights is mitigated sufficiently by additional measures that guarantee fairness and accountability in the execution of the measures. Finding proper mechanisms of oversight, transparency, and contestability is thus one of the important elements of the grand challenge of preserving human rights in a volatile context.

4 Conclusion

This Paper has sketched what are perceived as grand challenges of CC and CT research and policy, against the background of megatrends that have the potential to change society including crime, terrorism, and CC/CT policy in fundamental ways. The sketch shows the contours of the landscape in which the most pressing issues needing to be researched and addressed in policy can be positioned. To be sure, there are more challenges, including grand ones, included here: a landscape map, particularly of a broad field such as we are canvassing here, can always be made more fine-grained by zooming in on certain areas. Such zooming in is important, and should be undertaken by policy-makers and researchers. But let us not forget, while zooming in on more concrete challenges, to keep sight of the broader picture. Combating the challenges of CC and CT requires many different actions, involving a variety of pathways in policy and research. The real challenge, and the most daunting of all, is to combine these many actions into a coherent whole, based on a vision that takes on-going megatrends into account and that strives not, or not only, for individual policy measures and short-term research projects, but for tackling the challenges of CC and CT also in the long run.

Acknowledgement. The research leading to these results has received funding from the European Union Seventh Framework Programme (FP7-SEC-2013) as the COURAGE project under grant agreement no 607949.

References

1. Ayres, I., Braithwaite, J.: Responsive Regulation: Transcending the Deregulation Debate. Oxford Socio-Legal Studies. Oxford University Press, New York (1992)
2. Black, J., Lodge, M., Thatcher, M.: Regulatory Innovation: A Comparative Analysis. Edward Elgar, Cheltenham/Northampton (2005)
3. Brenner, S.W.: Cyber Threats: The Emerging Fault Lines of the Nation State. Oxford University Press, Oxford/New York (2009)
4. Brownsword, R.: Rights, Regulation, and the Technological Revolution. Oxford University Press, Oxford/New York (2008)
5. Gasson, M.N., Koops, B.J.: Attacking human implants: a new generation of cybercrime. Law Innov. Technol. **5**, 248–277 (2013)
6. Greenberg, A.: Hackers Remotely Kill a Jeep on the Highway With Me in It. Wired, 21 July 2015. http://www.wired.com/2015/07/hackers-remotely-kill-jeep-highway/
7. Koops, B.J.: Should ICT regulation be technology-neutral? In: Koops, B.J., et al. (eds.) Starting Points for ICT Regulation, pp. 77–108. T.M.C. Asser Press, The Hague (2006)
8. Koops, B.J.: Technology and the crime society: rethinking legal protection. Law Innov. Technol. **1**, 93–124 (2009)
9. Koops, B.J.: The internet and its opportunities for cybercrime. In: Herzog-Evans, M. (ed.) Transnational Criminol. Manual, vol. 1, pp. 735–754. Wolf Legal Publishers, Nijmegen (2010)
10. Koops, B.J., Goodwin, M.E.A.: Cyberspace, the Cloud, and Cross-Border Criminal Investigation: The Limits and Possibilities of International Law. WODC/TILT, The Hague/Tilburg (2014)
11. Koops, B.J., Leenes, R.E.: Code and the slow erosion of privacy. Mich. Telecommun. Technol. Law Rev. **12**, 115–188 (2005)
12. Koops, B.J., Leenes, R.E.: Privacy regulation cannot be hardcoded: a critical comment on the privacy by design provision in data-protection law. Int. Rev. Law, Comput. Technol. **28**, 159–171 (2014)
13. Naisbitt, J.: Megatrends: Ten New Directions Transforming Our Lives. Warner Books, New York (1982)
14. Ratcliffe, J.H.: Intelligence-Led Policing. Willan Publishing, Cullompton (2008)
15. Sykes, C.J.: The End of Privacy. St. Martins Press, New York (1999)
16. Wall, D.S.: Cybercrime. The Transformation of Crime in the Information Age. Polity, Cambridge (2007)

Towards a Systematic View on Cybersecurity Ecology

Wojciech Mazurczyk[1(✉)], Szymon Drobniak[2], and Sean Moore[3]

[1] Institute of Telecommunications, Warsaw University of Technology, Warsaw, Poland
wmazurczyk@tele.pw.edu.pl
[2] Institute of Environmental Sciences, Jagiellonian University, Kraków, Poland
szymek.drobniak@uj.edu.pl
[3] Centripetal Networks, Herndon, USA
smoorephd@gmail.com

Abstract. Current network security systems are progressively showing their limitations. One credible estimate suggests that only about 45 % of new threats are detected. Therefore it is vital to find a new direction that cybersecurity development should follow. We argue that the next generation of cybersecurity systems should seek inspiration in nature. This approach has been used before in the first generation of cybersecurity systems; however, since then cyber threats and environment have evolved significantly, and accordingly the first-generation systems have lost their effectiveness. A next generation of bio-inspired cybersecurity research is emerging, but progress is hindered by the lack of a framework for mapping biological security systems to their cyber analogies. In this paper, using terminology and concepts from biology, we describe a cybersecurity ecology and a framework that may be used to systematically research and develop bio-inspired cybersecurity.

Keywords: Bio-inspired security · Cybersecurity ecology · Bio-mimetic systems · Cyber-ecosystem · Nature-inspired cybersecurity

1 Introduction

It is estimated that current commercially available anti-virus products are able to detect only 45 % of the new threats that Internet users face each day [1]. Moreover, the number and functionality of malicious software utilised by cybercriminals, as well as its sophistication and complexity, is constantly increasing. As a result, the average length of time between initial injection of a threat into the network and its discovery is growing every year, and is now measured in months (according to Verizon's "2014 Data Breach Investigations Report"), if not years. Additionally, current defence systems are largely static and not sufficiently adaptable to cope with the attackers' changing tools and tactics.

B. Akhgar and B. Brewster (eds.), *Combatting Cybercrime and Cyberterrorism*,
Advanced Sciences and Technologies for Security Applications,
DOI 10.1007/978-3-319-38930-1_2

The inability to provide trusted secure services in contemporary communication networks could potentially have a tremendous socio-economic impact on both E2E and E2C global markets. Because currently available cyber defences are progressively showing their limitations, it is imperative to find a new direction for cybersecurity research and development to follow.

We propose that the network security community should look into nature for new approaches to cybersecurity, both offensive and defensive. Current and future cybersecurity solutions should be designed, developed, and deployed in a way that will fully leverage the experience, learning, and knowledge from ongoing biological evolution. Conversely, the community should also look to nature to anticipate how the threat may evolve, and respond accordingly.

The most notable pros and cons of the bio-inspired cybersecurity approach are detailed below.

First, nature has over 3.8 billion years of experience in developing solutions and adaptations to the challenges that organisms face living in extremely diverse environmental conditions. The estimated number of (largely undiscovered) species is tens of millions, and each of them possesses specific and unique traits facilitating survival and propagation of their own genes. The key process of living organisms that has led to the persistence of the most successful forms and behaviours is evolution. Evolution has developed optimal solutions for situations analogous to the threats faced by computer network systems.

Second, people have long sought inspiration from nature. Some relevant modern examples include biomimicry, which is the inspiration of such inventions as Velcro tape and "cat's eyes" (retroreflective road markings). Computer science has also taken a page out of nature's book by developing biologically inspired techniques like genetic algorithms, neural and sensor networks, etc. Although at first glance there may not appear to be a direct relationship between cybersecurity and the patterns present in nature, closer inspection reveals that the essence of most known Internet attacks and defence mechanisms have analogies in nature. For example the Kudzu vine is able to penetrate its ecosystem with an astounding speed of ca. 30 cm/day. Within a short time it can choke all other vegetation, including trees and shrubs, by blocking access to the resources necessary for survival – light and nutrients. The essence is just like in DDoS (Distributed Denial of Service) attacks for communication networks where legitimate users are deprived of the resources that they are entitled to like access to the service, bandwidth, CPU time, etc. Similar analogies can be drawn for other offensive techniques as well as for security solutions, as observed and described in [2].

Another powerful analogy is the "arms race" (a form of a co-evolution involving an aggressor developing its offensive mechanisms and a victim/host evolving countermeasures in the form of defensive barriers). "Arms race" is often observed between e.g. predators and prey in nature. Similar dynamics can be also found in interactions involving hosts and parasites, with the former constantly trying to invade host bodies and the latter constantly evolving countermeasures preventing the invasion. Both the above mentioned cases bear many resemblances with a "malware-security systems" scenario (or more generally "attackers-defenders") where there is a continual contention to develop offensive/defensive measures as fast as possible to, at least temporarily, dominate the other side. Thus, it is readily apparent that in both nature and cyber world, entities must evolve permanently and adapt to ever-changing environments. In biology this phenomenon – an organism's need to continually adapt and evolve to avoid extinction – is called the Red Queen hypothesis [17]. It was named after a character from Lewis Carroll's book "Through the Looking-Glass". In this book the Red Queen described her country as a place where "it takes all the running you can do, to keep in the same place". Exactly the same process can be observed in cybersecurity and in biological systems where there is a constant need for adaptation of offensive/defensive techniques to maintain a certain level of adaptation permitting survival and reproduction/propagation.

Bio-inspired cybersecurity is not a new idea. The first generations of cybersecurity research were bio-inspired, e.g., the immune system inspired defence methods based on signature analysis, as well as methods for handling polymorphic threats (which are analogous to, e.g., different influenza strains). Since then, however, the threats have evolved to make these first-generation defences less effective. In order to survive, cybersecurity must evolve and adapt accordingly to counter the new threats. A next generation of bio-inspired cybersecurity research is now emerging; however, we find the knowledge and achievements to be scattered because the field lacks a framework. This paper aims at filling this gap by defining, based on the terminology and concepts known from biology, the *cybersecurity ecology* (and related terms). This cybersecurity ecology will enable a rigorous analysis of the existing relationships between entities in the *cybersecurity ecosystem*. Such a systematic view of cybersecurity will allow the research community to analyse and compare biological organisms' interactions with those from the virtual world in order to identify differences, deficits and potentially new promising approaches to cybersecurity.

We need to be cautious, however, that the mappings from nature to the cyber world are not always "1-to-1", i.e., the analogies are not always perfect. Some of the reasons that exact mappings are not always possible include:

- Many mechanisms and relationships in nature are very complex and not yet understood sufficiently to correctly map them to the virtual world;
- In nature, individual organisms within a species are disposable, and death is a critical driver of evolutionary adaptation; but for many security-critical systems (e.g. military, utilities, and other critical infrastructure) any loss, compromise, or corruption is unacceptable;
- The main goal for any organism is to survive and reproduce, whereas our computers/networks have many different goals (specific tasks and functions).

Despite these imperfect mappings we strongly believe that there are still many important lessons from nature that can benefit and improve cybersecurity. Moreover, if we follow a Sapir-Whorf hypothesis [29], which states that language has a direct impact on thoughts, then finding analogies between cybersecurity and nature with its accompanying terminology, concepts and solutions can have a tremendous impact on the way we think about solving cybersecurity problems. New mechanisms and ideas may emerge. Therefore, the systematic view for bio-inspired cybersecurity that we are proposing should help to unveil new promising directions that could be pursued to discover and develop effective next-generation security solutions.

The rest of this paper is structured as follows. Section 2 summarises the state-of-the-art in bio-inspired cybersecurity. In Sect. 3 the analogy between the biology-based ecosystem and the cyber-ecosystem, including potential interactions, is drawn. Section 4 reviews most important concepts, interactions and models from the natural enemy ecology. Section 5 describes some promising research directions for cybersecurity. Finally, the last section concludes our work.

2 Related Work

The existing literature includes many attempts to map biological concepts to cybersecurity. And, many of these attempts have successfully transitioned to cybersecurity technologies and systems in common use nowadays, including anti-virus, intrusion detection, threat behaviour analysis, honeypots, counter-attack, etc. [2]. As already mentioned in the previous section, current research on bio-inspired cybersecurity is fragmented and lacks a systematic approach. A primary cause is the diversity of aspects from nature that can be used as inspiration for cybersecurity research. Current research may be broadly segmented into two groups, depending on how an inspiration is drawn:

- when the inspiration is drawn from a given organism's characteristic feature/defence mechanism (internal or external). Internal mechanisms include, for example, an immune system whereas external mechanisms include, for example, various camouflage and mimicry techniques;
- when the inspiration is drawn from various inter-organism interactions – this includes, for example, predator-prey associations.

2.1 Bio-inspired Cybersecurity Inspired by an Organism's Characteristic Feature/Defence Mechanism

In order to effectively avoid detection/observation an organism can hide or conceal its presence by using camouflage or mimicry techniques that modify the organism's external appearance [15].

Camouflage embraces all solutions that utilise individual's physical shape, texture, colouration, illumination, etc. to make animals difficult to spot. This causes the information about their exact location to remain ambiguous. Examples of animals that can easily blend into the background include the chameleon (family *Chameleonidae*) which can shift its skin colour to make it similar to ambient lighting and background colouration; stick and leaf insects (order *Phasmotodea*) that take the physical form of a wooden stick or a leaf; orchid mantis (*Hymenopus coronatus*) that resembles a tropical orchid which, although quite conspicuous, is difficult to detect against a background of developed flowers. Camouflage often occurs on levels other than visual recognition: e.g., many viruses code pathways and molecular signalling systems that mimic host cell transduction mechanisms – by doing so the virus can easily invade the cell and take control of the metabolism and immunological system of an individual [18]. In cyber space various information hiding techniques, e.g. steganography, can be utilised to provide means to hide the location of confidential data within an innocent-looking carrier or to otherwise enable covert communication across communication networks [16].

Patterns and/or colourations can be also used to confuse the predator, i.e., to make information about the prey hard to interpret. Such so-called "disruptive" camouflage is possible and can be seen in, e.g., a herd of zebras (*Equus quagga*) where it is difficult for an attacking lion to identify a single animal in a herd when they flee in panic. Patterns of contrasting stripes purportedly degrade an observer's ability to judge the speed and direction of moving prey, and they do so by exploiting specific mechanisms associated with the way brain processes visual information on movement [19]. An analogous idea is utilised by various moving target techniques/defence in cyberspace, which distribute the uncertainty between the attacker and the defender more fairly. For example, some first-generation solutions made periodic changes in a host's appearance from the network perspective, in order to mitigate the effectiveness of target reconnaissance [6]. Second-generation solutions include, e.g., an ant-based cyber defence which is a mobile resilient security system that removes attackers' ability to rely on prior experience, without requiring motion in the protected infrastructure [10].

Mimicry characterises the cases in which an organism's attributes are obfuscated by adopting the characteristics of another living organism. In particular, this means that the prey can avoid attack by making the predator believe it is something else, e.g., a harmless species can mimic a dangerous one. The prey hides information about its own identity by impersonating something that it is not. For example, harmless milk snakes (*Lampropeltis sp.*) mimic venomous coral snakes (*Micrurus sp.*) to confuse predators which are less likely to launch an attack in expectation of a venomous harmful bite. Cybersecurity solutions that utilise the same idea include various traffic type obfuscation techniques, e.g., traffic morphing [14].

Organisms' internal systems may also inspire new cybersecurity approaches. There are many recent studies attempting to map features and functions of the human immune system to cyber space [3,7–9]. Immune systems use a diverse range of receptors to detect external antigens (alien proteins). These variations are not inherited but instead are generated via recombination in the process of V(D)J (somatic) recombination, which generates repertoires of receptors undergoing clonal selection and reinforcement – preparing them for effective action against antigens, with the lowest possible level of autoagression (e.g. reaction against an organism's own proteins) [20]. The resultant Artificial Immune Systems (AIS) are designed to mimic certain properties of the natural immune system. In cybersecurity their main application is anomaly and misbehaviour detection. AIS typically rely on one of four major paradigms: *(i)* negative selection algorithm [3]; *(ii)* clonal selection algorithm [7]; *(iii)* dendritic cell algorithm [8] or (iv) idiotypic networks models algorithms [9]. The first generation AIS (*i* and *ii*) utilised only simple models of human immune systems, so the resulting performance was not comparable with its human counterpart. Recent AIS (*iii* and *iv*) are more rigorous and better correspond to natural immune systems.

2.2 Bio-inspired Cybersecurity Inspired by Organisms' Interactions

In nature, there are many interactions between organisms that potentially may serve as inspirations for cybersecurity.

For example, several studies focus on various aspects of predator-prey associations. In [11] the authors make the predator-prey analogy for the Internet and investigate how different levels of species diversification can serve as a defensive measure. They considered each type of a vulnerable device as a heterogeneous species and investigated what level of species diversification is necessary to prevent a malicious attack from causing a failure to the entire network. Subsequently, in [4] it was discovered that the cost to the predator in seeking its prey drastically impacts the predation process. In particular, it has been observed that even fairly simple strategies for raising the cost of predation can result in the significant reduction of the outbreak size. Other studies utilise biological models of epidemic spreading (a special case of antagonistic interaction between the pathogen and the victim) to predict or analyze malware outbreaks [12,13].

Finally, the relationships and interactions between existing malware (so called malware ecology) have been investigated in [5]. Numbers of interactions, both accidental and intentional, between different types of malware were analysed and the main conclusion was to seek ecologically-inspired defence techniques, because many ideas from ecology can be directly applied to all aspects of malware defence.

From the studies presented above we can conclude that bio-inspired cybersecurity is a wide, diverse, emerging, and evolving research field. However, from the research perspective, we see many "loose ends" that need to be tied by using a more systematic approach, which we next propose.

3 Cybersecurity Ecology

In this section, first we systematically review the key terms from biology related to ecology. Then by borrowing and adjusting the original biology-based definitions, we will describe the most important components of the cyber-ecosystem and then of cybersecurity ecology.

3.1 Cyber-ecosystem

In biology the term ecology is defined as the field of life sciences analysing and studying interactions among organisms and/or their environment. This means that it deals with the structure and functioning of ecosystems. An ecosystem is defined as a community of living organisms (biotic components) together with the non-living (abiotic) components of their environment that interact as a system. Apart from the biotic and abiotic components, interconnected by various interactions, the ecosystem is fuelled by energy, usually in the form of electromagnetic radiation (if production in an ecosystem is sun-driven, i.e., accomplished by green plants) and chemical energy (if an ecosystem relies on chemosynthetic bacteria). Both biotic and abiotic factors can influence an organism. For example, climate change or an atypically large number of predators can negatively impact some species [21].

In every ecosystem the energy flow is crucial as each ecosystem is energy-based and is capable of transforming, accumulating, and circulating energy. In nature the flow of energy is encapsulated in a food chain, and a concept of trophic levels is utilised to illustrate the position that an organism occupies in a food chain (Fig. 1, left). Depending on how energy is obtained, two groups of organisms can be distinguished: producers (that are able to manufacture their own food using inorganic components and chemical/radiation energy) and consumers (that feed on producers and/or other consumers) [22].

Ecology can be viewed as one of the approaches to study complex and dynamic systems. Thus, if we are able to understand how ecosystems and related concepts map to the cybersecurity field then the usefulness of various ecological methodologies can be evaluated. If such mappings are successful then application of many mathematical ecological systems models to cyber systems can be investigated.

Based on the above terms and definitions from ecology, we want to systematically recreate an analogous taxonomy for the cyber world.

Let us define *cyber-ecosystem* as a community of *cyber-organisms*, i.e., non-human actors, e.g., applications, processes, programs, defensive and offensive systems (analogues to the biotic components) that interact between themselves and with the environment (abiotic components). Let us also assume that the environment in which biotic components reside and interact is a communication network, e.g. the Internet, and it constitutes a non-living (abiotic) component with its hardware, links and interconnections.

In the cyber-ecosystem (the same as in nature) both biotic and abiotic factors can impact a cyber-organism. For example, malicious software can be utilised to compromise a user's device defences and steal their confidential data. On the other hand a failure of the link/networking device or network congestion influence a cyber-organism's ability to communicate and exchange information. In such a defined cyber-ecosystem we are particularly interested in the network of interactions among cyber-organisms, and between cyber-organisms and their environment. As mentioned above, in nature the key resource is energy. In communication networks, the analogous key resource is different kinds of information, including user personal or user-generated data, but also information about their behaviour. In such a cyber-ecosystem, information can be transformed, accumulated, and/or circulated (similar to energy in ecosystems).

To have more clear analogies between ecosystems and cyber-ecosystems the role of the humans in the present context is constrained to these roles:

- *Producers* which possess and generate information that forms a desirable resource for the consumers (e.g. the tools that attackers or digital marketing companies use to obtain desired information).
- *Components* of the offensive/defensive solutions. For example, a bot herder typically issues command to the bot that he controls so he is an inevitable "part" of the botnet. Another example is an ID/PS (Intrusion Detection/ Prevention System) which is configured and monitored by a security specialist.
- A part of *"evolutionary force"*. Humans influence cyber-organisms by changing their code, functionalities and applications. In this way an evolution is achieved. Typically, attackers try to outwit the defenders by developing malicious software that will be capable of overcoming existing defence mechanisms/systems. Conversely, defenders develop their defences to be "immune" to the existing threats. Thus, both sides are taking part in a cyber "arms race".

Considering the above, it is possible also for the cyber world to characterise certain "cyber food chains" and/or cyber-trophic levels (Fig. 1, right). Consumers can become cyber-predator (attacker) or cyber-prey (defender) depending on the location in the cyber food chain. Producers always take the role of cyber-prey.

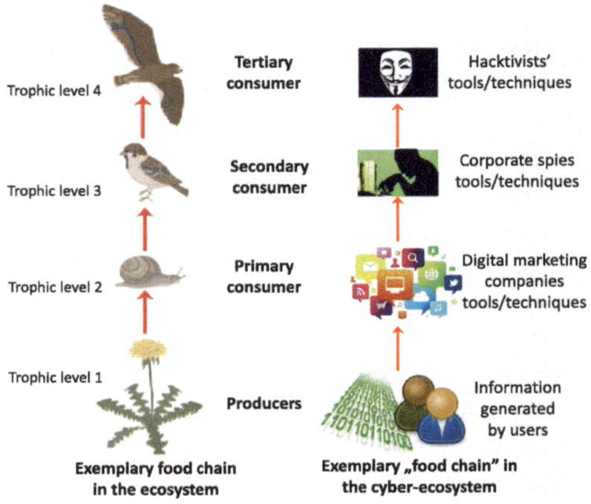

Fig. 1. Food chains and trophic levels in an exemplary ecosystem (left) and a cyber-ecosystem (right).

3.2 Cyber-ecology and Its Subtypes

By means of a simple analogy we can define the following terms that rigorously describe the toolbox of cybersecurity ecology:

- *Cyber Ecology (CE)* as a field that analyses and studies interaction among cyber-organisms and/or their environment.
- *Cybersecurity Ecology (CSE)* analyses and studies interactions among cyber-organisms and between cyber-organisms and their environment that influence their security. *CSE* is a sub-field of *CE*.
- *Attacker-Defender Ecology (ADE)* describes interactions between cyber-organisms which take roles of attackers and defenders in the specific cyber-ecosystem (e.g. in the Internet). As noted before such a relationship can be regarded not only as predation but also as parasitism. It is also worth noting that such interactions reside in different locations of the cyber food chain and depend on the trophic level (Fig. 1). *ADE* is a part of *CSE*.
- *Attackers Ecology (AE)* illustrates interactions between attackers (cyber-organisms) in a given cyber-ecosystem. The possible interactions encompass both antagonistic and non-antagonistic ones and depend on the context. Attackers can predate or parasite on each other, but the relationship can be of a symbiotic or a cooperative nature. *AE* is a part of *CSE*.
- *Defenders Ecology (DE)* provides insights into potential interactions between the defenders (cyber-organisms), and it incorporates mostly non-antagonistic ones. It includes both external defence mechanisms (interactions of malware and defence systems resulting in defence) and internal properties (analogous to animal immune systems). *DE* is a part of *CSE*.

The abovementioned terms like *AE* can be further divided into e.g. malware ecology, botnet ecology, etc. The relationships between the terms defined in this and in previous sections are illustrated in Fig. 2.

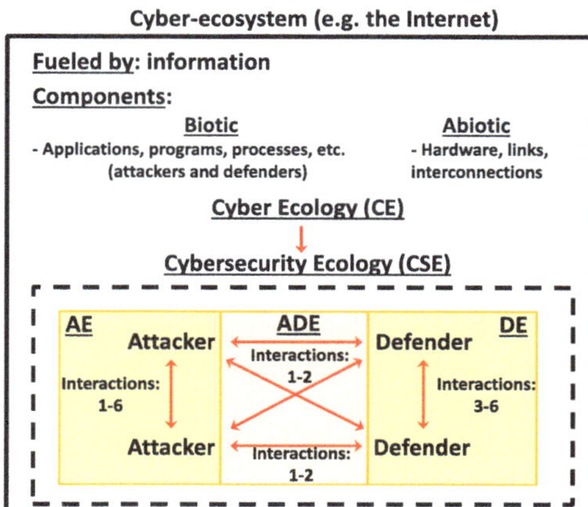

Fig. 2. Main components and interactions in a cyber-ecosystem (Interactions: 1-predation, 2-parasitism, 3-symbiosis, 4-cooperation, 5-sexual interactions, 6-competition).

3.3 Cyber-ecosystem Interactions

The structure and stability of an ecosystem in nature is determined by the set of interactions that interconnects different entities. Interactions can be roughly classified into antagonistic interactions (between species; mainly predation and parasitism), non-antagonistic interactions (between and within species; cooperation, symbiosis) and sexual selection-driven interactions (within species). In all three classes, interacting entities co-evolve, responding reciprocally to their current states in a positive/negative feedback loop mechanism (also known as the arms-race dynamics for antagonistic interactions) [21]. The interactions can be defined as follows:

- *predation:* a way of obtaining resources by killing/eating bodies of other organisms; results in the death of the prey; predation involves complex cycles of prey and predator abundances described by mathematical models such as the Lotka-Volterra equations system [21,23], which can be utilised to design the most optimal strategies of defence or attack, depending on which side of the predation-prey system the focal cyber-organism currently is. In communication networks ransomware can be treated as a predator as it is "killing" the host by encrypting vital information it stores and unless the ransom is paid this resource is "destroyed/lost" i.e. user's data cannot be retrieved;
- *parasitism:* interaction involving obtaining resources by eating other entities but not killing them [21,23]; it gave rise to a fruitful field of epidemiological parasitology, with mathematical models and defence systems that could be directly implemented in the context of cyber-epidemics. As already mentioned the current trend, especially for sophisticated malware such as Advanced Persistent Threats (APTs), is more similar to a parasite-host scenario than a predation-prey one. It means that it is more likely that the malicious software will be active on an infected host for a long time and obtaining its resources in a transparent manner;
- *symbiosis:* positive interaction involving obligatory interaction of two or more entities, necessary for all parties for survival and successful propagation. In cybersecurity this could include analysis of both attackers and defenders symbiosis. For example, for malware infection scenario it is common that the first infection is initially performed by exploiting some vulnerability on the host machine and this allows later for the second part of malware to be downloaded and executed in order to perform malicious actions for the cybercriminal;
- *cooperation:* facultative interaction of an individual within one species or members of different species, increasing the fitness and survival of other individuals (the acceptors of cooperation) often at the cost of the focal individual (the giver of cooperative behaviour) [21,24,25]; in communication networks cooperation should be recognised not only as a way of reinforcing defence mechanisms but also as a potential threat (a deceiver malware might exploit cooperating inclination of the system, wreaking havoc in its structures). A recent real-world example is the sharing of cyber threat indicators as prescribed in the US Cybersecurity Information Sharing Act of 2015;

- *sexual interactions:* occur exclusively within species and are channelled toward combining, in the most desired and effective way, the genes of females and males so that they maximise the fitness of offspring [26]; from the point of view of cyber-ecosystems the models of sexual selection based on compatible genes [27] are particularly interesting as they may serve as mechanisms for producing dynamic sets of the most optimal combinations of entities and their mutations that provide maximum protection against evolving malware. Moreover, using knowledge of how sexual selection works, it may be interesting to study how to become the most "unattractive" victim to the potential attacker.
- *competition:* this relationship is symmetrical and involves both organisms competing for the same pool of resources. Inherently the relationship between organisms can be broken without any harm to neither of the sides – as both influences are negative their cessation benefits both competitors. In communication network environment this interaction can occur e.g. between two types of malware trying to infect the same host – when one of them succeeds it tries to "secure" the host by patching the exploit used by the other type of malicious software. Competition can also occur between defenders when few similar defence systems (e.g. anti-virus software) are run together and they impact each other in a negative way.

A point of view of cyber-ecology may be to treat these interactions as purely mechanistic descriptions of cyber-systems – without looking at the consequences of interactions themselves and on the dynamics they describe. However, growing evidence suggests that the interactions not only influence the fitness and performance of entities but also significantly modify their physiology/performance in the interaction, altering the outcome of competition/synergy [28]. Such elastic responses of interacting entities to the interaction itself may have a significant role in cyber-ecosystems, as they may serve to design more efficient ways of controlling cyber-ecosystems and reacting to unknown, emerging threats.

As indicated in Sect. 2, existing work focuses mainly on predator-prey association. However, an interesting observation is that the relationship between the current malware and the host is in essence closer to parasitism than to predation. This means that the goal of the current malware is to live off the infected host (and the longer it remains undetected, the better) but not to immediately cause significant harm or permanent damage.

In the following section we will review the most important natural-enemy ecology models including parasitism models, and we will assess how this knowledge can be used for cybersecurity purposes.

4 Natural Enemy Ecology in Nature – Unifying Antagonistic Interactions

The field of antagonistic interactions in ecological studies has so far been dominated by a very sharp distinction between predator-prey interactions and parasite-host interactions. As pointed out recently such interactions are, however, much closer to each other, and together with a third class (competition) form a unified group of antagonistic interactions involving the aggressor, the victim and resources that are/may be available to one or both entities [30]. This has led to the emergence of a new field-of-study in ecology, which is broadly termed "natural enemy ecology", and encompasses all interactions involving detrimental effects of one organism on another, be it a direct or indirect (e.g. via shared resources) effect. In this section we discuss consequences of such a categorization and review the most prominent models of antagonistic interactions, while pinpointing their weaknesses [30].

4.1 Similarities Between Parasitic and Predatory Interactions

The strong distinction between parasitic and predatory relationships results mostly from an old methodology of categorizing nature [31]. In fact, all kinds of antagonistic ecological interactions (predation, parasitism and competition) share a common suite of components, which differs only in the strength/presence/direct character of the specific connections. All interactions involve conventionally at least two organisms (aggressor and victim, or two competitors in the competition model) that influence each other positively and/or negatively, and use each others' resources [30].

Competition: the least antagonistic of all interactions; the roles of the interacting organisms are indistinguishable and both exert mutually negative influence on the other. The relationship is symmetrical and involves both organisms competing for the same pool of resources. Inherently the relationship between organisms can be broken without any harm done to neither of the sides: as both influences are negative their cessation benefits both competitors [30].

Predation: occurs when the aggressor kills the victim directly and feeds on its tissue – therefore it is inherently asymmetrical; predation involves very short time-scales, much shorter than timescales necessary for the evolution of low-level (molecular, immunological) defence mechanisms and, thus, prey evolves defences in such system mostly at the higher, organismal (e.g. morphology and behaviour) level [32]. Instead of immunological mechanisms prey benefits more by evolving learning-like mechanisms that are much more flexible on one hand and can evolve within long generation times on the other hand. Because predators consume their victims, they are regarded as residing on a different, higher trophic level than prey [30].

Parasitism: in this form of interaction the aggressor feeds on the victim but does not kill it. Predatory interactions are inherently fatal whereas parasitic interactions have led to the phenomenon of intermediate virulence, which maximises parasite transmission to other hosts. The relationship between parasites and hosts is much more intimate and occurs at time-scales and generation times that allow the evolution of complex genetic (e.g. bacterial *Crispr-Cas* [33]) and immunological (e.g. vertebrate acquired immunity, invertebrate Toll receptors) defence mechanisms in victims/hosts.

It is clear that all three relationships are slightly different and involve different levels of inter-organismal contact. However they all draw from the same population processes related to the population growth and decline. Moreover, sometimes parasitism and predation are hard to delineate. For example, caterpillars feeding on plants could be regarded as predators, but they do not kill their victims and dwell on the surface of victim, as ectoparasites. Mosquitoes feed on the tissues of their victims (like parasites) but apart from this they display many properties of predators (longer generation time, short interaction timescale, high turnover rate of attacked victims). Recent literature has also pointed out that although seemingly different, parasitic and predatory interactions may give rise to similar ecological patterns. Some prominent examples include:

- *The evolution of inducible defences and attack anticipation* [34]: predation is often associated with behaviours and traits that are active and use resources only in the presence of predators – similar mechanisms may be present in the parasite-host systems where organismal systems (e.g. immunological) may optimise their activity window to match the activity window of aggressors,
- *Enemy-mediated facilitation* [35]: in the presence of more than one aggressor, host/prey communities may evolve mechanisms that make use of prey-specific resistance to aggressors and indirect ecological effects that result from variation in prey/host susceptibility to aggressors,
- *Managing the threshold of transmission*: in parasite-host systems there are specific host densities below which parasites are unable to effectively spread and persist; a similar concept might be applied to the predator-prey systems, where by managing the densities of particular predators ("superpredators" that affect prey densities the most) the population may be maintained at a desired level of prey density, avoiding extinction due to random fluctuations in predation pressures [30].

4.2 Models of Antagonistic Interactions

The ecological literature has developed a number of mathematical descriptions of the predator-prey or parasite-host interactions and not surprisingly, and in line with the abovementioned unifying considerations, all these models can be adjusted for the description of both predation and parasitism interactions. The most prominent and the oldest model is the Lotka-Volterra (L-V) model [31] that binds together aggressor and victim densities and models changes in these densities according to an assumed predation/parasitism rate. The model is defined using a system of two differential equations:

$$\frac{dx}{dt} = rx - ayx \tag{1}$$

$$\frac{dy}{dt} = -r'y + a'xy \tag{2}$$

where x and y denote prey and predator densities, r and r' describe population growth/decline of prey/predator populations, whereas a/a' quantifies the rate of encounters between prey and predators. The solution of this system describes the oscillatory behaviour of prey and predator densities. The L-V model was quickly considered simplistic (e.g. the assumption of constant encounter rates a/a' was considered as biologically unrealistic) and a number of other models have been developed. However, ecologists agree that all available models are just special cases of the L-V model, which in turn still remains the most important model for antagonistic interactions among organisms [31].

The models that followed the L-V system focused mostly on making some of its assumptions more realistic. For example, the Nicholson-Bailey model expanded on the results from the L-V system and generalised them to discrete generations of prey and predators (the L-V system was developed under the assumption of continuous overlapping generations). More advanced models, e.g. the Holling model [36], the Ivlev model [37], and the Watt model [38] remained in the reality set by the Lotka-Volterra model, changing and adjusting only the encounter function (i.e. the function that binds prey and predator densities together with time, providing the dynamics of the encounter rates between interacting individuals).

A proper integration of the existing models into the field of cybersecurity will likely involve a revision of the assumptions of different models of antagonistic interactions and relating them to the specific features of communication networks. Specific comparisons are necessary to elucidate the shared features and assumptions at the interface of biological and cyber systems – such comparative analysis can then identify models that are the most accurate in describing cyber reality with respect to the antagonistic interactions.

4.3 Antagonistic (Parasitic) Mimicry: Batesian Mimicry

Even without clear exploitation of material resources of the hosts, parasitism can be present if information content/reliability is being exploited by one organism at the expense of the costs born by the other organism [39]. One well-documented example of such behaviour is parasitic mimicry, which is relatively inexpensive to the mimicking organism as it is not associated with weapons/toxins this organism is pretending to have [40]. A well-known example is the *Chrysotoxum festivum* hoverfly that resembles toxic and stinging insects from the Hymenoptera group. By expressing warning colours the hoverfly avoids being attacked and eaten, and on the other hand it does not have to invest resources in actually having a sting.

Parasitic (Batesian) mimicry, due to its inexpensive nature, could readily be used in security applications in cyber systems. The mimic could be the security algorithm that could adopt some features of the actual hostile software to approach it and infiltrate without being detected [39]. Most existing models of Batesian mimicry operate on the balance between costs of being detected and the benefits of expressing certain masking phenotypes. Such models could be used to derive parameter ranges that ensure full masking in the cyber-ecosystem at the expense of the lowest possible resource allocation.

4.4 Non-antagonistic Interactions

Non-antagonistic interactions are more difficult to classify and organise, mostly because they combine intra- and inter-species processes. There exists no single model of synergistic interactions similar to the seminal Lotka-Volterra model; however, we have several ways of expressing the dynamics of such interactions mathematically. Non-antagonistic interactions that play major roles in development of cybersecurity solutions encompass all of the above sexual selection/mate choice processes, and symbiotic interactions. Both have the potential to substantially inform efforts to develop effective cybersecurity strategies; both also remain largely unstudied on a large, inter-species comparative level and thus are attractive targets of comparative biological research.

4.5 Symbiotic Interactions

Symbiosis is thought to underlie all life on Earth as, according to the endosymbiosis hypothesis, all eukaryotic cells are descendants of several prokaryotic organisms that merged together as symbionts, which gave rise to currently observed organelles such us chloroplasts and mitochondria [41]. Currently the most commonly known and well-studied examples of such interactions may serve as good models to derive mathematical parameters that can be used in developing cybersecurity solutions. From an evolutionary perspective, the symbiotic interactions can be readily modelled using the same mathematical reasoning as the one used in the Lotka-Volterra system, by modifying parameters of the equations so that interacting units benefit each other instead of harming [42].

From the point of view of cybersecurity applications, symbiotic interactions may potentially play roles in two scenarios. For one, symbionts in a cyber-ecosystem could be used to strengthen the protective/immunizing effects of applied techniques. Multiple symbiotic entities could enforce each others' defensive strategies and achieve fuller protection of the whole system. On the other hand, symbiotic interactions are intricately associated with other close interactions. In fact, the Lotka-Volterra-like model of symbiotic interactions [42] predicts that they can easily turn into parasitic interactions if conditions shift in the environment of symbionts (e.g. if available resources become more asymmetrically exploited by one of the symbionts). Thus, such models are also able to provide a testing space where a range of parameters that maintain the beneficial symbiotic interactions could be tested. In fact, such models can also be used to derive alternative scenarios of fighting cyber parasites – if it is possible to "mutate" them and modify their responsiveness to the environment – changing a parasitic interaction into a symbiotic one with an artificially introduced additional organism [43].

A special case of synergistic interactions occurs in cooperating organisms when individuals bear costs (often the highest fitness costs, i.e. by postponing/entirely abandoning reproduction) and benefit other individuals by helping them (usually in the form of raising their offspring) [25]. The dynamics of such interactions is best known in the altruistic forms of cooperation, where it is predicted and described by the Hamilton inequality [24] that binds costs of the donor, benefit of the receiver, and their coefficient of relatedness that defines how costs and benefits are balanced on both sides of the interaction [24, 44]. In the context of this paper, however, it is of a marginal importance – much more important kinds of cooperating interactions will be those encountered between non-related individuals. Such non-kin cooperation can easily be incorporated in our system (as reciprocal sharing of costs and achieved benefits), however this field of ecology is still strongly under-represented and no quantitative models exist that could be used and developed in the context of the cybersecurity.

4.6 Sexual Selection

From the point of view of cybersecurity, sexual selection may be the most difficult but also the most potent interaction that could be exploited [45]. The biggest difficulty comes from the fact that sexual selection operates through choice of the most suitable mates and thus would require creating and maintaining a population of sexually reproducing entities that would use cycles of selection in order to evolve new, more effective ways of fighting enemy software [26]. It is an important question how such selection would operate and currently evolutionary biology describes two major classes of sexual selection mechanisms. The first one, called "the good genes hypothesis" poses that selective individuals (in nature usually females) choose certain partners (usually males) because they provide them with "good genes" that increase offspring viability and fitness [46]. Such an indirect genetic benefits have been demonstrated in many animal studies and are a well-documented, although still weakly understood phenomenon [26, 27].

The second class of sexual selection drivers falls into the "Fisherian runaway" category, where the preference of one sex (females) evolves as a self-perpetuating mechanism that exploits certain male traits and is fuelled by a positive feedback loop generated by the strong genetic correlations between female preference and male display traits [45, 46]. This second form of sexual selection has also been suggested to occur in nature – however it is much more difficult to find its place in the cybersecurity reality as this form of sexual selection is not directly associated with any fitness benefits to females (apart from choosing males that can actually afford to have exaggerated and overgrown traits).

Both models of sexual selection are governed by one common mathematical model [47] that integrates female preference (P), male display (D) and residual fitness effects (F). If we denote variance and covariance of specific traits as V and C (e.g. $V(P)$ – variance in preference; $C(PD)$ – covariance between display and preference), b_s and b_n as respective selection gradients resulting from sexual (s) and natural (n) selection, the joint dynamics of these traits may be described as:

$$\Delta \begin{pmatrix} \overline{D} \\ \overline{P} \\ \overline{F} \end{pmatrix} = \begin{pmatrix} V(D) & C(PD) & C(FD) \\ . & V(P) & C(FP) \\ . & . & V(F) \end{pmatrix} \times \left(\begin{bmatrix} b_n(D) \\ b_n(P) \\ b_n(F) \end{bmatrix} + \begin{bmatrix} b_s(D) \\ b_s(P) \\ b_s(F) \end{bmatrix} \right) + \begin{pmatrix} u(D) \\ u(P) \\ u(F) \end{pmatrix} \tag{3}$$

where u denotes respective changes in phenotypes' values due to mutation. Different combinations of parameters of this model yield different modes of sexual selection, and exploration of these values within the ranges that are realistic to cyber systems will help uncover types of interactions that would be the most efficient in cybersecurity applications.

5 Potential Bio-inspired Research Directions for Cybersecurity

After defining key terms related to cybersecurity ecology, and describing most important models that characterise interactions between organisms in nature, the next step is to develop a "procedure" that will result in the potential new research directions. The steps of such a procedure related to interactions are illustrated in Fig. 3. First, it is important to map existing offensive/defensive measures as well as interactions in both types of ecosystems. From the biology perspective this includes performing rigorous meta-analyses describing comparatively and phylogenetically the diversity of defence/attack mechanisms present in nature and their complexity (e.g. their costs, the most optimal uses, their diversity at various level of life organization).

In the next step, the missing components in the virtual world that could be potentially ported from nature should be identified. All of the most promising candidates that do not have counterparts in cyberspace will form a list of most suitable bio-inspirations.

In the last step, it is also possible to identify security-related components that exist in cybersecurity but that are not sufficiently effective. Then, insights from mechanisms and relationships that exist in nature could provide important

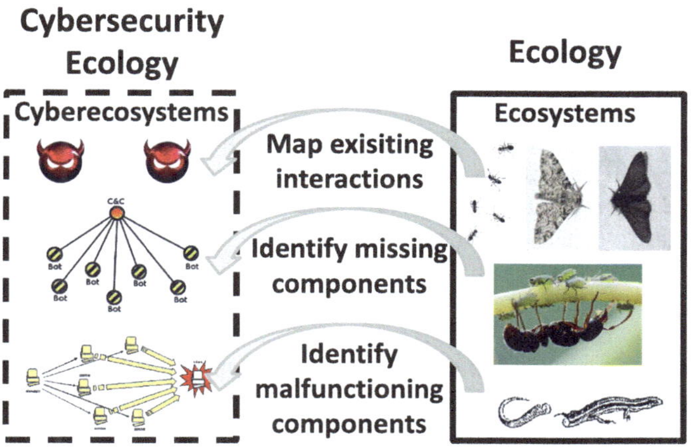

Fig. 3. Comparing interactions and components between ecology and cybersecurity ecology.

feedback on how these security techniques could be improved. To summarise, we believe that currently the most promising research directions include:

- Drawing further inspirations from the particular organism's characteristic feature/defence mechanism. For example, such features like *aposematism* (warning signal that is associated with the unprofitability of a prey item to potential predators) or *autotomy* (where an animal sheds or discards one or more of its own body parts to elude or distract the predator) could readily become an inspiration for future cybersecurity solutions.
- Careful investigation and applying knowledge from the mentioned nature-based interactions. As already observed the malware-host scenario is more similar to *parasite-host* than to predator-prey association. Therefore, more research attention should be turned to the models and achievements of biology in this field. This could provide many new, interesting insights. Another research direction that we believe has not been sufficiently explored is *sexual interactions* where, for example, the methods to become an attractive/unattractive target could be analysed.
- Comparative analysis of the features of parasitic and predatory systems that expose their common underlying mechanisms leading to their description within the natural enemy framework. Such common properties of these antagonistically interacting systems may be the most effective points (in a way identified by long evolutionary history of such systems) where new approaches to cybersecurity can be developed. The most promising avenues in this group of issues include *(i)* induced/anticipatory mechanisms that lower the costs of maintaining active defence mechanisms; *(ii)* enemy-driven facilitation – which, by exploiting multiple enemies, may lead to the establishment of reinforcement mechanisms that increase the effectiveness of enemy elimination; *(iii)* transmission threshold management which can provide tools to minimise the effort in eliminating threats, while maximizing the achieved security gain.

6 Conclusion

In this chapter we have presented a systematic ecology-based approach to cybersecurity. Based on the observation of the significant fragmentation of achievements and knowledge in the field of bio-inspired cybersecurity first we summarised the state-of-the-art in this field. Later, we drew the analogy between the biology-based ecosystem and the cyber-ecosystem, introducing terminology such as cyber-ecosystem, cybersecurity ecology, and other related terms before reviewing the most important concepts, interactions and models from the natural enemy ecology, making links as to how these can be used to study offensive/defensive mechanisms and interactions among cyber-organisms and/or between cyber-organisms and their environment. It is our belief that such an approach could help to reveal new potential future research directions which next generation cybersecurity solutions should follow.

References

1. Yardon, D.: Symantec develops new attack on cyberhacking. Wall Street J. (2014). http://www.wsj.com/articles/SB10001424052702303417104579542140235850578
2. Mazurczyk, W., Rzeszutko, E.: Security - a perpetual war: lessons from nature. IEEE IT Prof. **17**(1), 16–22 (2015)
3. Hofmeyr, S.A.: An immunological model of distributed detection and its application to computer security. Ph.D. thesis, University of New Mexico (1999)
4. Ford, R., Bush, M., Bulatov, A.: Predation and the cost of replication: new approaches to malware prevention? Comput. Secur. **25**(4), 257–264 (2006)
5. Crandall, J.R., Ladau, J., Ensafi, R., Shebaro, B., Forrest, S.: The ecology of malware. Proceedings of the New Security Paradigms Workshop (NSPW 2008), Lake Tahoe, CA, USA, pp. 99–106 (2008)
6. Okhravi, H., Hobson, T., Bigelow, D., Streilein, W.: Finding focus in the blur of moving-target techniques. IEEE Secur. Priv. **12**(2), 16–26 (2014)
7. de Castro, L.N., Von Zuben, F.J.: The clonal selection algorithm with engineering applications. In: Genetic and Evolutionary Computation Conference (GECCO), Las Vegas, USA, pp. 36–37 (2000)
8. Greensmith, J.: The dendritic cell algorithm. Ph.D. thesis, University of Nottingham, UK (2007)
9. Hart, E., Timmis, J.: Application areas of AIS: the past, the present and the future. Appl. Soft Comput. **8**, 191–201 (2008)
10. Fink, G.A., Haack, J.N., McKinnon, A.D., Fulp, E.W.: Defense on the move: ant-based cyber defense. IEEE Secur. Priv. **12**(2), 36–43 (2014)
11. Gorman, S.P., Kulkarni, R.G., Schintler, L.A., Stough, R.R.: A predator prey approach to the network structure of cyberspace. In Proceedings of the Winter International Synposium on Information and Communication Technologies (WISICT 2004), pp. 1–6. Trinity College Dublin (2004)
12. Kephart, J., White, S.: Measuring and modeling computer virus prevalence. In: Proceedings of the 1993 IEEE Computer Society Symposium on Research in Security and Privacy, Oakland, California, May 24-25, pp. 2–14 (1993)
13. Pastor-Satorras, R., Vespignani, A.: Epidemic spreading in scale-free networks. Phys. Rev. Lett. **86**, 3200 (2001)
14. Moghaddam, H.M., Li, B., Derakhshani, M., Goldberg, I.: SkypeMorph: protocol obfuscation for Tor bridges. In: Proceedings of the 2012 ACM Conference on Computer and Communications Security (CCS 2012), pp. 97–108. ACM, New York (2012)
15. Ruxton, G.D., Sherratt, T.N., Speed, M.P.: Avoiding Attack: The Evolutionary Ecology of Crypsis, Warning Signals and Mimicry. Oxford University Press, Oxford (2004)
16. Zielinska, E., Mazurczyk, W., Szczypiorski, K.: Trends in steganography. Commun. ACM **57**(2), 86–95 (2014)
17. Stenseth, N.C., Smith, J.M.: Coevolution in ecosystems: red queen evolution or stasis? Evolution **38**(4), 870–880 (1984)
18. Moore, P.S., Boschoff, C., Weiss, R.A., Chang, Y.: Molecular mimicry of human cytokine and cytokine response pathway genes by KSHV. Science **274**(5293), 1739–1744 (1996)
19. How, M.J., Zanker, J.M.: Motion camouflage induced by zebra stripes. Zoology **117**(3), 163–170 (2014)

20. Delves, P.J., Martin, S.J., Burton, D.R., Roitt, I.M.: Essential Immunology. Wiley-Blackwell, Hoboken (2011)
21. Krebs, C.J.: Ecology: The Experimental Analysis of Distribution and Abundance. Benjamin Cummings, San Francisco (2009)
22. Rooney, N., McCann, K.S.: Integrating food web diversity, structure and stability. Trends Ecol. Evol. **27**(10), 40–46 (2012)
23. Ings, T.C., et al.: Review: ecological networks - beyond food webs. J. Anim. Ecol. **78**(1), 253–269 (2009)
24. Axelrod, R., Hamilton, W.D.: The evolution of cooperation. Science **211**(4489), 1390–1396 (1981)
25. Riolo, R.L., Cohen, M.D., Axelrod, R.: Evolution of cooperation without reciprocity. Nature **414**, 441–443 (2001)
26. Andersson, M.: Sexual Selection. Princeton University Press, Princeton (1995)
27. Neff, B.D., Pitcher, T.E.: Genetic quality and sexual selection: an integrated framework for good genes and compatible genes. Mol. Ecol. **14**(1), 19–38 (2005)
28. Miner, B.G., Sultan, S.E., Morgan, S.G., Padilla, D.K., Relyea, R.A.: Ecological consequences of phenotypic plasticity. Trends Ecol. Evol. **20**(12), 685–692 (2005)
29. Whorf, B.L.: Language, Thought, and Reality: Selected Writings of Benjamin Lee Whorf. MIT Press, Cambridge (1956). Carroll J.B. (ed.)
30. Raffel, R., Martin, L.B., Rohr, J.R.: Parasites as predators: unifying natural enemy ecology. Trends Ecol. Evol. **23**(11), 610–618 (2008)
31. Royama, T.: Comparative study of models for predation and parasitism. Res. Popul. Ecol. **13**(Supp 1), 1–91 (1971)
32. Benard, M.F.: Predator-induced phenotypic plasticity in organisms with complex life histories. Annu. Rev. Ecol. Evol. Syst. **35**, 651–673 (2004)
33. Sorek, R., Kunin, V., Hugenholtz, P.: CRISPR - a widespread system that provides acquired resistance against phages in bacteria and archaea. Nat. Rev. Microbiol. **6**, 181–186 (2008)
34. Altizer, S., Dobson, A., Hosseini, P., Hudson, P., Pascual, M., Rohani, P.: Seasonality and the dynamics of infectious diseases. Ecol. Lett. **9**, 467–484 (2006)
35. Bruno, J.F., Stachowicz, J.J., Bertness, M.D.: Inclusion of facilitation into ecological theory. Trends Ecol. Evol. **18**(3), 119–125 (2003)
36. Holling, C.S.: Principles of insect predation. Annu. Rev. Entomol. **6**, 163–182 (1961)
37. Ivlev, V.S.: Experimental Ecology of the Feeding of Fishes. Yale University Press, New Haven (1955)
38. Watt, K.E.F.: Mathematical models for use in insect control. Can. Entomol. Suppl. **19**, 1–62 (1961)
39. Franks, D.W.: Modelling the Evolution of Warning Signals and Mimicry with Individual-Based Simulations. University of Leeds, Leeds (2005)
40. Pfennig, D.W., Harcombe, W.R., Pfennig, K.S.: Frequency-dependent Batesian mimicry. Nature **410**(323), 134–136 (2001)
41. Futuyma, D.: Evolution. Sinauer Associates, Sunderland (2015)
42. Neuchauser, C., Fargione, J.E.: A mutualism-parasitism continuum model and its application to plant-mycorrhizae interactions. Ecol. Model. **177**(3–4), 337–352 (2004)
43. Cheney, K.L., Cote, I.M.: Mutualism or parasitism? The variable outcome of cleaning symbioses. Proc. Royal Soc. B **1**(2), 12–19 (2005)
44. Nowak, M.A.: Five rules for the evolution of cooperation. Science **314**(5805), 1560–1563 (2006)

45. Prokop, Z.M., Michalczyk, L., Drobniak, S.M., Herdegen, M., Radwan, J.: Meta-analysis suggests choosy females get sexy sons more than "good genes". Evolution **66**(9), 2665–2673 (2010)
46. Drobniak, S.M., Arct, A., Cichon, M.: Extrapair paternity and genetic similarity - we are not quite there yet: a response to comments on Arct et al. Behav. Ecol. **26**(4), 973–974 (2015)
47. Kokko, H., Jennions, M.D., Brooks, R.: Unifying and testing models of sexual selection. Ann. Rev. Ecol. Evol. Syst. **37**, 43–66 (2006)

Challenges Priorities and Policies: Mapping the Research Requirements of Cybercrime and Cyberterrorism Stakeholders

Douglas Wells(✉), Ben Brewster, and Babak Akhgar

CENTRIC (Centre of Excellence in Terrorism, Resilience, Intelligence and Organised Crime Research), Sheffield Hallam University, Sheffield, UK
{D.Wells,B.Brewster,B.Akhgar}@shu.ac.uk

Abstract. The following chapter provides an in depth look at a broad selection challenges related to Cybercrime and Cyberterrorism, as identified through prolonged engagement with a multitude of horizontal and vertical cyber-security stakeholders. Out of six critical areas identified, the two leading causes, were through the evolving rate of technology, and, the subsequent lack of education, awareness and training. These two underlying factors further influenced and affected the severity of the additional four critical areas; the capability of investigators, cooperation and information sharing, legislative systems and data protection, and, organisational and societal resilience. Through the consultation and elicitation of information from over 90 individual domain experts, practitioners and security stakeholders, the research of this chapter is dedicated towards improving international awareness towards leading threats, vulnerabilities, and challenges to the continually evolving sphere of cybersecurity.

Keywords: Cybercrime · Cyberterrorism · Challenges · Priorities stakeholders · Recommendations · Delphi study

1 Introduction

The extent to which society now depends on information technology through all aspects of daily life has created new and extended vectors through which criminality can take place. As a result of this new environment, we are now inundated with media reports on a daily basis describing the latest, and often greatest, cybersecurity breaches, frequently exposing the confidential details of organisations' customers and employees. Such dialogue feeds the next wave of public and political debates surrounding topics such as the precarious balance between safeguarding privacy and liberty, against ensuring security and stability of society as a whole [1].

B. Akhgar and B. Brewster (eds.), *Combatting Cybercrime and Cyberterrorism*,
Advanced Sciences and Technologies for Security Applications,
DOI 10.1007/978-3-319-38930-1_3

In this chapter we present the results of a wideband Delphi study exploring the contemporary challenges that modern society faces in reference to the proliferating threat of cybercrime (CC) and the emergence of cyberterrorism (CT), towards establishing a number of priority areas to be targeted by future research and policy.

As if in accordance with Moore's Law itself, the concerns of cyberspace have increased exponentially throughout the 21st century [2, p. 11]. Each year the global flow of information brings new challenges to the fore. Criminals and terrorists appear to thrive in the anarchy of largely unlegislated and uncontrolled cyberspace, posing a clear and credible threat, not only to National Security, but to all levels of globally-connected society. Whilst technologies unfaltering progress continues to interconnect and transform the world bringing vast economic and social opportunities, it also accelerates the quantity and severity of threats. These new challenges are currently largely under-reported, under-acknowledged and often lack effective solutions, yet they critically threaten all sectors of European society [21]. The rapid rate of developing technology has continued to accelerate, creating large epistemic gaps. Within these gaps lie possible vulnerabilities for all European stakeholders, as their awareness and ability to adapt and take measures to prevent threats lags behind the ever-changing field of cyber-threats.

In total, over ninety unique stakeholders participated in the requirement extraction process representing a diverse range of sectors and professions, from government and civil society, to law enforcement, private sector organisations and beyond. A three phase process based upon the Delphi methodology [20] was employed in order to refine and establish priority areas for future research on issues related to Cybercrime (CC) and CT (CT). These requirements have been assessed and aggregated using a thematic analysis in order to identify significant challenges, trends and priorities, resulting in a set of key research themes each with a quantifiable scope and objectives. Six key areas of concern, based on the input of these stakeholders, emerged:

1. Increasing the capability of investigators
2. Improving the ways in which states and organisations cooperate at a public/private level and internationally.
3. Enhancing societies resilience capacity
4. Exploring issues related to the development, application and interpretation of legislative systems and policy
5. Expanding Awareness, Education and Training approaches
6. Challenges caused as a result of the pace of technological change and implications thereof.

In this chapter we first provide an overview of other existing work that contributes to or informs our study. Later sections provide an overview of the approach and methodology employed before moving on to discuss the results and their potential implications.

2 Related Work

Although it is not the purpose of this chapter to discuss the differences and potential ambiguities surrounding definitional issues associated with CC and CT, *as these are covered in Chap.* 16, a simple reference taxonomy of working definitions is utilised to establish a consistent context. For CC, we refer to Koops' definition [17, p. 737] that describes; '*a crime in which computer information networks are the target, or, substantial tool of an attack*', whilst acknowledging the existence of a range of definitions and frameworks based upon the European Convention against CC [9], also known as The Budapest Convention.

When focusing specifically on CT, arguably the most widely accepted and exchanged definitions are derived from Denning's Testimony before the 'Special Oversight Panel on Terrorism' [10]. In this definition, CT is described as; *"the convergence of traditional terrorism and cyberspace. It is generally understood to mean unlawful attacks and threats of attack against computers, networks, and the information stored therein when done to intimidate or coerce a government or its people in furtherance of political or social objectives"*. CT is further qualified as those attacks which *'result in violence against persons or property'*, that directly cause or lead to physical harm and/or damage, and induce fear. These are discussed in further depth in subsequent chapters.

The vector's through which threats associated with CC and CT are quantified, subsequently prioritised and communicated varies between different horizontal and vertical stakeholders. For example, Law Enforcement organisations have well established mechanisms for resource prioritisation through the production of threat assessments and other intelligence products such as baseline assessments. Such products are often used to identify problematic areas based on the analysis of existing crime records and intelligence reports. National and international threat assessments such as those provided by Interpol, Europol and the UK home office, build a picture of the serious and organised crime issues. Through identifying high level trends and core risks towards building an awareness of the actions needed to address them at national and EU levels. Overarching these artefacts are of course national and EU cybersecurity strategies, which target not only law enforcement, but are the underpinning basis for all national activity related to cybersecurity.

At the EU level, these priorities consist of actions aiming to increase levels of preparation, cooperation and information exchange between the public and private sectors regarding information security. Further actions include fostering the relevant environment to develop skills and expertise in relation to those in charge of the investigation and prosecution of CC, and improving issues related to the development and implementation of policy and cross border-issues towards the development of cybersecurity capacity both within and outside of the EU.

This is by no means a comprehensive or exhaustive picture of the information and intelligence landscape however; there are multiple active research communities, of multiple disciplines making significant contributions at all levels; from identifying system vulnerabilities, to informing strategy, management and EU policy. Subsequent sections of this chapter attempt to further detail these specific communities, presenting a snapshot of some of the most pressing challenges that society faces in response to CC and CT.

3 Approach

The approach utilised here adopts a wideband adaptation of the traditional Delphi method in order to integrate disparate perspectives on contemporary issues that are faced across different stakeholder groups in respect of CC and CT. The wideband approach was selected due to the renowned emphasis on the conversational aspects of the approach, enabling the participants to interact and debate with one-another, allowing for issues to be explored in greater detail. Typically, the Delphi approach is repeated until the responses reach a degree of consensus deemed appropriate to successfully satisfy the objectives of the research being undertaken, although in practice outside factors such as time and availability constraints often contribute to the exact nature and length of the process [20]. Traditionally, applications of the Delphi methodology take steps to avoid face-to-face interaction and thus disagreement between participants, instead relying on controlled statistical validation of responses in order to establish group consensus through rounds of iteration [19]. Subsequent adaptations of the Delphi approach, such as the method used here, have re-introduced the concept of face-to-face discussion as a value-adding-process in order to eliminate any biases introduced by individual participants. Such discussions provide opportunity for concepts to be questioned and elaborated upon in order to enhance both the original understanding and final response of items raised. Additionally, this may trigger further contribution and refinement of ideas in order to elicit a deeper understanding of the concepts under discussion. The Wideband adaptation of the Delphi method was originally conceived by Boehm was adopted using [5]; knowledge elicitation techniques -that incorporate iterative methods for wide-spread data capture, including surveys, interviews and workshops.

The approach made use of four individual stages, beginning with an anonymous survey propagated through the extended interdisciplinary stakeholder communities connected to the EU FP7 project COURAGE (CC and CT Research Agenda). The survey used a scoping mechanism to identify and signpost areas of interest, in a measure to help guide subsequent rounds of the process. Later rounds involved with the facilitation of interdisciplinary focus groups whereby participants initially ranked the identified interest areas and highlighted any potential knowledge gaps. The results of this process were then used to guide the discussions of the focus groups. The first round of focus groups targeted specific stakeholder areas. Participants were recruited based on their professional expertise and day-to-day engagement with matters of information security, CC and CT. In total seven focus-groups were held during this round, with each one thematically centred around a different group. These groups consisted of; academia, law enforcement, government and civil society, critical infrastructure, technology solutions, the private sector, and, legal, ethical and data protection.

The workshop transcripts were then thematically analysed, and the recurring topics and ideas extracted, before being categorised under six headings.

Data was initially coded in two steps. Firstly, open coding was using in order to examine the transcripts line-by-line in order to identify the nature of the core concepts being expressed by the research participants [14]. This process subsequently enabled the categorisation of concepts at each phase of the study. At the second step a selective approach was used in order to develop the aggregated concepts identified throughout the requirement elicitation process. The selective approach transformed aggregated concepts into genuine theory regarding actual requirements and challenges that, according to the data collected, should be addressed by research [8].

Initial results identified six core categories that the research should take steps to address:

1. Increasing the capability of investigators.
2. Improving the ways in which states and organisations cooperate at a public/private level and internationally.
3. Holistically enhancing the resilience capacity of organisations and society as a whole.
4. Issues related to legislative systems and policy
5. Expanding Awareness, Education and Training mechanisms
6. Challenges caused as a result of the pace of technological change and the implications thereof.

These categories were then used as the structure for further rounds of focus groups. In the third round, the thematic groups were abandoned in favour of mixed-discipline groups. This change aimed to facilitate the extraction of commonalities and differences between the various stakeholder groups. Careful consideration was taken to recruit participants in a manner to maximise the number of groups represented, whilst still retaining a group small enough to facilitate an environment conducive to fostering free expression and open conversation required to unpack the complexities of the domain. In total more than ninety stakeholders, from across fourteen EU member states were consulted in over the three phases of the engagement process. These later-stage focus groups were used to refine many of the concepts and ideas identified in previous sessions. The most prominent and relevant of these dialogues are discussed below, grouped under the six overarching categories identified earlier.

4 Challenge Paradigms

4.1 Enhancing the Capability of Investigators

The emergent threat of CC and the proliferation of the 'cyber-universe' into all forms of criminality, from online child exploitation and human trafficking, to fraud and the sale of weapons and illegal substances, has meant that the traditional skills and resources of law enforcement are coming under increasing pressure to evolve and adapt, in order to operate effectively in this emerging environment. Not only has the legitimacy of cyber threats grown to engulf matters of national security, it harbours a profound threat to all sectors and levels of socio-economic activity.

Although we're regularly confronted with reports prophesizing the sophistication and scale of national intelligence and law enforcement agencies data collection and analysis capability, in reality the specialist skills associated with data forensics and other aspects of cybersecurity are often lacking outside of specialised CC units, with frontline officers often lacking the general levels of awareness and competence needed to handle low level crime reports, as well as to offer advice to victims. Additionally, internal policies, structures and security cultures require review and possibly adaptation in order to prioritise the CC threat with contemporary policing.

Current investigative means appear to be primarily reactive, responding to threats and challenges once they have appeared, or have crossed a legal threshold. It is however, essential that investigators do not limit themselves to such approaches. In addition to pre-emptive capabilities, investigators must consider the known, or unknown, vulnerabilities that challenges stem from as they exploit *"pre-existing weaknesses in the underlying technology"* [16, p. 75]. Furthermore, methods such as the use of simulated threats to highlight vulnerabilities, such as the UK's; 'Cyberstorm III' simulations that were run across Europe, have been limited to just the public sector [11, 12]. Integration of the private sector into such simulations is essential as most elements of critical infrastructure remain privately owned [16, p. 84].

These challenges present a number of areas where research and further work can help law enforcement, not just at a national and EU wide level, to improve capability and raise capacity. Models used by organisations such as Europol, Interpol and well-funded national agencies that are better equipped to manage, respond and prevent the cyber-threat should be analysed to identify cost effective measures to address gaps in the capability and capacity of police in this regard. Of course, the skills and tools needed to cope effectively with CC are not static, therefore it is vitally important for research to assist and inform of changes in the behaviours and tools used by criminals. Alongside this, steps must be taken to ensure the retention and dissemination of this specialist knowledge to alternative departments and sectors in a digestible manner.

4.2 Cooperation and Information Exchange

The importance of cooperation between the various internal sectors of a nation is essential to meet the challenges presented by CC and CT. The borderless, transnational nature of cyberspace acts as an enabler of other forms of criminality as well as being the dependent factor in conducting attacks that seek to disrupt, destroy or steal from digitally-interlinked systems. This has resulted in increased pressure on vertical and horizontal stakeholders to more effectively collaborate and cooperate in response. Thus, the requirement for inter-agency and international cooperation in terms of law enforcement, and collaboration with and between public and private-sector organisations and citizens has grown exponentially in the 21st century. Political, economic and business pressures however all pose barriers to these relationships being established and maintained.

One particular consideration can be observed regarding Critical Infrastructures; "information sharing between the public and private sectors is even more important, with the added difficulty of devising methods for civilians and the military to collaborate in times of peace" [6, p. 129]. Indeed, the current pan-European climate shows some countries to feature potential public-doubts, misconceptions, and mistrust towards the information gathering and processing of government and military affiliates, particularly following the information leaks of Julian Assange and Edward Snowden. This gap in cooperation is often further exacerbated by a lack of understanding towards the specifics of cybersecurity, surveillance, investigation, and so forth, creating what Beck refers to as the 'politics of insecurity' [4]. Examples of insecurity can be seen as citizens often point the finger of blame at key individuals, such as politicians, military figures, business and organisations CEO's, for problems for which they may not be responsible for, nor hold any direct influence over. This processes of misunderstanding responsibilities and distorting accreditations and blame is clearly evident throughout most cybersecurity concerns, highlighting a clear epistemic gap. Therefore, initial research suggests the need for proper education and awareness mechanisms to be propagated across all facets of society provide a stable foundation for other stakeholder requirements such as cooperation and information exchange to be more effectively addressed.

In response, attempts should be made to further unpack the underlying problems that prevent effective cooperation and information exchange between stakeholders, both in terms of international and public/private sector approaches. Consequently, this process should involve evaluating the effectiveness and sustainability of existing cooperation and information sharing approaches between different groups.

4.3 Legislative Systems and Data Protection

In section 2 of this volume the legal and ethical issues associated with conducting research in the field of CC and CT have been discussed at length. Specifically, issues of legislation and policy, have a direct and focused impact on these issues. Considering the integral role of the internet on the ways in which we communicate, conduct business and carry out many other facets of our lives, also means that it holds significant implications in terms of data protection, privacy and fundamental human rights. Furthermore, criminal law is an essential component in enabling the prosecution of criminals, and thus must be validated and reviewed to ensure its continued effectiveness against the continually evolving challenges, vectors, and vehicles of cybercrime and cyberterrorism.

Critics of the increasingly globalised and inter-connected Earth such as Bauman and Arendt, suggest that new technologies are increasingly distancing moral responsibility from action and consequence. Indeed, Bauman a survivor of the holocaust and condemner of its technological and bureaucratic formation would likely see his greatest fears manifest themselves in the sophistication, complexity, and emotional distance of botnets and other criminal networks [3]. Furthermore, the new hierarchies and emotional, moral and physical distance of internet technologies is likely to have also raised concerns with Arrendt, one of the most influential behavioural and socio-political scientists of the 20th century. Using Arrendt's theoretical considerations of power as; power to, and, violence as; power over: *"Cyberspace is thus both, a modern space of appearance and political freedom and an un-explored context for Arendt's conception of power as well as an anti-space of appearance, a space filled with Arendt's conception of violence that denies the positive attributes of a space when filtering and control techniques are implemented"* [18, p. xiii]. Today, large divides are clearly evident within Europe and across the world, between the legal rights for freedom, anonymity and freedom of speech, and, between trying to collect data, build profiles, and censor or prevent illegal, undemocratic, 'unconstitutional', and immoral activities. For example, the absence of public trust in intelligence agencies such as the NSA (National Security Agency) in the US and GCHQ (Government Communications Headquarters) in the UK has led to activist (and hacktivist) groups campaigning for internet anonymity as a fundamental human right. This process, for right or wrong, creates barriers to efforts carried out by law enforcement and national-security agents to prevent other serious human rights grievances, such as human trafficking and child abuse.

Due to the ways in which personal data is handled evaluation is needed towards the effectiveness and applicability of existing criminal law, and the way in which it is interpreted through enabling criminals to be prosecuted appropriately. Furthermore, a review of current judicial systems and personnel capabilities is needed to increase efficiency in preventing and prosecuting CC, especially regarding abilities to comprehend technical and complex cyber issues. Particular attention should be paid as to what mechanisms would be both effective and realistic to implement as a centralised pan-European framework, to transcend borders and improve cooperation and consistency.

4.4 Awareness, Education and Training

Inadequate knowledge has proven to be an underlying issue contributing towards many, if not all, of the concerns highlighted across vertical and horizontal stakeholder groups. Building stakeholder requirements such as legal and policy frameworks, as well as investigative and resilience policies on top of an ill-formed foundation of epistemic knowledge, society runs the risk of entering into an age of *'organised irresponsibility'* [4].

Part of the misunderstanding comes not only from the identified research gaps, but from the removal of the perception of consequence in cyberspace. Indeed, as [15] would point out; risk awareness is closely linked to human agency, trust and understanding. This lack of understanding may be related to the widespread inadequacies of specialised and generic training, as well as a historical lack of education in computer and information technologies, therefore leaving most to forfeit their risk awareness responsibilities to subject matter experts.

Despite the ubiquity at which network devices and digital services have been adopted into society, general levels of awareness regarding best practices and security hygiene are relatively low, despite widespread attempts to encourage and disseminate such principles. Indeed, traditional perceptions paint cybersecurity as an 'IT department' concern within organisations. Such attitudes, proliferate the human element remains as an expanded vulnerability in our pursuit of increased societal resilience. These factors define scope for the identification and development of mechanisms to improve this awareness through focusing on the eliciting the training requirements of target audiences and providing mechanisms that reach out and meet them.

Research aimed towards addressing requirements to improve awareness and education should seek to achieve greater identification of target audiences, from citizens with low levels of technological capability, to organisational executives to IT infrastructure providers in order provide tailored relevant materials and mechanisms to increase proficiency. Greater emphasis needs to be placed on assessing and evaluating the impact and efficiency of existing initiatives at raising awareness and training so that best and effective practices can be better understood and more widely applied.

4.5 Technological Evolution

The continued development and evolution of technology and the ways in which it is utilised breeds an unstable and uncertain environment, in which the threats, vulnerabilities and dangers are consistently evolving. However, when we consider the full spectrum of issues associated with CC, and the emerging threat of CT, the impact and importance of these changes extends across all aspects commonly associated with cyber-resilience. In some instances the same technologies being used to enable or further research bring with them new opportunities and capabilities that can enhance efforts to increase security, or aid investigators to better respond to crime. However, technological change is not only a matter of assessing the impact of new technology, but also an appreciation of the widening gap between old and new, with legacy systems potentially presenting their own security risks. The demand for ever progressive technology however, may lead to further insecurities and vulnerabilities, as the quality control of systems, software and service providers aiming to secure and safeguard these developments do not necessarily increase proportionately.

The current inability to predict the threats from evolving cyber-technology has created what [22] call, the 'Risk Paradox'. The paradox states that the vulnerabilities created from emerging technologies, for developing countries, outweigh the benefits of joining the technological advancement. Microsoft's report considers this by assessing the challenges of existing threats to the defensive and economic capabilities of developing nation-states. For many developing states, the risks of cyber progression are too high, yet paradoxically all states are forced into modernising as a result of competition from globalisation [7, p. 8]. This concern is even more severe considering that the Microsoft report only takes into account the threats of established cyber-challenges and not horizon dangers, and undiscovered vulnerabilities. The acceleration of information technologies has, and will continue to, produce 'known unknown' threats [23]. Such threats are perhaps best demonstrated by the propagation and potential exploitation of zero day vulnerabilities, such as the disclosure of the 'Heartbleed' vulnerability in 2014 which identified an issue in the OpenSSL cryptographic library used by many popular websites [25].

Unless Europe is able to overcome the challenges as a result of the continued pace of technological evolution it may find, like developing countries that continued online integration will create an increasingly fertile risk landscape. It is essential not just to meet the known challenges, but to create pre-emptive capabilities for the emergence of known-unknown threats. Additionally, stakeholders must be vigilant of a state of constant insecurity, whereby vulnerabilities may exist, yet remain undiscovered.

Research surrounding the impact, potential utility, and, appreciation of new technology should be 'future facing'. Through utilising effective horizon scanning to identify and assess new risks, research communities must continue to drive the identification of new threats and novel security applications and the utility of many of the new and emerging technologies and products discussed in other chapters of this volume.

4.6 Organisational and Societal Resilience

The concept of resilience encompasses the full spectrum preparation, response, and capacity to recover from any given threat. Extending this, cyber-resilience has been defined as the *"capability to withstand negative impacts due to known, predictable, unknown, unpredictable, uncertain and unexpected threats originating from cyberspace"* [13, p. 4].

The established proliferation of connected devices and information systems across society has created an environment where the potential impact of successful cyber-attacks is now greater than ever. This trend will continue to grow as the lines between the physical and virtual worlds are increasingly blurred and the levels of integration and reliance upon communication networks and computer systems continue to rise. These trends carry implications for all citizens and organisations big and small, making the responsibility of ensuring our resilience to such occurrences an overall, collective one.

The rate of technological advancement creates and exposes vulnerabilities in existing technologies. Perhaps most notably, industrial systems are often built on legacy infrastructure and subsequently exposed to the internet via the demands of the modern supply chain, this series of interdependencies therefore offers great potential risks, as outdated and unsupported software and hardware may fall prey to contemporary cyberattacks. Holistic and transparent frameworks are essential for the successful development of stable resilience programs that seek to encompass all aspects of cybersecurity and risk mitigation, throughout the different horizontal and vertical stakeholders [24, p. 35]. In today's ever-interconnected world it is essential to increase the notions of responsibility and awareness for all stakeholders individually, thus reinforcing overarching societal and organisational collective resilience responsibilities.

Research addressing these challenges should seek to identify and address the abilities of society as a whole, and that the actors within it take responsibility for ensuring the sufficient practices, managements, technologies, and, awareness is provided to ta CC and CT. Specifically, research should address the development and implementation of cyber focused resilience strategies for critical infrastructure, taking a holistic approach that appreciates the full spectrum of risk across the broader supply of stakeholder chains.

5 Concluding Remarks

In this chapter we have discussed a number of broad priority areas that, in the eyes of a diverse range of domain experts and stakeholders could, and should, be assisted through ongoing and future research. The research priorities have been discussed under six thematic headings; Enhancing the capability of investigators, cooperation and information exchange methods, organisational and societal resilience, legislative systems and data protection, awareness education and training, and technological evolution. Later chapters in this volume will further refine issues identified in this chapter towards road-mapping and defining the specific actions. The primary focus was found to be across six dimensions; (1) enhancing the capability of investigators, (2) cooperation and information exchange, (3) legislative systems and data protection, (4) awareness, education and training, (5) technological evolution, (6) organisational and societal resilience.

Of the six interlinked areas identified, we can clearly draw a nexus between the rate of technological change and the (lack of) awareness, education and training as the fundamental, underlying issues that infiltrate all six areas of concern. Furthermore, these two of the leading problems are intrinsically linked to one another, as technology rapidly evolves it creates great gaps in knowledge and understanding, these gaps are dangerous as within them exist vulnerabilities to be exploited, or, lead to misinformation that then produces ineffective frameworks that inform practices across sectors. By considering that all stakeholder concerns are routed from; a lack of Awareness, Education and Training, and, the rate of Technological Evolution, we are able to consider not just existing threats, but known areas of specific concern that require attention. Furthermore, it allows all end-users to consider the threat of unknown vulnerabilities, thus promoting the need to carefully investigate all existing systems, frameworks, legislature and policies.

Acknowledgements. The research leading to these results has received funding from the European Union Seventh Framework Programme (FP7-SEC-2013) as the COURAGE project under grant agreement no 607949.

References

1. Akhgar, B., Staniforth, A.: Tackling Cyber Crime and Cyber Terrorism Through a Methodological Approach. Freedom From Fear 2015 (2015). http://f3magazine.unicri.it/?p=621. Accessed 25 Apr 2016
2. Aspray, W.: Chasing Moore's Law. SciTech Pub., Raleigh (2014). http://univebooks.com/file/C/Chasing-Moores-Law.pdf. Accessed 25 Apr 2016
3. Bauman, Z.: Modernity and the Holocaust. Polity Printing Press, Cambridge (1991)
4. Beck, U.: Risk Society: Towards a New Modernity. Sage Publications, Munich (1992)
5. Boehm, B.W.: Software Engineering Economics (1984). https://extras.springer.com/2002/978-3-642-59413-7/4/rom/pdf/Boehm_hist.pdf. Accessed 25 Apr 2016
6. Brenner, S.: Cyberthreats: The Emerging Fault Lines of the Nation State. Oxford University Press, New York (2009)
7. Castells, M.: Information Technology, Globalization and Social Development. United Nations research Institute for Social Development (1999). http://www.unrisd.org/80256B3C005BCCF9/(httpAuxPages)/F270E0C066F3DE7780256B67005B728C?OpenDocument. Accessed 25 Apr 2016
8. Corbin, J., Strauss, A.: Grounded theory research: procesures, canons and evaluative criteria. Zeitschrift fuer Soziologie **19**(6), 418–427 (1990). http://link.springer.com/article/10.1007/BF00988593 Accessed 25 Apr 2016
9. Council of Europe, Convention on CC. Counceil De L'Europe, Budapest (2001). http://conventions.coe.int/Treaty/EN/Treaties/Html/185.htm. Accessed 25 Apr 2016
10. Denning, D.E.: "Cyberterrorism," Testimony Before the Special Oversight Panel on Terrorism, Committee on Armed Services, U.S. House of Representatives (2000). http://faculty.nps.edu/dedennin/publications/Testimony-Cyberterrorism2000.htm. Accessed 25 Apr 2016
11. ENISA, FAQ Cyber Europe 2010 Final (2015). www.enisa.europa.eu/media/news-items/faqs-cyber-europe-2010-final/view?searchterm=cyber%20europe. Accessed 25 Apr 2016
12. ENISA, CERT Cooperation and its further facilitation by relevant stakeholders (2015). https://www.enisa.europa.eu/activities/cert/background/coop/files/cert-cooperation-and-its-further-facilitation-by-relevant-stakeholders. Accessed 25 Apr 2016
13. EU Executive Report, Smart Cities (2015). https://eu-smartcities.eu/sites/all/files/blog/files/Transformational%20Smart%20Cities%20-%20Symantec%20Executive%20Report.pdf. Accessed 25 Apr 2016
14. Glaser, B.G.: Theoretical Sensitivity: Advances in the Methology of Grounded Theory (1978)
15. Giddens, A.: The Consequences of Modernity (1990). http://ewclass.lecture.ub.ac.id/files/2015/02/Giddens_-_Consequences_of_Modernity_17388b4f6c76919ffe7817f7751c61fa.pdf. Accessed 25 Apr 2016
16. Guinchard, A.: Between Hype and Understatement: Reassessing Cyber Risks as a Security Strategy (2011). http://scholarcommons.usf.edu/cgi/viewcontent.cgi?article=1107&context=jss. Accessed 25 Apr 2016
17. Koops, B.J.: The Internet and its Opportunities for CC (2010). http://papers.ssrn.com/sol3/papers.cfm?abstract_id=1738223. Accessed 25 Apr 2016

18. Kremer, J.-F., Müller, B.: Cyberspace and International Relations: Theory, Prospects and Challenges. Springer, Heidelberg (2014)
19. Landeta, J.: Current validity of the Delphi method in social sciences. Technological Forecasting and Social Change **73**, 467–482 (2005). http://www.sciencedirect.com/science/article/pii/S0040162505001381. Accessed 25 Apr 2016
20. Linstone, H., Turoff, M.: The Delphi method: Techniques and applications. Addison-Wesley (1975). https://www.ncjrs.gov/App/Publications/abstract.aspx?ID=256068. Accessed 25 Apr 2016
21. Medland, D.: U.K. Study reveals Serious Underreporting of Cyber Atacks by Business. Forbers (2016). http://www.forbes.com/sites/dinamedland/2016/03/02/u-k-study-reveals-serious-under-reporting-of-cyber-attacks-by-business/#79392e1c392e. Accessed 25 Apr 2016
22. Microsoft, The Cybersecurity Risk Paradox: Impact of Social, Economic, and Technological Factors on Rates of Malware. Microsoft Security Intelligence Report, Special Edition (2014). http://download.microsoft.com/download/E/1/8/E18A8FBB-7BA6-48BD-97D2-9CD32A71B434/Cybersecurity-Risk-Paradox.pdf. Accessed 25 Apr 2016
23. Rumsfield, D.: Known and Unknown: A Memoir. Penguin Press, London (2011)
24. Salazar Ortuno, A.: Collective Intelligence. Crisis Response 10(3) (2015)
25. Schneier, B.: Heartbleed. Schneier on Security (2014). https://www.schneier.com/blog/archives/2014/04/heartbleed.html. Accessed 25 Apr 2016

A (Cyber)ROAD to the Future: A Methodology for Building Cybersecurity Research Roadmaps

Davide Ariu[1(✉)], Luca Didaci[1], Giorgio Fumera[1], Giorgio Giacinto[1], Fabio Roli[1], Enrico Frumento[2], and Federica Freschi[2]

[1] Department of Electrical and Electronics Engineering, University of Cagliari, Piazza d'Armi, 09123 Cagliari, Italy
{davide.ariu,didaci,fumera,giacinto,roli}@diee.unica.it
[2] CEFRIEL - ICT Institute Politecnico di Milano, Via Fucini 2, 20133 Milano, Italy
{enrico.frumento,federica.freschi}@cefriel.com
http://pralab.diee.unica.it

Abstract. We describe the roadmapping method developed in the context of the CyberROAD EU FP7 project, the aim of which was to develop a research roadmap for cybercrime and cyberterrorism. To achieve this aim we build on state-of-the-art methodologies and guidelines, as well as related projects, and adapt them to the specific characteristics of cybercrime and cyberterrorism. The distinctive feature is that cybercrime and cyberterrorism *co-evolve* with their contextual environment (i.e., technology, society, politics and economy). This poses specific challenges to a roadmapping effort. Our approach could become a best practice in the field of cybersecurity, and could also be generalised to phenomena that exhibit a similar, strong co-evolution with their contextual environment. In this chapter, we define our route to developing the CyberROAD research roadmap and contextualise it with an example of Enterprise 2.0.

1 Introduction

CyberROAD[1] is a project funded by the European Commission under the 7th Framework Program. It aims to develop a research roadmap for cybercrime (CC) and cyberterrorism (CT) by providing a categorisation of CC and CT, identifying the major challenges, gaps and needs, and finally proposing desirable solutions and methods to evaluate them in practice. Such points are addressed by providing a thorough and comprehensive analysis which will encompass the technological, social, economical, political, and legal aspects of CC and CT. The project spans 24 months (June 1st, 2014–May 31st, 2016) and is implemented by a consortium of 20 members from 10 different EU countries representing all the key players involved in the fight against CC and CT (defence, law enforcement agencies, research, academia, private and public companies).

The task of CyberROAD is known as "science and technology roadmapping" (S&TRM). S&TRM has been adopted since mid-1980s by corporations

[1] http://www.cyberroad-project.eu/.

B. Akhgar and B. Brewster (eds.), *Combatting Cybercrime and Cyberterrorism*, Advanced Sciences and Technologies for Security Applications,
DOI 10.1007/978-3-319-38930-1_4

and industries as a tool for strategic planning of S&T resources toward a well-defined goal, which usually consists of supporting the development of new products or technologies, with a focus ranging from a single product to an entire technological sector. Since mid-1990s, S&TRM has been increasingly exploited also by research institutions and think-tanks for providing intelligence to policy-makers, with the aim of optimising public R&D investments and ensuring their relevance to society [5,12]. The CyberROAD roadmap belongs to the category of policy-oriented roadmaps.

It is commonly acknowledged that a successful S&TRM project requires a principled methodology [5,11–13]. So far, several roadmapping methods have been proposed in the literature; several guidelines are also available from public and private organisations that promoted roadmapping efforts in fields as different as industry and government, as well as many useful case studies. This means that a novel roadmapping effort can exploit and build on a considerable body of knowledge, possibly adapting existing methods to the characteristics and needs of the specific project. Accordingly, we started from a thorough analysis of S&TRM literature, focusing on policy-oriented roadmapping, and analysed recent S&TRM projects in the cybersecurity and related fields. We then developed a method that takes into account the specific application field of our project (the fight against CC and CT), as well as its contextual environment which encompasses societal, political and economic issues beside technological ones.

After a survey of the relevant literature on S&TRM and of related projects in Sect. 2, in Sect. 3 we describe the specific roadmapping method we developed for CyberROAD. We then give an example of its application in Sect. 4, taken from the outcomes of CyberROAD. We finally discuss the proposed method in Sect. 5; in particular, we argue that it can be exploited not only in a cybersecurity context, but also in other S&TRM projects that, analogously to CyberROAD, involve different fields, and thus require the integration of different domain expertise.

2 State of the Art on S&T Roadmapping

S&T roadmaps can be broadly categorised as either *normative* (*goal-oriented*) or *exploratory* [5,12]; although, hybrid roadmaps also exist [1]. The choice between these two kinds of roadmaps is among the first ones to be made in a roadmapping project, based on its context, goal and target audience. Normative roadmaps are commonly used by corporations and industries. They define the paths to attain a well-defined, desired future state from the present one on a relatively short time horizon (usually, 6 months up to 5 years). The desired state is defined in detail by high-level decision makers such as end users or policymakers. Exploratory roadmaps aim instead at enhancing future outlook or foresight of the evolution of an industrial, technological or social landscape over a usually longer time span (up to 20 years). These roadmaps take into account various alternative futures including rupture scenarios and major technological breakthroughs. Accordingly, *scenario building* (see below) plays a key role in this kind of roadmap, as well as the investigation of non-technical fields of influence [7].

In particular, exploratory roadmaps are believed to be a useful tool for providing intelligence for policymakers in areas where science and technology play a prominent role, e.g., to highlight emerging S&T issues and to anticipate long-term needs. Policy-oriented roadmaps, the category in which CyberROAD roadmap belongs, are considered to be still emerging [9]. They exhibit several distinctive features from corporate/industry-oriented roadmaps: (i) Their scope and goals are wider and less well defined; e.g., they can address far-reaching societal challenges. (ii) They usually involve also social, cultural, political, legal and economical dimensions, and cover a longer time span. (iii) Their target audience is made up of "generalists" rather than "experts". (iv) They are built by *multiple* organisations, and are aimed at an an *external* target audience (usually government, and often different organisations/departments). (v) Their main goal is *political persuasion* about actions to be implemented toward some objective.

Another crucial issue is the definition of a principled roadmapping method. To this aim, different resources are currently available. So far, several roadmapping methods have been proposed in the academic literature, as well as guidelines for successfully constructing and *implementing* roadmaps, in many different contexts such as company, industry and government [11,13]; in particular, policy-oriented roadmaps have been analysed in [5,9,12]. Guidelines and best practices have also been defined by private and public organisations. A relevant example in the policymaking context is the roadmapping process developed by the International Energy Agency for the energy technology sector [8]. From our analysis of S&TRM, focused on policy-oriented roadmapping, the following five key issues emerged (see also Fig. 1).

(1) Identifying the target audience. Since policy-oriented roadmaps are not aimed at the same organisation than produces them, a wide set of target stakeholders from different domains has to be effectively and evenly considered.

(2) Data sources. The main data sources are the scientific literature in the field of interest, the stakeholders, and the domain experts. Their careful selection is critical due to the wide scope of policy-oriented roadmaps and the involvement of a number of stakeholders and domain experts from different fields, including non-technological/scientific ones. Effective and efficient information/knowledge elicitation techniques must also be defined.

(3) Roadmap representation and visualisation. Policy-oriented roadmaps are targeted to the generalist view of policymakers. A clear, focused synthesis and presentation of their core issues is thus crucial. This can be attained by suitable graphical representations which allow decision-makers to focus on the most relevant elements and relations in complex systems involving scientific, technological, economic, political and social dimensions, rather than on low-level details.

(4) Roadmap validation and quality assessment. Early actions must be carried out to achieve this aim, starting from the roadmapping planning stage. It is widely acknowledged that evaluating the quality of a roadmap during its construction is not sufficient: clear criteria and metrics have to be defined to evaluate a roadmap during its *implementation*. For instance, the following issue related to guaranteeing the roadmap reliability and replicability is particularly relevant in the context of CyberRoad: "To what degree would a roadmap be replicated if a completely different development team were involved in its construction?"

(5) Roadmap construction technique. Last but not least, a sound method for developing the roadmap should be applied. As mentioned above, several methods have been proposed so far, due to the widespread usage of S&TRM. Therefore, no unique paradigm or standard for roadmap construction exists, neither a single definition of S&TRM, even in the specific case of policy-oriented ones. Nevertheless, as argued in [5], defining a unique, general roadmapping method is not a practical nor a desirable goal: instead, "the approach should be based on a light and modular process using a 'methodological toolbox' with different modules depending on the roadmapping areas, issues, context and objectives." This is witnessed by recent, policy-oriented S&TRM projects in fields related to CyberROAD, such as:

- Time2Learn[2], Sept. 2002 – Nov. 2003, FP5
- eGovRTD2020: Roadmapping eGovernment Research – Visions and Measures towards Innovative Governments in 2020, January 2006 – May 2007, FP7 [4]
- iCOPER[3]: Interoperable Content for Performance in a Competency-driven Society, 2008–2011, *e*Content*plus*
- EHR4CR[4]: Electronic Health Record for Clinical Research, 2011–2015, partially funded by Innovative Medicines Initiative (IMI)

Their roadmapping methods are similar at a high level but their implementation has been devised ad hoc according to the specific characteristics and goals of the project.

In the rest of this section we focus on key issue five: which is the subject of this paper. In our survey we identified some specific, potentially useful roadmapping approaches as a starting point to head towards a definition of a method suitable to CyberROAD. In particular, two interesting examples of normative and exploratory roadmap construction methods are explained in [10] and [7], respectively. The normative approach of [10] is tailored to the implementation of government policies which define the high-level, future vision for a given public service. As a case study, the Royal Australian Navy's fleet plan along the time horizon of 2010–2030 was considered, based on the 2009 Australian Government's Defence White Paper. In this kind of application, high-level objectives exists, defined by policymakers, and the setting is mainly static and under their control. The goal of roadmapping is to prescribe actions to reach such objectives. The proposed roadmapping approach consists of the following steps (see also Fig. 2). (i) Defining the **Context**, i.e., the trends and drivers that govern the overall, high-level goals of the roadmapping activity. For example, in the case study mentioned above, they include the defence policy, the strategic interests,

[2] http://www.cordis.europa.eu/project/rcn/64013_en.html.

[3] http://nm.wu.ac.at/nm/icoper.

[4] http://www.ehr4cr.eu.

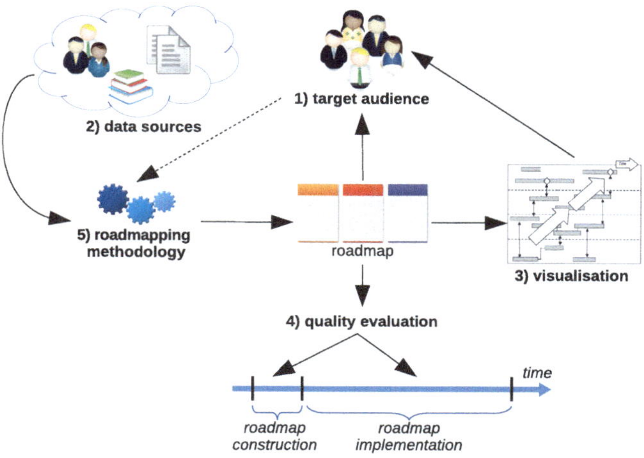

Fig. 1. Five key issues to be addressed to guarantee a successful roadmap.

and the military capabilities. (ii) Based on the Context, a **Backcasting** process is applied to define in detail the **Desired state** at the end of the roadmap time span; then, reasoning backwards in time up to a medium-term period, the actions to be carried to attain the Desired state must be defined. (iii) Since the **Current state** influences what can be attained in the future, it is necessary to define the **Path dependency**, i.e., the actions to be carried out from the current state to enable the ones identified through Backcasting.

In [7], an approach for scenario-based, exploratory TRM is proposed. It is based on the observation that technology is often influenced not only by *endogenous* factors, like market trends and standards, but also by *exogenous*, non-technical factors related to the evolution of society, economy and politics. As a consequence, technology does not follow an evolutionary path, making it very difficult to predict its development, and preventing the use of a normative roadmapping approach. In this context, exploratory roadmapping is useful as an instrument of technology forecasting, i.e., to understand how a technology may evolve, and forms the basis for subsequent planning activities. The approach of [7] consists of the following main steps (see Fig. 3, bottom): (i) identifying the exogenous and endogenous influencing factors of the technology under investigation (see, e.g., Fig. 3, top); (ii) projecting the possible evolution of the most relevant exogenous factors in one or more time steps during the roadmap time span (several alternative projections are usually possible); (iii) combining alternative projections into a few, consistent and alternative scenarios (even just two "extreme" scenarios); (iv) analysing how the influencing factors interact with each other, to identify the "driving factors" exhibiting the highest impact on the considered technology; (v) envisioning how the latter may evolve under each scenario; (vi) developing a roadmap for each scenario.

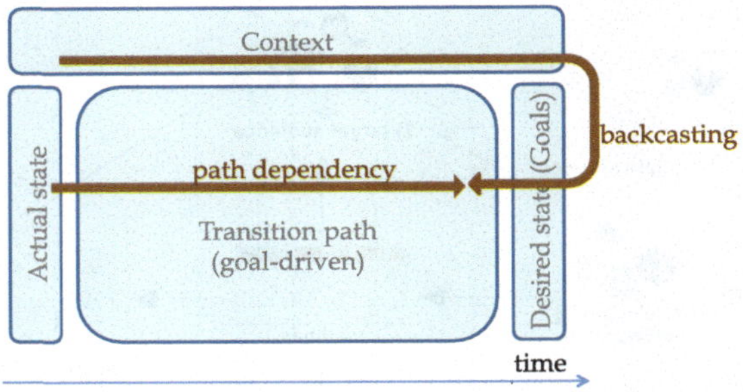

Fig. 2. Sketch of the normative roadmapping approach of [10].

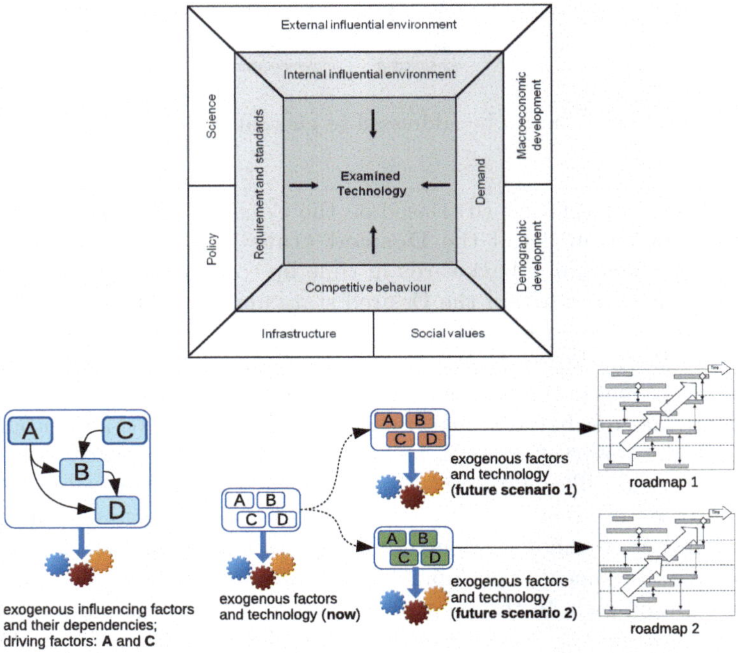

Fig. 3. Top (taken from [7]): high-level view of the exogenous and endogenous factors influencing a given technology. Bottom: sketch of the exploratory roadmapping approach of [7].

We finally discuss scenario building (aka scenario thinking or planning), which is a key component of exploratory roadmaps. It was introduced in a corporate R&D context in the 1950s [2], and is nowadays a strategic planning tool for supporting decision-making in complex and rapidly changing environments. It is widely used in business, industry and government. Its main purpose is to explore *different* potential evolutions of a given field (including non-technological issues) under the influence of some *driving forces*, to support proactive development and planning, and to cope with future challenges [16]. For instance, scenario building can enable the recognition of technological discontinuities or disruptive events. Consequently, they can include them into long-range planning and facilitate an organisation to be better prepared to handle new situations as they arise [15].

Broadly speaking, a *scenario* can be defined as a coherent and concise description of a possible future, often in a narrative form, in which the underlying driving forces are pointed out. In practice, a number of different scenario building methods have been proposed so far, and, as pointed out in [14], they still lack of a solid conceptual foundation, and are usually adapted by the users to suit their needs. This is exemplified by the ad hoc scenario building techniques used in the roadmapping projects mentioned above.

In three main methodological "schools" [2] are identified and analysed: Probabilistic Modified Trend (PMT), *La Prospective* (LP), and Intuitive Logics (IL). The PMT methodology mainly provides probabilistic forecasting tools, involving the analysis of historical data. The LP approach is more complex and mechanistic and heavily relies on computer-based mathematical models and simulations. Both the above approaches aim at producing the *most probable* scenarios. The IL methodology is more flexible, subjective and qualitative. This makes it suited to a wider range of scenario purposes, including CyberROAD. Another feature of such a methodology is that it produces a small number of scenarios which are considered to be *equally probable*. We point out that the scenario building approach used in [7] (see above) mainly follows the IL methodology. The main steps of the IL methodology are the following: (i) determining two main driving forces affecting the subject of scenario building characterised by the highest impact and the highest uncertainty in their evolution; (ii) defining two extreme but possible outcomes for both driving forces; (iii) developing a scenario for each of the four combinations of outcomes.

3 The Proposed Method

The choice of a suitable roadmapping method has been guided by the characteristics of the CC and CT phenomenon. Under this viewpoint, the main feature of CC and CT is that they *co-evolve* with their contextual environment, i.e., technology, society, politics and economy, beside being also driven by internal forces. This in sharp contrast with the *independent* evolution of crime and information security before 2000. In particular, the emergence of new technologies, as well as novel social habits and issues (like social networks and privacy issues) can generate new opportunities for CC and CT, enabling novel kinds of attacks. In turn, CC and CT are among the forces that influence the evolution of technology (in the broadest sense of the word) and society. At the same time, the evolution of CC and CT is also driven by *internal* forces, which recently mostly coincided with market trends and laws. A clear example can be seen in the evolution of marketing and consumer profiling techniques, and in the corresponding evolution of social engineering techniques, both based on the same methods; other similar examples can be found in linked open data, psychology and personality profiling, cyber sociology, modern sentiment analysis techniques, and anonymising techniques (see, e.g., [6][5]).

The above considerations imply that the evolution of CC and CT can not be understood by considering them as "black boxes" influenced only by their contextual environment; instead, their peculiar, internal driving forces must be taken into account as well, like the cyber-logic and cyber-economy (see, e.g., the Hacker Profiling Project). Accordingly, a project like CyberROAD requires a specific roadmapping method; in particular, it must be different from methods adopted in projects like those mentioned in Sect. 2, whose subjects (e.g., e-government and health services) are related to phenomena that mainly evolve under the influence of external driving forces, and do not exhibit any significant co-evolution behaviour.

3.1 Toward the CyberROAD Roadmapping Method

As pointed out in Sect. 2, the first choice about the development of a roadmapping method is between a normative and an exploratory approach. Given the characteristics of CC and CT discussed above, this choice is not straightforward. On the one hand, since the contextual environment, including long-term government policies, influence the evolution of CC and CT (e.g., enabling new attacks), a policy-oriented, normative approach can be used, like the one of [10]. This would allow one to apply a Backcasting process to define a Desired state, and the main actions required to attain it; for instance, one could exploit existing, high-level EU policy objectives (e.g., white papers), to derive more specific, technical goals, such as hypothesising specific policies against CC and CT at the end of the roadmap time span.

On the other hand, the peculiar co-evolution of CC and CT with its contextual environment makes it infeasible to predict their development with a degree of certainty as the one required by the Path dependency step of [10], even in the short term. Therefore, even if specific objectives can be defined in the Backcasting step, defining specific actions to reach them in the Path dependency is not possible under such a dynamic setting. This implies that a purely normative approach is infeasible for analysing the evolution of CC and CT from the Current state. Accordingly, a scenario-based, exploratory approach appears better suited to define the Path dependency. To this aim, the approach of [7] is appealing. In particular, we point out that this approach is based on analysing the evolution of the roadmapping subject as a function of two distinct kinds of influencing factors: the "exogenous" and "endogenous" ones; in the context of CyberROAD, such a distinction closely resembles the one between the external driving forces of CC and CT (e.g., their contextual environment) and the internal ones.

[5] Available at http://www.sicherheitsforschung-magdeburg.de/uploads/journal/MJS_033_Frumento_Assessment.pdf.

Based on the above rationale, we initially developed a hybrid normative-exploratory approach by combining the ones of [10] and [7]. This approach is sketched in Fig. 4. The Backcasting step starts from a Context to be derived from long-term, high-level EU policies. For instance, they can refer to strategic interests and assets (like critical infrastructures), and to future EU roles in the cybersecurity field. This should lead to hypothesising more specific goals (the Desired state), as explained above. Subsequently, starting from the Current state of CC and CT and of their contextual environment, their possible evolution has to be envisioned in the Path dependency step through a scenario-based, exploratory approach. In particular, several "vertical" roadmaps can be developed to investigate the evolution of different, specific environment/business scenarios of interest, like social networks and mobile workforces. In the end, the outcomes of these exploratory roadmaps will be compared with the desired state, which allows one to address questions like the following ones: What goals can be achieved, given the transition path? How to change the scenarios (technology, legislation, etc.) so that also the other goals can be achieved? What are the research priorities during the transition path?

The above hybrid solution is coherent with methods proposed in the roadmapping literature, as well as conceptually elegant. However, its normative component turned out to be infeasible in the specific CyberROAD context. The main reason is that CC and CT are worldwide phenomena. This implies that defining a normative "desired state", limited to EU, is nearly infeasible. Moreover, in a field like cybersecurity, a cooperation between research teams from very different fields (such as social sciences, economics, computer security, etc.), as well as government, law enforcement agencies and private companies, is necessary. We therefore chose to retain, and further develop, only the explorative part of the above approach. In particular, we let the final scenarios emerge in a bottom-up fashion from an aggregation of distinct, "vertical views" of the contextual environment of CC and CT; each view is autonomously developed by experts in the different domains involved, without reference to a desired state. This approach is described in the rest of this section.

3.2 Outline of the Proposed Method

The roadmapping method we finally developed builds on the one of [7] and, partly, on the method followed in the eGov2020 project [4]. It is based on scenario analysis, coherent with the chosen exploratory approach. In particular, in the CyberROAD context the final aim of scenario building consists of identifying the resulting CC and CT threats, and the corresponding desired defences. To this aim, the wide contextual environment of CC and CT has to be taken into account, i.e., the technological, social, economical, political, and legal aspects that can influence the evolution of CC and CT. Accordingly, we defined a *scenario* as a concise, internally consistent and coherent sketch of a possible future state of CC and CT and of their context. In particular, the state of CC and CT consists of the threats that may arise under a given scenario, and of the corresponding desired defences. The roadmap is then obtained after a *gap analysis*

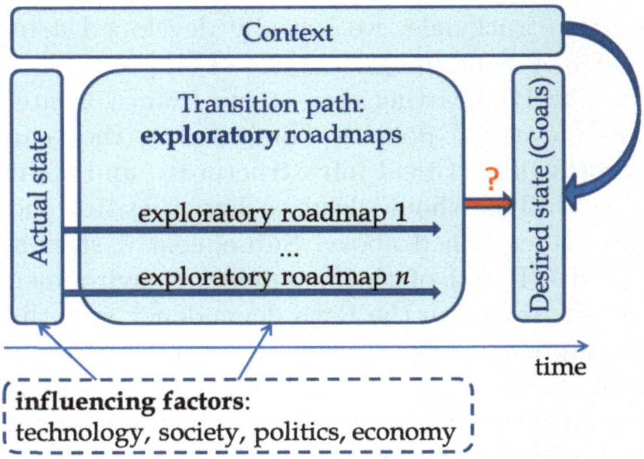

Fig. 4. Sketch of the preliminary roadmapping method developed for CyberROAD, as a hybrid normative-exploratory approach that combines the ones of [10] (see Fig. 2) and [7] (see Fig. 3).

step, which consists of identifying *research gaps* emerging from the comparison between the threats and the defences in the actual state, and the ones in each future scenario.

Accordingly, our roadmapping approach consists of the four main steps summarised in Fig. 5, and described in more detail in the following:

1. Representing the actual state as a scenario, to allow a direct comparison with future scenarios
2. Scenario building
3. Gap analysis
4. Roadmap construction

Actual State Scenario. The actual state has to be described as a scenario, using the template shown in Table 1. It consists of a short summary of the contextual environment, followed by the existing CC and CT threats and the available defences. In particular, each threat has to be characterised by the following information, in order to allow quantifying its *risk* in the subsequent roadmapping steps: the assets targeted by the threat, its likelihood, and its consequences.

Given the multidisciplinary nature of this subject, the actual state scenario can be subdivided into several coherent, vertical *views* of the contextual environment. Each view focuses on a specific, sectoral aspect, like payment systems, driver-less vehicles, mobile devices and services. This allows each view to be defined by *different* domain experts.

Finally, the *key driving factors* of each view must be identified, i.e., the ones that are expected to exert the highest influence on the evolution of future scenarios.

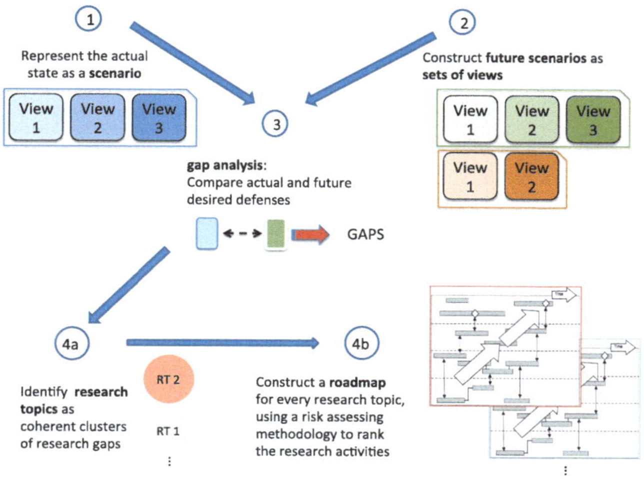

Fig. 5. Outline of the proposed rodmapping method.

Table 1. Scenario/view template

View title
Summary (one page)
Key driving factors (only for the actual state)
Threats: – description   – targeted assets   – threat likelihood   – consequences
Defences

Scenario Building. The goal of this step is to produce a set of possible future scenarios, which should explore a range of potential evolutions of CC and CT and of their contextual environment as wide as possible, highlighting the threats that can emerge, and the corresponding, desirable defences. For the same reason above, we chose a bottom-up scenario building approach, in which the final scenarios emerge by aggregating several vertical views of the contextual environment. This can be attained with three sub-steps (see Fig. 6):

1. Domain experts on each of the subjects that compose the contextual environment (society, politics, economy, and technology), build a set of *initial* views.
2. Coherent initial views are then combined to obtain a small set of broader, *final* views of the contextual environment, which must be *alternative* to each other

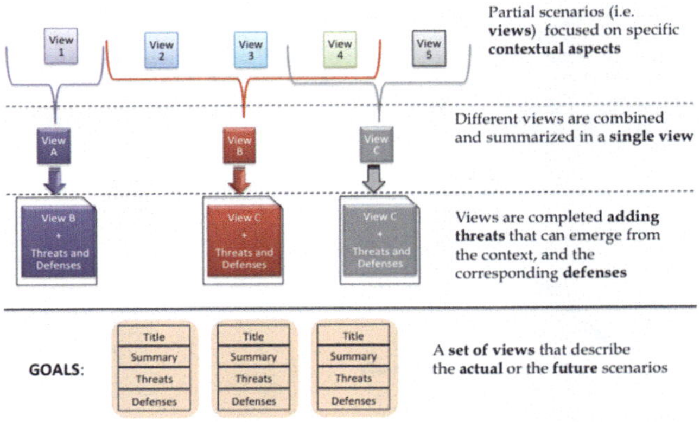

Fig. 6. Scenario building.

(i.e., contradictory). To this aim, the most relevant and interesting groups of initial views should be identified, using the following guidelines:
- a final view can be obtained by merging initial views that are coherent (non-contradictory), and contain elements which can interact, resulting in specific CC and/or CT threats;
- the same initial view can be included into more than one final view, provided that such final views are alternative to each other (i.e., they must contain also contradictory initial view, as explained above).

3. Each final view has to be completed by adding the specific aspects related to CC and CT, i.e., by envisioning the possible, corresponding threats and defining the desired defences.

Each final view must be described according to the same template used for the actual state scenario, excluding only the key driving factors (see Table 1).

Gap Analysis. The goal of this step is to identify the research gaps that emerge from the comparison of each of the future views with the actual state views (see Fig. 7). We define a research gap as a specific research issue that needs to be addressed in the context of a EU project to enable a desired defence against a specific threat. Research gaps have thus to be identified by tracking the changes of the threats from the actual to the future scenarios and comparing the corresponding existing and desired defences. In particular, a given threat in the actual state can increase, decrease, remain unchanged, or disappear in a future scenario. Novel threats can also appear in a future scenario. The outcome of gap analysis must be summarised in a table in which each row contains a single threat from a future scenario (either a known or a novel threat), the defence existing or pursued in the actual state (only for known threats), the desired defence in the future view, and the identified research gaps (see the example in Table 3).

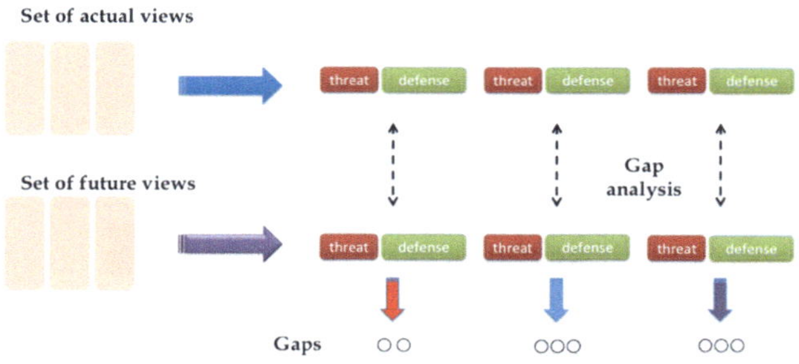

Fig. 7. Gap Analysis.

Roadmap Construction. The final roadmap, aimed at addressing the identified research gaps, is defined through the following sub-steps (see Fig. 8):

(a) Defining a set of broad *research topics*, as coherent clusters of related research gaps, that can be addressed by a suitable sequence of research actions, i.e., EU projects.
(b) The identified research topics are prioritised using a suitable risk assessment method defined as a part of CyberROAD, taking into account the relevance of the threats they address. In particular, this will be attained by evaluating the **risk** of each threat (using the information mentioned in Sect. 3.2), as well as the following **non-risk** (cost) factors that have to be defined for each research action: distance to the market (in terms of Technology Readiness Level[6]), cost of the action, estimated in terms of the number of projects the EU should fund for getting proper results, and availability of competences in Europe.
(c) A distinct, "vertical" roadmap is defined for each research topic. This is attained by identifying the specific research actions required to address the corresponding gaps, and then putting the actions into a clear time frame, taking into account their interdependencies.

4 Scenario Definition and Gap Analysis: An Example

The application of the above roadmapping method by the partners of the CyberROAD consortium has lead to the definition of ten scenarios (final views) of broad interest, each one related to a specific aspect of the technological, social, economic or political landscape that defines the context of CC and CT. Each scenario is made up of several vertical views that in turn focus on a specific subject inside the corresponding scenario. In total, twenty-four views have been

[6] See the definition of TRL proposed by the European Commission in the Horizon 2020 context: http://ec.europa.eu/research/participants/data/ref/h2020/wp/2014_2015/annexes/h2020-wp1415-annex-g-trl_en.pdf.

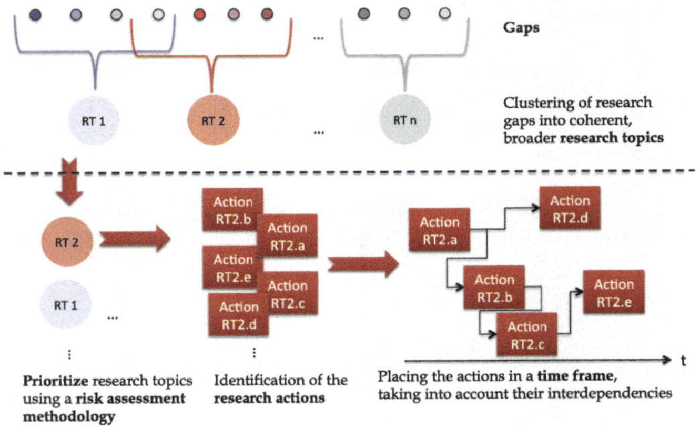

Fig. 8. Roadmap construction.

defined. For each vertical view, the current state has been depicted, and a possible future state has been envisaged. The list of the scenarios and of their vertical views is reported in Table 2. Note that some scenarios are made up of a single vertical view. Threats and defences related to CC and CT (both existing ones, and possible, future threats) have then been identified for each vertical view, and gap analysis has been carried out on them.

In the following we give an example of the outcome of the above process, focusing on the definition of the current state and of a possible future state of a single vertical view, and on the subsequent gap analysis. For our example we chose the "Enterprise 2.0" view of the "Workforce" scenario (note that this is a single-view scenario, as shown in Table 2). In the next sections we first report a description of this view. We then report the current and a possible future state of the "Enterprise 2.0" view that have been devised by the CyberROAD partners, including the key driving forces, threats and defences, for both CC and CT. We finally report a subset of the research gaps identified from the considered scenario.

4.1 View Description

The "Enterprise 2.0" view of the "Workforce" scenario is related to the evolution of the *workforces*, i.e., how people are accustomed to work. This is one of the aspects strongly influenced by the wide adoption of mobile technologies. The digital devices have strongly shaped the way people are working and collaborating. Figure 9 reports a simplified user-centric model of the modern way of working. This schema has four directions surrounding a worker, that impact his/her working habits: Dataspace, Enabling Technologies, Use Cases and Context. In general, we could define a worker as a person that owns (i.e., has legit rights to access edit and modify it) a dataspace (also called a personal

Table 2. The scenarios, and the corresponding views, that have been defined in the CyberROAD project.

Scenarios	Views
Social Sharing	Social Network
	Life Logging
	Wearable Devices
Building Automation	Smart Building
Utilities	Water Utilities
	Gas Utilities
	Smart Grids
Transportation	Rail Transport
	Aviation
	Maritime Transport
	Road Transport
	Freight
Healthcare	e-Health
	P4 medicine
Security and Safety	Cybercrime as a Service
	Attribution of Cybercrime
	Trusted Components
Workforce	Enterprise 2.0
Industry	Industry 4.0
	Just in Time Production
Financial Services	Cryptocurrencies
	Online Banking
Data Driven Economy	Big Data
	Control Over Data

information space[7]) where all their data are stored. What they do is to extend, elaborate and create new elements in this dataspace, even with the collaboration of other workers (shared dataspaces) or objects (internet of things). Synthesising, for the sake of clarity, the everyday working activity is as a continuous process of updates to the personal dataspace. A simple definition of a working dataspace, useful for understanding, is a virtual place where to store and access the data, that could either be strictly personal, shared or both. Nowadays trends are moving toward a complete dematerialisation of the personal dataspace on centralized

[7] VV.AA., "The Future of Identity — Personal information space – The future of identities in a networked world," 1st ed., Giesecke & Devrient, 2013, http://mcaf.ee/l209yu.

cloud services (no more disjoint data islands) and toward an intersection of the personal and working dataspaces.[8]

To access the dataspace, a worker can use several enabling technologies with different usability characteristics. Choosing any of these technologies is in general just a matter of usability and easiness for the worker. By "easiness" we define how easy is to perform a task, or a use case, in a specific place (context) with an enabling technology. Presently, the market is constantly offering new "methods" to access a user's own dataspace: Google Glasses are just the newest one, and others are following behind, for example the expected revolution of the wearable electronics.[9,10,11]

Summing up, so far a user accesses his own dataspace utilising an enabling technology, selected among several based on usability and personal preferences (that are indeed also an usability issue).

With reference to Fig. 9, a Use Case is the "invariant" portion of this scenario that the technologies and social trends do not affect. For example, along the years a user could have written a commercial letter in different ways: using a typewriting machine, a video terminal with a word processor, more recently a tablet, and in a future a wearable smart glasses that understands speech or thought.[12] What always remains the same, is the way of writing a commercial letter.

Thanks to mobile and ubiquitous terminals, a user can complete a task from any location, home, public spaces or company office. It does not matter where they perform the work: only ergonomics matter (for example doing tasks with a laptop does not have the same ergonomy if traveling on public transportation, such as a train, compared to working at a desk). Therefore, sensing the Context of a user is of enormous importance in order to adapt the enabling technologies' usability.[13] The Context is also important to help define which data from the personal dataspace a user can access in a specific place: to protect their identity, privacy or to respect some security policies. For example, consider a situation where a user wants to access a secured document, from a crowded place, over a data network: the system might prevent the access since in a crowded place someone else might spy over their shoulder while they type the access password.

[8] "Gartner: 10 critical IT trends for the next five years," http://www.networkworld.com/news/2012/102212-gartner-trends-263594.html.

[9] Canina, M. and Bellavitis, A.D. (2010) "IndossaMe: il design e le tecnologie indossabili." Milano: FrancoAngeli (in Italian).

[10] Talk to my shirt blog, http://www.talk2myshirt.com/blog/.

[11] Crunchwear, http://www.crunchwear.com/.

[12] "Control Your Mobile Phone or Tablet Directly from Your Brain," NextNature.net, http://www.nextnature.net/2013/05/control-your-tablet-directly-from-your-brain/.

[13] "Context-Aware Computing: Context-Awareness, Context-Aware User Interfaces, and Implicit Interaction," http://www.interaction-design.org/encyclopedia/context-aware_computing.html.

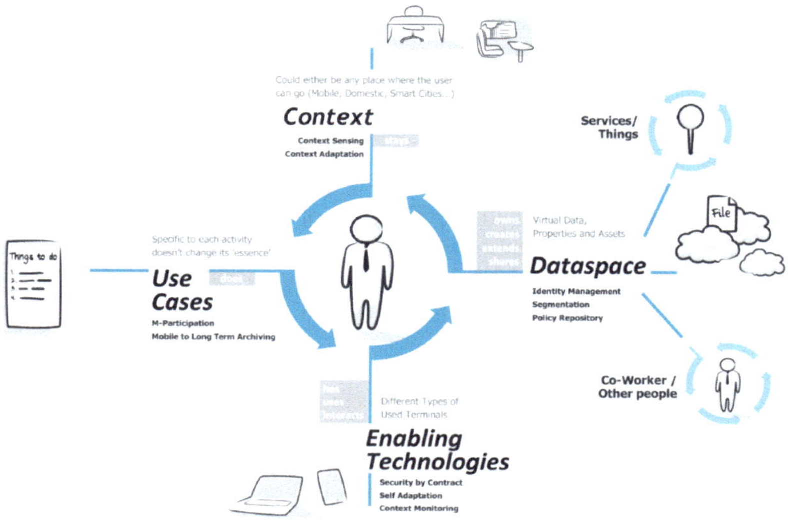

Fig. 9. Schematization of modern mobile work forces (source: CEFRIEL).

4.2 Actual State View

Title: Enterprise 2.0

Summary: The recent global recession directly influences labour market adding new paradigms, more flexibility and more mobility. Thanks to mobile and ubiquitous terminals, a user could complete a task in any possible place, home, public spaces or company office. Today we assist to a blending between private and professional lives due to the flexibility to work at any time from different locations and, as a consequence, physical and virtual encounters seamlessly merge. The widespread distribution of social platforms is another key element within the current workforces scenario. The market is constantly offering new "methods" to access a user's own dataspace, like for example the expected revolution of the wearable electronic and IoT. Nowadays, more and more services gather personal data (different services collect different data). In order to verify users' identity (and decide whether to grant access) machines collect personal data from users who want to have access to services. Users want to use those services and are therefore willing to give away personal data, following a data-for-(free)-services logic. At the same time, our identity, trust and privacy constraints are different in the contexts of different environments we live in (business identity, cultural identity, administrative identity, etc.). From a technological point of view, we are faced to the presence of a digital ecosystem: a community of people who interact, exchange information, combine, evolve in terms of knowledge, skills and contacts, in order to improve their lives and meet their needs. New dataspace services are available moving toward a complete dematerialization of the personal dataspace on centralized cloud services. Among cloud services is emerging the concept of "federated" cloud where there are common standards for both hardware and software companies. An important issue emerging from this scenario is the change in trust chains. They are growing in number and are influenced by logical and physical contexts.

Possible key driving factors:

- **Mobile devices:** widespread distribution of mobile and wearable devices. Thanks to mobile and ubiquitous terminals, a user could complete a task in any possible place, home, public spaces or company premises.
- **Blending life:** a world where physical and virtual encounters seamlessly merge.
- **Social platforms:** widespread distribution of social networking platforms.
- **Ubiquitous workforces:** user wants to complete a task in any possible place, home, public spaces or company office.
- **Usability:** to access the dataspace a worker can use several tools with different usability characteristics in order to accomplish easiness of use purpose.
- **New data space:** moving toward a complete dematerialisation of the personal dataspace on centralized cloud services.
- **Payment system:** diffused online payment systems in every environment.
- **Communication service provider:** large and long bandwidth.
- **Sensing the context:** sensing the Context of a user is of enormous importance in order to adapt the enabling technologies' usability. The Context is also important to help defining which data of the personal dataspace a user can access, in a specific place.

Threats (CC). Nowadays the traditional concept of a corporate trust zone is not valid any more. While in the past it was relatively easy to separate between "personal" and "corporate" information space, nowadays there is an overlap of these two spheres. The continuous evolution of tools and services, indeed, enabled access to corporate information systems from almost everywhere (and no more limited to the internal perimeter), through different devices that probably are not owned by the company itself. This is a problem from information security point of view, mainly because the risk mitigation processes and techniques may not be so effective outside the company perimeter. This context is going to expand again, enabled by a multitude of new technologies such as Internet of Things and wearable devices. This disappearance of the Trust Zones introduces some weaknesses, exposing enterprises to a new series of threats. Perimeter break-ins are diminished, because, from an attacker perspective, it is enough to obtain access to one of the devices or services outside the perimeter, which might be successfully targeted. In modern Advanced Persistent Threat schema, it is enough to establish remote access to corporate network through any of the connected devices, in order to allow exfiltration of critical information. The essence of CC is to abuse the trust chains to steal assets. Hence, changes in trust models and importance of assets implies changes in cybercrime. In general, we assist to the enhanced importance of the human element in the enterprise processes. The enterprise offers an increased and heterogeneous surface of attack (e.g., on social networks via "bring-your-own-device", BYOD) within a context characterized by legislation inconsistencies, which make it difficult to regulate employee behaviours. CC is increasingly adopting models taken from marketing, conforming itself as "crime as a service". Cyberespionage goes corporate: the dark market for malware code and hacking services could train cyberespionage malware used in public sector and corporate attacks to be used for financial intelligence-gathering.

Threats (cyberterrorism). As today, most of the enterprise offering services relies on cloud computing resources to store their critical data. Moreover, today, financial incentives for companies to invest in greater information security are low. It is easily guessed how this phenomenon of data management "delegation" could constitute an inherent vulnerability in cybersecurity. If we consider cyberwar scenario, we can imagine how cyberterrorists, for ideological or religious reasons, could decide to cause an economic crisis by breaking down the production system of a country. As mentioned above, the target to attack could no longer be the single enterprise, but compromising cloud services (not offering sufficient security countermeasures) on which many big companies rely on.[14]

Defences (CC and CT).

- Legal and Law Enforcement issues:
  - Privacy and data legislation is important to help defining which data of the personal dataspace a user can access, in a specific place to protect his identity, privacy or to respect some security policies.[15]
  - Relevance of the Cybersecurity insurance and connection with the active defence systems.[16]
- Technological issues:
  - New authentication methods (no password, behavioural, fuzzy security, etc.).
  - New Counterattack and prevention technologies (see for example the discussion at "Will DPM 5GL Save Cybersecurity?"[17]).
- Inclusion of human elements inside an holistic strategy of protection.
- The spread of IoT and wearable have reached a significant level of market penetration. It occurs as well the necessity to have user safety guidance and precise industry best practice in order to accomplish information security needs appropriate for the devices.

---

[14] Sullivan, D. (2015) "How is cloud penetration testing different for AWS, Google, azure?" Available at: http://searchcloudcomputing.techtarget.com/answer/How-is-cloud-penetration-testing-different-for-AWS-Google-Azure (accessed: 13 January 2016).

[15] https://www.ftc.gov/system/files/documents/reports/federal-trade-commission-staff-report-november-2013-workshop-entitled-internet-things-privacy/150127iotrpt.pdf (accessed 18-02-2016).

[16] http://www.govtech.com/dc/articles/Will-DPM-5GL-save-cybersecurity.html (accessed 18-02-2016).

[17] http://www.govtech.com/dc/articles/Will-DPM-5GL-save-cybersecurity.html (accessed 18-02-2016).

4.3 Future State View

Title: Enterprise 2.0

Summary: The future scenario approximately located in 10 years is characterized by the radicalization of "Blending life" concept, the "Immersed Human". Humans are constantly surrounded by a technological environment in every aspect of their life. Persistent interference by the service providers in providing suggestions (covering every sphere of life) in line with the person profile. The trend for social platforms is oriented to more decentralized networks. It seems there is no more need to be member of the same social network to share the information with one's own friends because event streams are transferred between social networks. Some networks even take money for their event stream, e.g., because they host all the stars and celebrities.

The integration of service largely uses a peer-to-peer decentralized approach; it is in general possible to have isolated service providers and isolated peers. Their business is to be disconnected from others for several reasons (privacy, independency, or hiding themselves from the others).

People are less dependent on one service provider; interoperability is forcing services and platforms to compete in offering the best user experience. However, interoperability remains a challenge and it is likely that in an interoperability-friendly environment, consequently the personal information space will probably include more parts than people are aware of.

This kind of society has moved toward a complete dematerialisation of the personal dataspace on cloud services. Thanks to the use increasing performance of big data analysis on anonymised data sources it is also possible since few years to create repositories of social and transactional data. These data allow shaping the average habits of most of users: purchasing habits, media consumption, travel plans.[18]

The public services (e.g., health) can exchange the data they need to deliver proactive personalized alerts and reminders. All these elements combine to create an idea of growing service customization.

Another important aspect to be considered in this scenario is the revolution in automation field: diffusion of automatic transports (e.g., electric cars).

This is how private and working lives are blending into a unique stream of services and habits without a solution of continuity.

Threats (CC).

- New forms of abuses/new targets (Human, IoT, Infrastructure, linked open data, social, connected things, etc.).
- Minor perception of information security risk because of people, finding themselves living in blending life, starts to take for granted the technological infrastructure and it becomes somehow "transparent" to the user.

[18] Fan, W., and Bifet, A. (2013) "Mining Big Data: Current Status, and Forecast to the Future," Association for Computing Machinery (ACM). Available at: http://kdd.org/exploration_files/V14-02-01-Fan.pdf.

- Detection evasion: it means the attackers' attitude in trying to avoid detection targeting new surfaces and using sophisticated attack methods and actively evading security technology. Difficult-to-detect attack styles will include "fileless" threats, encrypted infiltrations, sandbox evasion malware, and exploits of remote shell and remote control protocols.
- Below-the-OS attacks: applications and operating systems are hardened against conventional attacks so attackers could look for weaknesses in firmware and hardware. The consequence could be the broad control performed by the attackers.
- Abuse of unnoticed trust chains also due to the increasing of disappearing computing or immersed human paradigms.
- We will assist to the increase of volume and value of personal digital data. The availability of this amount of extremely attractive data (in cybercriminals perspective) will likely promote extreme data broker, i.e. fake identity trading.

Threats (CT). The modern way of working interferes dramatically with the inner organization of the enterprise digital backbones, completely changing the consolidated trust zones (e.g., the concept of demilitarised zone meant as the most secure core of an enterprise information system is not realistic anymore). The result is an increased and diffused vulnerability of the enterprises to new threats, coming from different sources (e.g. employees and more easily targeted by modern social engineering attacks). Phenomenon like "blending lives", "spread of BYOD terminals" and "social networks abuses" changed the risks landscape. Actually, this diffused and blended way of working and living forms a digital ecosystem into which users and enterprises, personal and professional services coexists and exchange data. The concept could just become more pervasive including also the fragmentation of the productive processes and the consumerization and externalization of several parts of an enterprise (e.g., external cloud services, disaster recovery and mail systems). The prevalence of digital ecosystems that offer a multitude of services opens up new possibilities of attack provoking denial of critical services. For example, a cyberterrorist may decide to cripple the emergency services that come into play during a terrorist attack, directly compromising the ecosystem.

Defences (CC).

- The security industry will develop more effective tools to detect and correct sophisticated attacks. Behavioural analytics could be used to detect irregular user activities that could indicate compromised accounts. Shared threat intelligence is likely to deliver faster and better protection of systems. Cloud-integrated security could improve visibility and control. Finally, automated detection and correction technology promises to protect enterprises from the most common attacks, allowing IT security staff to focus on the most critical security incidents.
- Threat intelligence and detection of new opportunities before they are exploited; emulate human behaviour and creation of "human honey pots".

Table 3. Example of gap analysis on the view "Enterprise 2.0" of the "Workforce" scenario. The symbols next to the gap number denote whether the corresponding threat is increasing (↑), decreasing (↓), unchanged (=), or a new one (!), going from the actual to the future view.

Gap #	Threat (future view)	Defence (actual view)	Defence (future view)	Research gap
1 (↑)	Abuses on new targets: Human, IoT, Infrastructure, linked open data, social, connected things, etc. (CC and CT)	Statistics and detection of preferred attacks patterns	Threat intelligence and detection of new opportunities before they are exploited; emulate human behaviour and creation of "human honey pots"	Threat and attack intelligence, attack simulation infrastructures
2 (=)	Abuse of unnoticed trust chains also due to the increasing of disappearing computing or immersed human paradigms (CC and CT)	Identification of trust chains; extended testing; arm race with attackers in finding exploits	Identification of *new* trust chains before attackers with proper testing and developing CMMs	Automated ways to identify existing trust chains, increasing of threat management models
3 (!)	Term-of-service (ToS) are becoming more invasive (CC)	NA	Market is becoming extremely aggressive in terms of what it can be done with released data	Monitor the ethical and legislative infrastructure for the ToS of non-EU entities

- Improved awareness methodologies for citizens; security by design; law protecting e-citizen against "bad" design.
- Identification of *new* trust chains before attackers with proper testing and developing CMMs.

Defences (CT). One of the most interesting changes in the secure governments of such complexities will rely the regulatory bodies (e.g. standards de facto or de jure and best practices). The regulation bodies will be in the position to take the responsible step of looking to the greater threats in manufacturing industry and ensure that all the software meets a minimum-security standard with new legislation activities. It will have the effect to remove the typical "moral-hazard" approach for industries.[19,20,21,22]

4.4 Gap Analysis

From the gap analysis process (carried out on the whole set of scenarios and views in Table 2), seven research gaps related to the "Enterprise 2.0" view of the "Workforce" scenario have been identified. Four of them are in common with other scenarios and views. As an example, in Table 3 we report a subset of the outcome of the gap analysis process, organized according to the template defined by CyberROAD partners. We show three research gaps, each of them characterized by a numeric identifier (Gap #), a future threat, the actual defence (if any) against such a threat, the desired defence in the future, and the description of the corresponding research gap. In particular, for each research gap we point out whether the corresponding threat is believed to increase, decrease, remain unchanged, or even be a new one, going from the actual to the future view.

5 Conclusions

We described the method we developed in the context of the CyberROAD EU FP7 project for constructing a policy-oriented research roadmap for cybercrime and cyberterrorism, at the EU level. In the preparatory phase, we analysed the state-of-the-art of S&TRM methodologies, as well as the available guidelines and related projects, focusing on policy-oriented roadmaps. We also considered the peculiarities which distinguish CC and CT from other fields: one is that they require a multidisciplinary approach involving very different domains; the other, and most relevant one, is that they co-evolve with their contextual environment. This makes a roadmapping effort particularly challenging, and a normative approach infeasible. Accordingly, we chose an exploratory approach based on a bottom-up scenario building step, in which the possible, future scenarios are obtained by aggregating vertical views of the contextual environment, obtained by combining the contributions of the different domain experts involved.

We believe that the scope of our method, as well as its rationale, is not limited to the CyberROAD project, nor to its specific subject. On the one hand, it can become a best practice for future roadmapping projects in the cybersecurity

[19] Help Net Security (2016) "Most companies do nothing to protect their mobile apps." Available at: http://www.net-security.org/secworld.php?id=19318 (Accessed: 13 January 2016).

[20] Farrington, P. (2015) "Driving an industry towards secure code." Available at: http://www.net-security.org/article.php?id=2431 (Accessed: 13 January 2016).

[21] Kassner, M. (2015) "Data breaches may cost less than the security to prevent them." Available at: http://www.techrepublic.com/article/data-breaches-may-cost-less-than-the-security-to-prevent-them/ (Accessed: 13 January 2016).

[22] Karisny, L. (2015) "Will DPM 5GL save Cybersecurity?" Available at: http://goo.gl/A2iSQ6 (Accessed: 13 January 2016).

field; on the other hand, it can be generalised to phenomena that exhibit a similar, peculiar behaviour as CC and CT, i.e., a strong co-evolution with their own contextual environment.

Acknowledgement. The research leading to these results has received funding from the European Union Seventh Framework Programme (FP7-SEC-2013) as the Cyber-ROAD project under grant agreement no 607642.

References

1. Beeton, D.A., Phaal, R., Probert, D.R.: Exploratory roadmapping for foresight. Int. J. Technol. Intell. Planning **4**(4), 398–412 (2008)
2. Bradfield, R., Wright, G., Burta, G., Cairns, G., Van Der Heijden, K.: The origins and evolution of scenario techniques in long range business planning. Futures **37**, 795–812 (2005)
3. Carvalho, M.M., Fleury, A., Lopes, A.P.: An overview of the literature on technilogy roadmapping (TRM): Contributions and trends. Technol. Forecast. Soc. Chang. **80**, 1418–1437 (2013)
4. Codagnone, C., Wimmer, M.A., (Eds.): Roadmapping eGovernment Research - Visions and Measures towards Innovative Governments in 2020. Results from the EC-funded Project eGovRTD2020, IST-2004-027139 (2007)
5. Da Costa, O., Boden, M., Friedewald, M.: Science, technology roadmapping for policy intelligence. lessons for future projects. In: Proceedings of the 2nd Prague Workshop on Futures Studies Methodology, pp. 146–161 (2005)
6. Frumento, E., Puricelli, R.: An innovative and comprehensive framework for Social Driven Vulnerability Assessment. Magdeburger Journal zur Sicherheitsforschung **2**, 493–505 (2014)
7. Geschka, H., Hahnenwald, H.: Scenario-based exploratory technology roadmaps - a method for the exploration of technical trends. In: [13], pp. 123–136. Springer, Heidelberg (2013)
8. Energy Technology Roadmaps – a guide to development and implementation, International Energy Agency, Paris (2014). http://www.iea.org/roadmaps/
9. Jeffrey, H., Sedgwick, J., Robinson, C.: Technology roadmaps: an evaluation of their success in the renewable energy sector. Technol. Forecast. Soc. Chang. **80**, 1015–1027 (2013)
10. Kerr, C.I.V., Phaal, R., Probert, D.R.: Roadmapping as a responsive mode to government policy: a goal-orientated approach to realising a vision. In: [13], pp. 67–87 (2013)
11. Kostoff, R.N., Schaller, R.R.: Science and technology roadmaps. IEEE Trans. Eng. Manage. **48**(2), 132–143 (2001)
12. Londo, H.M., More, E., Phaal, R., Wütenberger, L., Cameron, L.: Background paper on technology roadmaps, Report for United Nations Framework Convention on Climate Change (UNFCCC) (2013)
13. Moehrle, M.G., Isenmann, R., Phaal, R. (eds.): Technology Roadmapping for Strategy and Innovation. Non-series, vol. 125. Springer, Heidelberg (2013)
14. Wright, G., Bradfield, R., Cairns, G.: Does the intuitive logics method - and its recent enhancements - produce 'effective' scenarios? Technol. Forecast. Soc. Chang. **80**, 631–642 (2013)

15. Mietzner, D., Reger, G.: Advantages and Disadvantages of Scenario Approaches for Strategic Foresight (2005)
16. Ratcliffe, J.: Scenario building: a suitable method for strategic property planning. Property Manag. **18**(2), 127–144 (2000)

Part II: Legal, Ethical and Privacy Considerations

Data Protection Law Compliance for Cybercrime and Cyberterrorism Research

Arnold Roosendaal[1], Mari Kert[2], Alison Lyle[3](✉), and Ulrich Gasper[4]

[1] TNO, The Hague, Netherlands
arnold.roosendaal@tno.nl
[2] European Organisation for Security, Brussels, Belgium
Mari.Kert@eos-eu.com
[3] Office of the Police and Crime Commissioner for West Yorkshire, Wakefield, UK
alison.lyle@westyorkshire.pnn.police.uk
[4] Cybercrime Research Institute, Cologne, Germany
gasper@cybercrime.de

Abstract. Data protection is perhaps the most important area in which legal requirements determine whether and how research into cybercrime and cyberterrorism may take place. Data protection laws apply whenever personal data are processed for the purposes of research. There are legal risks of non-compliance with data protection regimes emanating from strict legal frameworks and from rules on data security and data transfer. Researchers are strongly recommended to explore the possibilities of anonymisation as well as all obligations relating to notification and consent, which affect the legitimacy of data processing. The presentation of findings, with implications for research carried out in the area of cybercrime and cyberterrorism, begins with exploring definitions of data protection and privacy. We introduce the most relevant aspects of data protection for cybercrime and cyberterrorism research before an overview of the applicable legal and regulatory frameworks is presented. The way in which data protection interacts with other fundamental rights, namely freedom of speech, academic freedom and security, is considered in order to highlight important issues which may affect researchers. Another key feature of data protection law is the difference between countries in the way it is applied; member states have a degree of autonomy in this respect which is summarised and an overview provided. General conclusions are drawn from all findings and implications of the research undertaken for this chapter and key recommendations for those involved in research are presented.

Keywords: Data protection · Data transfer · Data security · Privacy · Anonymisation · Notification · Consent · Data processing

1 Introduction

Any research in the area of cybercrime and cyberterrorism (CC/CT) takes place within the framework of society at large, which has an impact on how this

B. Akhgar and B. Brewster (eds.), *Combatting Cybercrime and Cyberterrorism*,
Advanced Sciences and Technologies for Security Applications,
DOI 10.1007/978-3-319-38930-1_5

research may be carried out. This legal section aims to identify and analyse some of the main legal and ethical issues that may, or will, arise when carrying out this type of research.

In addition to the more general issues of social cohesion and discrimination against gender, religion and minorities that are quickly revealed by traditional methods of research,[1] further, more specific topics are also relevant. Data protection issues are of central importance; they address both privacy and personal data protection and are the subject of legislative reform at European level.[2] Illegal content issues may arise and affect the legality of CC/CT research and the fundamental rights of victims and suspects must also be at the forefront of any CC/CT research considerations.

In order to address and examine these crucial areas, this legal section is divided as follows:

1. *Data Protection Law Compliance for CC/CT Research*
2. *Non-discrimination and Protection of Fundamental Rights in CC/CT Research.*
3. *Risks Related to Illegal Content in CC/CT Research*

Each chapter contains the results of expert and thorough research into the main issues; the overview and recommendations presented for each are drawn from the in-depth analysis of legislation, case law and practical examples. The structure of each chapter facilitates explanation of key issues such as definitions and relevant legal standards, with special focus on freedom of speech and academic freedom. An overview of country studies illustrates the importance of understanding different approaches across national jurisdictions when carrying out this type of research. The general conclusion to each chapter presents the main findings and includes specific recommendations.

2 Definitions

Defining privacy is not an easy task; so far there have been no successful attempts and there is no obvious or universally accepted answer.[3] Some state that privacy is an important fundamental right because it underpins values such as human

---

[1] For example a PESTLE or STEP approach.

[2] EU data protection reform consists of a General Data Protection Regulation and a Data Protection Directive for the area of police and criminal justice, both of which received final agreement on 14 April 2016, will come into force 20 days after appearing in the Official Journal, and Member States have a further two years to achieve compliance.

[3] Dan Svantesson (2010), A Legal Method for Solving Issues of Internet Regulation; Applied to the Regulation of Cross-Border Privacy Issues. EUI Working Papers LAW No. 2010/18. Via: http://papers.ssrn.com/sol3/papers.cfm?abstract_id=1785421.

dignity and freedom of speech.[4] However, the concept of privacy is evolving, particularly in the networked society which allows large-scale data processing and aggregation. How privacy is understood varies within different contexts and encompasses ideas related to the right to be left alone, a right to confidentiality of communications, the right to determine how to live one's life and a right to personal data protection. The scope and reach of privacy are un(der)determined; judges will decide when privacy interests are at stake and when their protection can rightfully be invoked.

Data protection is broader and more specific than privacy, but the relationship between them is important. Data protection also incorporates the protection of freedom of expression, freedom of religion and conscience, the free flow of information and the principle of non-discrimination, albeit conditional upon competing rights and interests of others. Data protection is more specific, since it applies every time personal data are processed. The application of data protection rules does not require an answer to the question of a violation of privacy; data protection applies when the conditions stipulated by legislation are fulfilled, they are not prohibitive by default but channel and control the way data are processed.[5]

3 Relevance of Aspects of Data Protection to Research

As technology continues to grow, and new technologies emerge, research related to CC/CT is crucial. However, the same rapid evolution of technology means that research will usually involve the automatic processing of the electronic personal data of individuals participating in, or the subject of, the research.

The primary consideration for researchers is to establish a legitimate basis for the data processing, which includes collection, processing, storing and dissemination. At all stages it is imperative that measures are taken to ensure the security of any personal data; this is of greater importance when sensitive personal data is processed.

Engaging in CC/CT research will inevitably engage data protection laws and those involved will have obligations and duties that they must be aware of. Although this might be seen as restrictive, the same rules contain exemptions for those carrying out research; this could perhaps be seen as creating a balance between academic freedom and the right to data protection, both fundamental and protected rights.

---

[4] Michael Friedewald, David Wright, Serge Gutwirth & Emilio Mordini: Privacy, data protection and emerging sciences and technologies: towards a common framework - Innovation: The European Journal of Social Science Research, Volume 23, Issue 1, March 2010, page 61–67, via: http://www.sciencedirect.com/science/article/pii/S0267364909001939.

[5] Michael Friedewald, David Wright, Serge Gutwirth & Emilio Mordini: Privacy, data protection and emerging sciences and technologies: towards a common framework - Innovation: The European Journal of Social Science Research, Volume 23, Issue 1, March 2010, page 61–67.

Research related to data protection is more important now than ever before. Google's and Facebook's policies, the NSA scandal, fast-moving cross-border data flows and the heated discussions on 'the right to be forgotten' mean that the call for the protection of data and privacy is all the more urgent. The relevance of aspects of data protection with regard to research is highly important; it will provide insights into what kind of research has already been done and what still needs to be conducted, particularly with the increase in CC such as large-scale attacks on companies such as Facebook and Google, aimed at retrieving personal data.[6] Since fundamental rights need to be protected, it is essential to ensure that data protection is at the centre of cyber research, where particular challenges to the right appear.

Specifically with regard to research, it must be determined what type of research is at stake and what type of data is required. A distinction between personal and sensitive personal data must be made, as the latter requires more attention. Any data used must be relevant for the research. Decisions must be made about specific retention periods for the data as this requires consent from the data subjects. Additionally, whether data should be anonymised for archiving and the impact this might have on future research must be considered. In relation to criminal investigations, questions arise in respect of legitimate use and whether this justifies infringement of rights.

Several complications have been identified in legal literature:

Legislation

- Disparities in national legislation and a lack of harmonisation across different EU member states (fragmentation);[7,8]
- Uncertainties and complexities associated with the definition of personal data, particularly in the UK, have given rise to practical difficulties.[9,10] The definition of explicit consent and the situations in which it is required need further explanation. There is also uncertainty about when consent can be implied or when it may be waived on grounds of public interest. Other issues that need to

[6] Stefan Savage, Collaborative Center for Internet Epidemiology and Defenses (CCIED), "An Agenda for Empirical LCybercrime Research", 2011 USENIX Federated Conferences Week, June 14–17, 2011, Portland-OR. via: https://www.youtube.com/watch?v=ILOtIMShi9s.

[7] OECD Guidelines on the Protection of Privacy and Transborder Flows of Personal Data – OECD Website. Via: http://www.oecd.org/internet/ieconomy/oecdguidelinesontheprotectionofprivacyandtransborderflowsofpersonaldata.htm.

[8] Judith Strobl, Emma Cave and Tom Walley (2000) Data protection legislation: interpretation and barriers to research. Via: http://www.ncbi.nlm.nih.gov/pmc/articles/PMC1118686/ .

[9] Judith Strobl, Emma Cave and Tom Walley (2000) Data protection legislation: interpretation and barriers to research. Via: http://www.ncbi.nlm.nih.gov/pmc/articles/PMC1118686/.

[10] Christopher Millard & W. Kuan Hon (2011), Defining 'Personal Data' in e-Social Science, Information, Communication and Society, 2012 Vol 15(1) p 66.

be clarified include anonymisation and its effects and those relating to access of confidential data.

Technology

- Encryption technologies are frequently used to protect users' privacy, however not all data about a user will be under their control.[11]
- Items of interest (IOIs) can be linked, allowing an attacker to distinguish whether they are related or not within a system.
- Identifiability of a subject, means an attacker can sufficiently identify the subject associated with an IOI, for example the sender of a message.
- Information disclosure threats may expose personal information to individuals who do not have legitimate access.
- Policy and non-compliance refers to the lack of guarantee that a system complies with its advertised policies.[12]

Other

- The use of anonymisation may raise ethical problems such as when DNA reveals a propensity to certain diseases.[13]
- There may be a lack of awareness about data protection rights, by both data subjects and legal experts.[14]

Several solutions to problems have also been identified:

- The implementation of data protection principles in a cyber-security policy may act as a proxy to reduce cyber threats.[15]
- International harmonisation of data protection. For example, the Organisation for Economic Co-operation and Development (OECD) member countries developed a set of guidelines to help harmonise national privacy legislation which would uphold human rights and prevent interruptions in international data flows.

[11] ITU (2006), Research on legislation in data privacy, security and the prevention of cybercrime, via; http://www.itu.int/ITU-D/cyb/publications/2006/research-legislation.pdf.

[12] M. Deng, K. Wuyts, R. Scandariato, B. Preneel and W. Joosen (2010): a Privacy Threat Analysis framework: supporting the Elicitation and Fulfilment of Privacy Requirements. Via: https://www.cosic.esat.kuleuven.be/publications/article-1412.pdf.

[13] FP7, Privacy and emerging fields of science and technology: Towards a common framework for privacy and ethical assessment – PRESCIENT (2012).

[14] European Union Agency for Fundamental Rights (2014) Access to data protection remedies in EU member states, via: http://fra.europa.eu/en/publication/2014/access-data-protection-remedies-eu-member-states & http://fra.europa.eu/sites/default/files/fra-2014-access-data-protection-remedies_en_0.pdf.

[15] Porcedda, Maria Grazia (2012), Data Protection and the Prevention of Cybercrime: The EU as an area of security?

- Research efforts such be directed towards increasing privacy protection, particularly in respect of location privacy.[16]
- In order to promote future internet trustworthiness, research should address security requirement engineering and users' security awareness.[17]
- Identification of the risks and issues related to the extensive use of profiling and the counter measures that have been adopted across EU member states.[18]

4 Relevant Standards

Data protection at European level is afforded by both the Council of Europe, which focuses primarily on protecting human rights and fundamental freedoms, and by the European Union which regards data protection as a fundamental right at Treaty level[19].

The Council of Europe (CoE) was established to promote human rights in the states of Europe and so adopted the **European Convention on Human Rights (ECHR)**[20] which came into force in 1953 and has influenced all data protection law. Article 8 ECHR protects the privacy of all citizens.

In recognition of the importance of protecting privacy in developing societies, the CoE adopted the **Convention for the protection of individuals with regard to the automatic processing of personal data (Convention 108)**.[21] This is still the only international legally binding document in force and has been ratified by all the European member states as well as the EU itself.[22] Convention 108 applies to all data processing carried out by public and private organisations and seeks to protect citizens against violations of their rights in respect of the collection, processing, storage and dissemination of their data. The principles of fairness, lawfulness and proportionality are enshrined in this legislative instrument.

Transborder data flows of personal data were later focused on by the CoE when it adopted an **Additional Protocol to Convention 108 (Additional Protocol 181)**.[23] This recognised the development of exchanges of personal

[16] Network of Excellence on Engineering Secure Future Internet Software, see www.nessos-project.eu/.

[17] ibid.

[18] Dan Svantesson (2010), A Legal Method for Solving Issues of Internet Regulation; Applied to the Regulation of Cross-Border Privacy Issues. EUI Working Papers LAW No. 2010/18. Via: http://papers.ssrn.com/sol3/papers.cfm?abstract_id=1785421.

[19] Since the Treaty of Lisbon 2009.

[20] Council of Europe, European Convention for the Protection of Human Rights and Fundamental Freedoms, as amended by Protocols Nos 11 and 14, 4 November 1950, CETS 5.

[21] Council of Europe Convention for the Protection of Individuals with regard to Automatic Processing of Personal Data, CETS 108 1981.

[22] Art. 23 (2) of Convention 108 amended allowing the European Communities to accede, adopted by the Committee of Ministers on 15 June 1999.

[23] Council of Europe, Additional Protocol to the Convention for the protection of individuals with regard to automatic processing of personal data, regarding supervisory authorities and transborder data flows, CETS 181 2001.

data across national borders and sought to further the protection of citizens in this respect.

In 2010 the Committee of Ministers of the CoE adopted the **Profiling Recommendation**[24] in respect of automatic processing of personal information and which recognised the new challenges created by technological advancements, in particular the practice of 'profiling' whereby data processors are able to obtain different information about an individual by using various software applications. The Recommendation sought to afford protection to citizens' right to privacy and data protection and recognised the potential for violations of the rights relating to non-discrimination and dignity enshrined in the ECHR.

In the European Union, the **Charter of Fundamental Rights of the European Union (The Charter)**[25] achieved Treaty status with the Treaty of Lisbon in 2009, and affords respect for private and family life in Article 7 as well as specific protection of personal data in Article 8, thereby placing these concerns at the core of the EU.

The **EU Directive 95/46/EC (the Data Protection Directive)**[26] currently forms the most important legal framework. Under EU law, personal data can only be gathered legally under strict conditions, for a legitimate purpose. Furthermore, persons or organisations which collect and manage personal information must protect it from misuse and must respect certain rights of the data subjects, which are guaranteed by EU law. Common EU rules have been established to ensure that personal data enjoys a high standard of protection everywhere in the EU. Citizens have the right to complain and obtain redress if their data is misused anywhere within the EU.[27]

Applicants must follow within their project the EU legal framework; these standards should also apply to participants in third countries meaning Non-EU Member States. Prior to any transfer of data outside the EU Member States, applicants should make sure that the place where the data is to be sent has a data protection regime in place that is at least as solid as that required in the EU, or at least conform to the Data Protection Directive's requirements. Data storage must be secured so as for the data not to become accessible to unauthorised third parties and to be protected against disaster and risk.[28]

---

[24] Recommendation of the Committee of Ministers to member states on the protection of individuals with regard to automatic processing of personal data in the context of profiling, CM/Rec (2010)13.

[25] Charter of Fundamental Rights of the European Union [2010] OJ C 83/02.

[26] Directive 95/46/EC of 24 October 1995 on the protection of individuals with regard to the processing of personal data and on the free movement of such data [1995] OJ L 281/31.

[27] European Commission website, Justice, Data protection. Via: http://ec.europa.eu/justice/data-protection/.

[28] European Commission, Experts Working Group on data protection and privacy (2009), EU - Data protection and Privacy Ethical Guidelines. Ethical review in FP7.

Directive 2002/58/EC (the E-Privacy Directive)[29] is set out to provide for protection of citizens' privacy in respect of personal data processed in the electronic communications sector. It translates the provisions in the Data Protection Directive and accounts for the development of the electronic communications services and the risks this may pose for the user in respect of their personal data and privacy.

The new **General Data Protection Regulation** which will replace Directive 95/46/EC was agreed by the EU Parliament on 14 April 2016. Twenty days after its appearance in the Official Journal it will become law, and member states have two years in which to comply. Issues that are relevant for personal data processing in relation to research activities can be recognised. First, the fact that it is a Regulation instead of a Directive means that there will be more harmonisation at the EU level. Thus, national differences in the implementation of the law will be limited (if not eradicated). This is supported by the one-stop-shop principle which is included in the Regulation, and means that if data are processed in different Member States, there will be only one responsible data protection authority for all processing, which ensures a uniform approach.[30]

Another important change concerns consent from the data subject. If consent is to be obtained for research purposes, the Regulation indicates that this consent not only has to be free, informed, specific, and unambiguous, but also explicit.[31] So there has to be an explicit act of the data subject indicating that he or she provides consent for the data processing.[32]

5 Data Protection vs. Academic Freedom

Several regulations have been established to protect personal data in research, however what falls within this scope is still uncertain particularly in countries such as the UK. The balancing exercise between the rights of data protection and academic freedom is illustrated in Article 9 of Convention 108[33] which allows for the rights protected in Article 8 to be restricted when automated data files are used for research, as long as data subjects' privacy is not violated.

The rights relating to data protection and academic freedom need not be competing ones; if data protection principles are followed when carrying out research using personal data an appropriate balance can be achieved. Some measures which will ensure this can be identified:[34]

[29] Directive 2002/58/EC of 12 July 2002 concerning the processing of personal data and the protection of privacy in the electronic communications sector (Directive on privacy and electronic communications) [2002] OJ L 201/37.

[30] Article 51 of the proposed Regulation.

[31] See also: https://secure.edps.europa.eu/EDPSWEB/webdav/site/mySite/shared/Documents/EDPS/Publications/Speeches/2014/14-09-15_Article_EUI_EN.pdf.

[32] Article 4(8) of the proposed Regulation.

[33] Council of Europe, Convention for the Protection of Individuals with Regard to the Automatic Processing of Individual Data, 28 January 1981, ETS 108.

[34] J. Ritchie et al. Qualitative Research Practice (2nd edition): A Guide for Social Science Students and researchers (2013).

- Personal information stored separately from research data and archived by use of a serial number;
- Electronic files or any document linking serial numbers to participants kept in a separate location from research data;
- Storing of paper documents and electronic files securely in lockable cabinets or on password-protected or encrypted electronic devices. Restricting access to data and documents to members of the research team;
- Qualitative data secured in transit from meetings with participants using encrypted digital records. Removing memory cards from the recording device to minimise the risk of data being lost or stolen;
- Consideration given to securely transfer data being shared. Mailing or e-mailing may risk interception or delivery to an unintended recipient. Electronic folder or use of encryption constitute more secure means of transfer.

6 Privacy vs. Security

The difficulty of balancing the apparently competing rights of privacy and security may be present in research on CC/CT. Some positive aspects can be found already, however. For instance, the EU, in its Cybersecurity Strategy, "goes beyond the traditional approach of opposing security to privacy by providing for the explicit recognition of privacy and data protection as core values which should guide cybersecurity policy in the EU and internationally."[35] At the same time, the EDPS argues that the definition of cybercrime in the Strategy is too overarching, instead of specific, which brings a risk for inappropriate balancing of security and privacy.[36] In the context of criminal investigations, the Council of Europe is of the opinion that data protection principles have to be respected, in line with Convention 108.[37] The challenge is to balance security and privacy and to take both into account when executing research on cybersecurity and cyberterrorism.

7 Country Studies

7.1 United Kingdom

This section provides a general overview of data protection in the United Kingdom that are relevant to research activities on CC/CT.

[35] Opinion of the EDPS on the Joint Communication of the Commission and of the High Representative of the European Union for Foreign Affairs and Security Policy on a 'Cyber Security Strategy of the European Union: an Open, Safe and Secure Cyberspace', and on the Commission proposal for a Directive concerning measures to ensure a high common level of network and information security across the Union (2013).

[36] Ibid.

[37] van den Hoven van Genderen, R. (2008), Discussion paper: Cybercrime investigation and the protection of personal data and privacy, p. 49.

The UK **Data Protection Act 1998** implements the European Data Protection Directive[38] and contains certain rules relating to the processing of data for research, which are set out in section 33. The definition of 'personal data' in the Act is similar to that in the EU Directive, however the UK definition contains two additional elements relating to identifiability of an individual to a controller and proximity of data:

"'[P]ersonal data' means data which relate to a living individual who can be identified

(a) from those data, or

(b) from those data and other information which is *in the possession of, or is likely to come into the possession of, the data controller*,

and *includes any expression of opinion about the individual and any indication of the intentions of the data controller or any other person in respect of the individual.*"[39]

The difference between (a) and (b) is that under (a) the data themselves include unique identifiers, such as name and address or other information that uniquely identifies a person. Clause (b) covers the situation in which the data themselves do not contain unique identifiers but which allow identification to take place if the data are combined with other data sets that the controller has, or is likely to acquire. The other data sets need not necessarily contain unique identifiers—if the combination of different anonymous data sets (in the possession or likely to be acquired by the data controller) allow identification of an individual, then the data in each data set are also to be considered personal data, as per clause (b).

The broad definitions in the EU Directive allows for this national divergence,[40] but may also lead to uncertainty. However, the characteristic of the EU Directive is an instrument aiming at harmonisation which is not minimal but "generally complete".[41] A degree of deviation is inherent in the nature of a European Directive which has to be implemented by national legislation.

Processing of personal data of research purposes has to comply with the UK Data Protection Act and only limited exceptions are allowed in section 33. The Act does not define what is meant by 'research'. However, the Information Commissioner's Office (ICO), which is the Data Protection Authority in the UK, "uses an ordinary meaning of 'research' [...], [namely,] research is a systematic investigation intended to establish facts, acquire new knowledge and reach new conclusions."[42] That is, forms of research other than statistical or historical research mentioned in the Act (such as market, social, commercial or opinion research) could be subject to the exemption.[43] The concept of "research" covers

[38] http://www.legislation.gov.uk/ukpga/1998/29/contents.

[39] S1(1)(b) Data Protection Act 1998.

[40] Article 29 Working Party opinion 4/2007 on the concept of personal data, 20 June 2007, p. 3 (WP 136).

[41] CJEU C-101/01 Bodil Lindqvist [2003] ECR I-1297, paras 96, 97.

[42] Ibid.

[43] Ibid.

research carried out in the public or private sector and it can be commercial or academic.[44] There are conditions relating to these exemptions which are understood as safeguards[45] and stipulate that the data are not used for profiling or in a way that causes, or is likely to cause, substantial damage or distress to the data subject.[46] When these safeguards are met, then research can be carried out on the personal data even if the data were not collected originally for research purposes, and can be retained for an indefinite period of time.[47]

Importantly, for the purposes of the research exemption, personal data continue to be treated as processed for research purposes also when "the data are disclosed (a) to any person, for research purposes only, (b) to the data subject or a person acting on his behalf, (c) at the request, or with the consent, of the data subject or a person acting on his behalf, or (d) in circumstances in which the person making the disclosure has reasonable grounds for believing that the disclosure falls within paragraph (a), (b) or (c)"[48]

The processing of personal data for research must satisfy one of the grounds for legitimate processing, which will usually be consent or because it is necessary for the legitimate interests of the data controller.[49] However, research in a specific area such as the evaluation of the effectiveness of a certain measure in law may satisfy the legitimate ground of necessity for a legal obligation to which the controller is subject.[50]

The processing of sensitive data is subject to greater restrictions and what constitutes sensitive data is set out in s2 of the Act. Grounds for the processing of this data are specified in Schedule 3 to the Act. Research might rely on either explicit consent[51] or a substantial public interest,[52] depending on the purpose of the work. Either would still engage the conditions previously referred to. Ground 9(1) of Schedule 3 regulates the processing of sensitive data relating to racial or ethnic origin. This is only allowed if the processing "is necessary for the purpose of identifying or keeping under review the existence or absence of equality of opportunity or treatment between persons of different racial or ethnic origins, with a view to enabling such equality to be promoted or maintained, and (...) is carried out with appropriate safeguards for the rights and freedoms of data subjects".

Although the Data Protection Act defines 'personal data' narrowly, it is interesting to note that the case law suggests a wider interpretation. In the case

[44] Rosemary Jay, Angus Hamilton, Data Protection – Law and Practice, London, Sweet & Maxwell, 2003, section 18-09 (p. 414).

[45] ICO' s Anonymisation: managing data protection risk code of practice, p. 44.

[46] S33(1)(a) and (b) Data Protection Act 1998.

[47] Rosemary Jay, Angus Hamilton, Data Protection – Law and Practice, London, Sweet & Maxwell, 2003, section 18-09 (p. 413).

[48] S33(5) Data Protection Act 1998.

[49] Schedule 2 to the Data Protection Act 1998.

[50] ibid.

[51] Ground 1 of Schedule 3 to the Data Protection Act 1998.

[52] Para. 1(2)(a) Data Protection (Processing of Sensitive Personal Data) Order 2000.

of *Common Services Agency v. Scottish Information Commissioner*[53] the court considered whether anyone else would be able to identify the data subject if they came into contact with other data sets which, when combined with anonymised data, would allow identification of an individual. If so, then the data would be classed as personal. This approach is more in line with the European Data Protection Directive.

Other regulatory initiatives in the UK include guidance on anonymisation by the ICO[54] which may assist CC/CT researchers using anonymised data sets. The **Digital Economy Act 2010**[55] may be relevant for CC researchers studying copyright infringements, to the extent that the research might use personal information about subscribers' online activities or information about copyright infringement reports.

7.2 Belgium

The **Belgian Data Protection Act** (DPA)[56] implements the EU Data Protection Directive and aims at protecting the fundamental rights and freedoms of the person, especially the right to protection of privacy, with regard to the processing of personal data.[57] The Belgian Commission for the protection of privacy[58] (Commission de la Protection de la Vie Privée, hereinafter referred as CCVP) is the authority that oversees and enforces the DPA.

In general, the DPA defines personal data as "any information relating to an identified or identifiable natural person". The general obligations on data controllers to ensure fair and lawful processing are as follows:

- Data controller can only process personal data with the data subject's consent;
- The data processor has the obligation to notify the CCVP of any wholly or partly automatic operation or set of operations;
- Data can be collected for specified, explicit and legitimate purposes and should not be further processed in a way that is incompatible with those purposes;
- Personal data must be adequate, relevant, not excessive in relation to the purpose for which it is collected and/or further processed, accurate, up-to-date, kept in a form permitting identification of data subjects for no longer than necessary;

[53] House of Lords, Common Services Agency v. Scottish Information Commissioner [2008] UKHL 47.

[54] Anonymisation: managing data protection risk code of practice, available at http://ico.org.uk/for_organisations/data_protection/topic_guidelines/anonymisation.

[55] Sections 3-16 of DEA amended the Communications Act 2003 by introducing the new sections 124A-N.

[56] Loi relative à la protection de la vie privée à l'égard des traitements de données à caractère personne" of 8 December 1992 As amended by the Law of 11 December 1998 and the Royal Decree of 13 February 2001.

[57] Ibid Art. 2.

[58] http://www.privacycommission.be/fr.

- Data controllers must provide certain information to the data subjects concerned and grant the data subjects concerned the rights to access, object, rectify, block and/or delete the personal data relating to him;
- Data controllers must implement appropriate technical and organisational security measures to protect personal data.

The DPA allows processing of data for historical, statistical or scientific purposes.[59] The DPA specifically mentions data processing requirements for scientific purposes on health related personal data[60] and personal data related to litigation that has been submitted to courts and tribunals.[61] There are no specific requirements on research carried out on CC/CT. Article 9 DPA imposes an obligation for the data controller to inform the data subject about the processing, but limits this in cases where doing so would involve disproportionate effort, in particular for statistical purposes or for historical or scientific research.

The **Royal Decree**[62] provides more details about rules on data processing involving anonymised or encoded data, which are relevant for those engaging in research:

- When using anonymous[63] data for research the DPA does not apply so informed consent is not required;
- When using encoded (pseudonymised) data, informed consent is not required but there is an obligation to inform the data subject, unless this proves impossible or involves a disproportionate effort.[64]
- The data subject must give his explicit consent to the processing of non-encoded personal data relating to him for historical, statistical or scientific purposes prior to the processing.[65]
- When encoded data are further processed, informed consent or re-consent needs to be obtained from the data subject, unless the further processing is restricted to non-encoded personal data that has been made public as a result of steps deliberately taken by the data subject or that are closely related to the public character of the data subject, or proves to be impossible or involved disproportionate effort.[66]

No self-regulatory initiatives in the field of cybercrime research were found relating to Belgium and no case law on the matter was discovered.

59 Article 4 (2), Loi relative à la protection de la vie privée à l'égard des traitements de données à caractère personne" of 8 December 1992.
60 Ibid Art. 7.
61 Ibid Art. 8.
62 Royal Decree implementing the Act of 8 December 1992 on the protection of privacy in relation to the processing of personal data.
63 'Anonymous' shall be construed as relating to data that cannot be related to an individual or identifiable person.
64 Article 15, Royal Decree implementing the Act of 8 December 1992 on the protection of privacy in relation to the processing of personal data.
65 Ibid Article 19.
66 Ibid Article 20(2).

7.3 The Netherlands

The relevant legislation in the Netherlands on data protection is the **Wet bescherming persoonsgegevens (Wbp)** which is the national implementation of the EU Data Protection Directive. In the Netherlands, personal data is defined as "elk gegeven betreffende een geïdentificeerde of identificeerbare natuurlijke persoon" ["any data concerning an identified or identifiable natural person"].[67] This is in line with the definition in the EU Data Protection Directive (95/46/EC), which states that: "'personal data' shall mean any information relating to an identified or identifiable natural person ('data subject')".[68]

Depending on the type of research and the means used for the research, the privacy provisions of the Telecommunicatiewet (Telecommunications Act), which implement the ePrivacy Directive[69] may also apply. In particular, this can be the case when it concerns processing of traffic or location data, or the use of cookies or other means to recognise devices. Article 9(3) of the Wbp indicates that further processing of personal data for scientific or statistical purposes is allowed, as being not incompatible with the original purposes for which the data have been acquired, if the data controller has taken appropriate measures to guarantee that the data shall only be used for these specific purposes. In these cases, it is also allowed to store the data for a longer term than usual.

Article 23(2) Wbp states that sensitive personal data may be processed for statistical purposes if:

- The research serves a public interest;
- The processing is necessary for the research or statistical analysis;
- Obtaining explicit consent from the data subjects is impossible or requires a disproportionate effort;
- Sufficient guarantees are taken so that the privacy of data subject is not disproportionately harmed.

Furthermore, according to Article 44(1), organisations or agencies for academic research (e.g., universities) or statistics (e.g., National Bureaus of Statistics) are exempted from the obligation to notify data subjects of the processing, and they can refuse requests from data subjects for information, provided that the necessary measures have been taken to ensure that the personal data will only be used for statistical and academic research purposes.

In respect of non-regulatory approaches in relation to data protection for CC/CT research purposes, it is possible to make use of binding corporate rules or codes of conduct to ensure the appropriate protection of personal data and to limit the use for scientific or statistical purposes. At a national level, such a code of conduct exists for the Association of Universities (VSNU).[70]

[67] Article 1(a) of the Wet bescherming persoonsgegevens (Dutch Data Protection Act).
[68] Article 2(a) of the Data Protection Directive.
[69] Directive 2002/58/EC of 12 July 2002 concerning the processing of personal data and the protection of privacy in the electronic communications sector (Directive on privacy and electronic communications) [2002] OJ L 201/37.
[70] http://www.vsnu.nl/files/documenten/Domeinen/Accountability/Codes/Bijlage%20Gedragscode%20persoonsgegevens.pdf.

The following case illustrates the crucial importance of applying data protection rules and principles when carrying out research. The role of data controller involves obligations and responsibilities for individuals' personal data. The case in question, from 2012, involved a botnet attack named Pobelka. The botnet was used to harvest data from numerous computers and systems, including computers from universities in the Netherlands.[71] Even though it remained unclear what data exactly had been compromised, the case created quite some stress at universities, since they were unable to guarantee that the data they had collected for research was safeguarded. The compromised data sets also included personal data, such as email addresses.

We have found no other specific issues in the Netherlands concerning the processing of personal data for research on CC/CT.

8 General Conclusion

The legal framework at EU and CoE level, as well as at national level, sets a number of strict requirements that have to be met in order to make processing of personal data legitimate. In general, certain exemptions apply to the legal requirements at European level when data are processed for scientific research or statistical purposes. These exemptions are applied at national law level and may apply to requirements such as notifying data subjects of processing or obtaining consent.

Research projects and those carrying out research in the area of CC/CT may enjoy some privileges in respect of data protection obligations, however the principles of data minimisation and purpose limitation will still apply. Moreover, when research results are published, these have to be anonymous.

Any international transfer of personal data from a member state of the EU has to ensure that the receiving jurisdiction provides an equivalent and adequate level of data protection. Data transfers to jurisdictions within the EU are unproblematic, but transfers to non-EU states have to ensure that the data protection measures in the receiving jurisdiction are at least as high as in the EU. This is an area of law which is rapidly evolving, so researchers must be aware of up to date requirements in this respect.

9 Recommendations

Harmonisation of data protection requirements will be finally achieved in 2018 when all EU member states will have to apply with the new General Data Protection Regulation. Until this time, CC/CT researchers who invariably engage in cross-border studies will have to ensure compliance with different national and supra national data protection regimes. In furtherance of this aim, the following measures are recommended:

[71] http://www.realphantom.com/content/botnet-kaapt-16-miljoen-e-mailadressen-en-wachtwoorden. Also report 'NSCS Cybersecuritybeeld Nederland 2013', p. 66.

- The use of anonymised data; data protection rules only apply to information which is capable of identifying an individual, therefore effective anonymisation will avoid compliance requirements. Anonymisation should take place at the earliest stage and in any case prior to the publication of results;
- Notification of data subjects: in some jurisdictions the requirement to notify participants of the processing of their data may be subject to exemption if this involves a disproportionate effort, or if the nature of the research does not require it. Researchers need to be aware of the difference in requirements across jurisdictions;
- Obtain consent: researchers must verify whether consent must be obtained from the data subjects prior to the research. This may vary across jurisdictions and may not be required if it involves a disproportionate effort or if the research is of a certain nature;
- Legitimate data processing: each research project must specify the purpose of the research and have a legitimate reason. Research must comply with the principles of data minimisation and purpose specification. In line with the purpose of processing, it has to be decided for how long the data will be used, how long it will be stored and when it will be deleted. The data subjects, the data controller and the data processor will have to be identified. Stricter rules apply to the processing of sensitive data;
- Specified purpose: researchers must make sure that the personal data are only processed for scientific or statistical purposes and must not be further processed for a different purpose.

Acknowledgement. The research leading to these results has received funding from the European Union Seventh Framework Programme (FP7-SEC-2013) as the COURAGE project under grant agreement no 607949.

Non-discrimination and Protection of Fundamental Rights in Cybercrime and Cyberterrorism Research

Francesca Bosco[1], Elise Vermeersch[1], Vittoria Luda[1], Giuseppe Vaciago[2], Ulrich Gasper[2], and Alison Lyle[3](✉)

[1] UNICRI, United Nations Interregional Crime and Justice Research Institute, Turin, Italy
{bosco,vermeersch,luda}@unicri.it
[2] Cybercrime Research Institute, Cologne, Germany
gasper@cybercrime.de
[3] Office of the Police and Crime Commissioner for West Yorkshire, Wakefield, UK
Alison.Lyle@westyorkshire.pnn.police.uk

Abstract. This chapter presents and explores the legal issues surrounding the fundamental human rights of victims and in relation to non-discrimination, in the context of cybercrime (CC) and cyberterrorism (CT) research. In relation to non-discrimination, the focus is on social inclusion, minimising disparities and avoiding marginalisation of groups, particularly when presenting results of studies involving identified sections of society. The importance of victims' rights in relation to CC/CT research is then explored and the most relevant aspects as a possible limiting factor in this area are outlined. The infinite value of awareness of these considerations as well as independence and neutrality of research is emphasised.

Keywords: Victims' rights · Non-discrimination · Gender equality · Minority protection · Bias · Neutrality · Independence · Privacy · Re-victimisation · Data protection · Right to be forgotten

1 Introduction

In order to create understanding of issues surrounding the fundamental human rights of victims in relation to non-discrimination, definitions are presented and their relevance to cybercrime (CC) and cyberterrorism (CT) research outlined. The way in which rights relating to victims and non-discrimination are protected by law and other regulatory frameworks is set out and examples of how these are applied by the courts facilitates further understanding. The cross-border nature of much of the research in this area requires researchers to be aware of how different countries deal with these issues, so a brief overview and comparison of selected European Member States is included.

B. Akhgar and B. Brewster (eds.), *Combatting Cybercrime and Cyberterrorism*,
Advanced Sciences and Technologies for Security Applications,
DOI 10.1007/978-3-319-38930-1_6

Finally, general conclusions are drawn from the findings and implications of the research undertaken for this chapter, and key recommendations for those involved in research projects are developed.

2 Definitions

It is important to clarify what is understood by the terms and issues referred to within the context of the present discussion.

The term 'victim' is universally used to identify those who, either individually or collectively, have suffered harm. This includes physical, mental, emotional, economic loss or substantial limitations or harm to their fundamental rights through actions, or failures of action, that subsequently breach criminal laws.[1] Irrespective of the crime type, victims have fundamental needs which include being treated with respect and dignity, receiving support, enjoying protection and having access to justice. These needs have been recognised at the EU level as being worthy of greater consideration[2]. The focus here is on issues concerning victims' rights in relation to CC/CT and what distinguishes them from victims of other types of crime.

Gender has been defined as: "*... the set of qualities and behaviours expected from men and women by their societies and forms their social identity; an identity that differs from culture to culture and at different periods in history.*"[3] Gender equality can be defined as considering, valuing and favouring the different behaviour, aspirations and needs of men and women equally.[4]

The concepts of Religion and belief are not defined by most EU and international documents targeted at protecting them, however the Council of Europe[5] has stated that the concept of religion includes the holding of theistic or non-theistic beliefs and recognition of formal worship as well as expressing religious views and forms of conduct based on religious belief. 'Belief' can be understood

1 This definition of victim of crime can be found in article 1 of the "Declaration of basic principles of justice for victims of crime and abuse of power", approved by resolution number 40/34 of 29 September 1995 by the General Assembly of the United Nations (https://www.unodc.org/pdf/compendium/compendium_2006_part_03_02.pdf).

2 Speech made on 18 May 2011 in Brussels by the vice president Viviane Reading, Vice-President and Commissioner responsible for Justice during the presentation of the measures known as the "Victims Package" to protect the victims of crime (http://europa.eu/rapid/press-release_IP-11-585_en.htm).

3 EUROPEAN COMMISSION (2004), EQUAL Guide on gender mainstreaming. Employment & European Social Fund. Available online at: http://ec.europa.eu/employment_social/equal_consolidated/data/document/gendermain_en.pdf.

4 The European Institute for Gender Equality at: http://eige.europa.eu/gender-mainstreaming/concepts-and-definitions accessed August 2015.

5 COUNCIL OF EUROPE (2004), Council Directive 2004/83/EC of 29 April 2004 on minimum standards for the qualification and status of third country nationals or stateless persons as refugees or as persons who otherwise need international protection and the content of the protection granted. Available online at: http://eur-lex.europa.eu/LexUriServ/LexUriServ.do?uri=CELEX:32004L0083:en:HTML.

as a system of interpretation consisting of personal convictions concerning the basic structure, modality and functions of the world; it is not a scientific system. Beliefs can include perceptions of humanity, views of life and morals.[6] Religious discrimination can be defined as "a disadvantageous consideration or distinction of people on the basis of their religious affiliation, their personal belief (or non-belief), their faith-based appearance or behaviour or their assumed religious affiliation."[7]

Minority can refer to a national minority, ethnic minority groups, a religious minority, linguistic minority or sexual minority. According to the Oxford English Dictionary, minority refers to "a small group of people within a community or country, differing from the main population in race, religion, language, or political persuasion".[8] These characteristics can be important elements of identity and are often the source of discrimination.

Social cohesion is a very wide term which can include many elements, however for the present purposes it is understood as being aligned with the Council of Europe's definition of "*the capacity of a society to ensure the well-being of all its members, minimising disparities and avoiding marginalisation.*"[9]

3 Research Issues

As a result of the development of the Internet, most 'traditional' or 'offline' crimes are now also committed online; meaning that the crime in itself does not change but the Internet becomes a tool used as a facilitator of the crime. This makes the crimes much more dangerous, potentially enabling them to affect a larger number of victims and interfere more directly with their personal privacy. The anonymity afforded to perpetrators may increase the sense of impunity while committing cybercrime. All these considerations have been highlighted and developed in the recent Internet Organised Crime Threat Assessment (iOCTA) prepared by the European Cybercrime Centre (EC3) of the European Police Office (EUROPOL).[10] This often leaves victims of these crimes with little

6 EUROPEAN NETWORK OF LEGAL EXPERTS IN THE NON-DISCRIMINATION FIELD (HUMAN EUROPEAN CONSULTANCY, MIGRATION POLICY GROUP (MPG) 2008), Explanatory notes of the amended Equal Treatment Act, Country Report Austria. Available online at: http://www.non-discrimination.net/content/media/2008-AT-Country\%20Report\%20final.pdf.

7 EUROPEAN NETWORK AGAINST RACISM (2007), Religious Discrimination and Legal Protection in the European Union. Fact Sheet N°34 Available online at: http://www.cie.ugent.be/documenten/ENAR_religiousdiscrimination_oct2007.pdf.

8 OXFORD ENGLISH DICTIONARY. Available online at: http://www.oxforddictionaries.com/definition/english/minority.

9 COUNCIL OF EUROPE (2008), Report of High-level task force on social cohesion towards an active, fair and socially cohesive Europe. Available online at: http://www.coe.int/t/dg3/.

10 EUROPOEAN POLICE OFFICE (EUROPOL) (2014), The Internet Organised Crime Threat Assessment (iOCTA). Available online at: https://www.europol.europa.eu/sites/default/files/publications/europol_iocta_web.pdf.

opportunity to seek justice, whilst the damaging effects on reputation or the violation of rights can be longer lasting and more difficult to resolve.

Engaging in CC/CT research related to gender, religion, minority and social cohesion issues allows researchers to obtain vital sets of information that zero-in on the effects of a particular policy or phenomenon, providing a better understanding of the research question being raised at the macro-level and determining which societal actors are being disproportionately affected. Research that creates an understanding of how individuals and groups are perceived by their communal peers and the effects of CC/CT on particular segments of society in relation to social cohesion is of fundamental importance to the development of any CC/CT research methodology.

Researching victims' rights in these and other cybercrimes is also highly relevant; it will provide insights into what kind of measures have already been taken and what kind of research still has to be conducted, particularly with the aim of improving assistance to victims of CC/CT attacks and developing programmes at an EU level that protect their rights.

4 Aspects of Non-discrimination and Victims' Rights as a Possible Limitation of Research

- Lack of consensus on the definition of the terms 'gender', 'minority', 'religion' and 'social cohesion';
- Lack of coherent legal framework and legal certainty;
- Unintentional outcome of groups being isolated, prejudices being reinforced or stigmatisation patterns as a result of focusing on specific ethnic, linguistic, religious or other minorities;
- Indirect discrimination caused by researching characteristics attributable to particular ethnic or other minorities;
- Questions of independence (from government, financing organisation, peers, media etc.) and neutrality of researchers;
- Difficulty conducting transnational investigations;
- Different legislative frameworks applying to various aspects of research;
- Lack of a common legal framework to protect victims of crime that is committed in other jurisdictions;
- Difficulty accessing assisted protection systems;
- Difficulty in identifying victims of the same crime in different countries

5 Relevant Standards

There are several major international and European legislative instruments protecting fundamental rights, which include the areas outlined above that are frequently subject to discrimination.

The Universal Declaration of Human Rights (1948)[11]

[11] Universal Declaration of Human Rights (adopted 10 December 1948 UNGA Res 217 A(III) (UDHR) art 5. Available online at: http://www.un.org/en/documents/udhr/.

Article 2 provides entitlement to all rights and freedoms set out in the Declaration without distinction of any kind. This includes: "*..race, colour, sex, language, religion, political or other opinion, national or social origin, property, birth or other status.*"[12] This entitlement is irrespective of the status of the country or territory to which the person belongs, thereby encompassing the 'universal' element of the instrument.

Article 18 addresses beliefs and religions and sets out the right of freedom for all people in respect of manifestation of these in "*..teaching, practice, worship and observance.*"[13]

European Convention on Human Rights (1950)[14]

Article 9(1) sets out the same right to freedom of thought, conscience and religion as the Universal Declaration of Human Rights. This right is limited in Article 9(2) in respect of overriding interests which have a legal foundation and are necessary for the protection of public safety, public order or other people's rights and freedoms. However, this limitation is subject to strict interpretation by the European Court of Human Rights.

Article 14 again reflects the Universal Declaration of Human Rights, on which it is based, in respect of the right to non-discrimination on any grounds.

Protocol No. 7 to the Convention for the Protection of Human Rights and Fundamental Freedoms[15]

Article 5 provides for equality between spouses in respect of rights and responsibilities towards each other and their children. This right is limited only in the case of overriding interests of the children.

The Charter of Fundamental Rights of the European Union (2000)[16]

Article 10 of the Charter echoes the rights set out in previous instruments in respect of '*Freedom of thought, conscience and religion*'.

Article 21 sets out the right to non-discrimination "*...on any ground such as sex, race, colour, ethnic or social origin, genetic features, language, religion or belief, political or any other opinion, membership of a national minority, property, birth, disability, age or sexual origin..*". Discrimination on grounds of nationality is specifically prohibited in Article 21(2).

Article 22 affords respect for diversity in relation to "*..cultural, religious and linguistic diversity.*"

Article 23 reiterates the principle of equality between men and women in all areas and allows for measures providing advantages for under-represented sexes.

---

[12] Ibid Art. 2.

[13] Ibid Art. 18.

[14] Convention for the Protection of Human Rights and Fundamental Freedoms (European Convention on Human Rights, as amended) (ECHR) art 3, 1950.

[15] Protocol No. 7 to the 1950 European Convention for the Protection of Human Rights and Fundamental Freedoms, (ETS No. 117), entered into force Nov. 1, 1988.

[16] European Union, Charter of Fundamental Rights of the European Union, 26 October 2012, 2012/C 326/02. The Treaty of Lisbon 2009 conferred on the Charter the same legal status as the European Treaties.

There are several other Conventions, Protocols, Declarations, Recommendations and Directives, again at the international and European levels, defending gender, religion and minorities against various forms of discrimination.[17] This reveals the importance placed on protecting against all forms of discrimination, which is at the core of all the activities of the Council of Europe and is integrated into the founding Treaties of the European Union. Social cohesion is also of central importance to the Council of Europe who recognise that it is an essential complement to the promotion of human rights and dignity.[18]

Because the Internet is used as a tool to commit cybercrimes related to gender, religion and minorities, the legal instruments which apply to "offline crimes" can also apply to online crimes. However, given the potentially much larger scale of crime occurring on Internet, specific instruments have been developed. In particular, the Additional Protocol to the Convention on Cybercrime, concerning the criminalisation of acts of a racist and xenophobic nature committed through computer systems,[19] is relevant here. According to the Additional Protocol, "*racist and xenophobic materials*" covers "*any written material, any image or any other representation of ideas or theories, which advocates, promotes or incites hatred, discrimination or violence, against any individual or group of individuals, based on race, colour, descent or national or ethnic origin, as well as religion if used as a pretext for any of these factors.*"

Since the beginning of the 1980's the Council of Europe and the United Nations have issued recommendations and resolutions as well as agreed on conventions for their Member States in order to improve the care for the victims and the development of a criminal policy focused on the interests and needs of victims, while the EU has adopted EU legislation both in relation to all victims and in relation to specific groups of victims. All these legal instruments have been and are adopted to eventually ensure that victims can participate actively, have adequate rights and are being treated fairly within criminal proceedings.

[17] For examples see: UN OFFICE FOR THE COORDINATION OF HUMANITARIAN AFFAIRS, IRIN humanitarian news and analysis. Available online at: http://www.irinnews.org/indepthmain.aspx?InDepthId=20&ReportId=62846; EUROPEAN UNION AGENCY FOR FUNDAMENTAL RIGHTS AND THE EUROPEAN COURT OF HUMAN RIGHTS, Handbook on European non-discrimination law. Available online at: http://fra.europa.eu/sites/default/files/fra_uploads/1510-FRA-CASE-LAW-HANDBOOK_EN.pdf; and UNITED NATIONS INTERREGIONAL CRIME AND JUSTICE RESEARCH INSTITUTE (UNICRI) - LIGHT ON Training Manual: Investigating and Reporting Online Hate Speech, p. 50.

[18] COUNCIL OF EUROPE (1997), Final Declaration of the Second summit of heads of State and Government. Available online at: https://wcd.coe.int/ViewDoc.jsp?id=593437.

[19] COUNCIL OF EUROPE (2003), Additional Protocol to the Convention on Cybercrime, concerning the criminalisation of acts of a racist and xenophobic nature committed through computer systems. Available online at: http://conventions.coe.int/Treaty/en/Treaties/Html/189.htm.

Council of Europe

Recommendation No. R (83) 7 of the Committee of Ministers to Member States on Participation of the Public in Crime Policy[20]

It is recommended that the governments of Member States promote participation of the public in the drawing up and implementation of a crime policy aimed at the prevention of crime, the use of alternatives to custodial sentences and the provision of assistance to victims.

Recommendation no. R (85) 11 of the Committee of Ministers to Member States on the Position of the Victim in the Framework of Criminal Law and Procedure[21]

It is provided that, amongst other things, Judges should have the authority to oblige the convicted to restore the victim and to link probation to effective restoration.

Recommendation No. R (87) 21 of the Committee of Ministers to Member States on Assistance to Victims and Prevention of Victimisation[22]

This Recommendation, together with the n. 85 of 1985 and the UN Declaration on the basic principles of justice for victims of crime and abuse of power - adopted by the UN General Assembly November 29, 1985 - shows a list of rights to be granted to the victims not only from the profile of compensation, but above all in terms of service, privacy assurance, information and participation in the criminal process, and possibly protection.

Recommendation Rec (2003) 20 of the Committee of Ministers to Member States Concerning New Ways of Dealing with Juvenile Delinquency and the Role of Juvenile Justice[23].

Regarding the victims of juvenile crime, it is recommended that to address serious, violent and persistent juvenile offending, Member States should develop a broader spectrum of innovative and more effective (but still proportional) community sanctions and measures. Examples of these are to directly address offending behaviour as well as the needs of the offender, and to evaluate whether to involve the offender's parents or other legal guardian and, if possible, deliver mediation, restoration and reparation to the victim. In addition, the UN Committee of Ministers addresses the issues related to negative perceptions, affirming that information strategies on juvenile delinquency and the work and effectiveness of the juvenile justice system should be developed to inform public opinion and increase public confidence.

European Convention on the Compensation of Victims of Violent Crimes, Strasburg, 24 November 1983[24]

[20] http://www.coe.int/t/dghl/standardsetting/victims/Rec(1983)7.pdf.

[21] http://www.coe.int/t/dghl/standardsetting/victims/recR_85_11e.pdf.

[22] https://wcd.coe.int/com.instranet.InstraServlet?command=com.instranet.CmdBlobGet&InstranetImage=608023&SecMode=1&DocId=694280&Usage=2.

[23] https://wcd.coe.int/ViewDoc.jsp?id=70063.

[24] http://apav.pt/apav_v2/images/pdf/pk06032_031.pdf.

This Convention regards a specific group of victims: victims of intentional crimes of violence who have suffered bodily injury or impairment of health and of dependants of persons who have died as a result of such crimes, and establish general rules about contribution to the compensation of them by sources of the Member States.

United Nations

Declaration A/RES/40/34 of Basic Principles of Justice for Victims of Crime and Abuse of Power, 29 November 1985[25]

The UN General Assembly adopts the Declaration of Basic Principles of Justice for Victims of Crime and Abuse of Power, annexed to the resolution, which is designed to assist Governments and the international community in their efforts to secure justice and assistance for victims of crime and victims of abuse of power, mostly by regulating the access to justice and fair treatment, the restitution and compensation processes.

European Union

Council Framework Decision of 15 March 2001 on the Standing of Victims in Criminal Proceedings[26]

It provides for the possibility for Member States to "promote" the mediation as a search for a negotiated solution between the victim and the offender, in the context of criminal proceedings. Such activities must be counterbalanced by a support activity for the victim, covering the processing of their experiences prior to the offense right away (fear, anger, confusion, anxiety, and so on), as well as an accompaniment during the course of mediation/repair.

Council Directive 2004/80/EC of 29 April 2004 Relating to Compensation to Crime Victims[27]

The Directive sets up a system of cooperation to facilitate access to compensation to victims of crimes in cross-border situations, which should operate on the basis of Member States' schemes on compensation to victims of violent intentional crime, committed in their respective territories. Therefore, a compensation mechanism should be in place in all Member States.

Directive 2011/36/EU of the European Parliament and of the Council of 5 April 2011 on Preventing and Combating Trafficking in Human Beings and Protecting its Victims, and Replacing Council Framework Decision 2002/629/JHA[28]

Member States must ensure that assistance and support are provided to victims before, during and after criminal proceedings (e.g. with the provision of

---

[25] http://www.un.org/documents/ga/res/40/a40r034.htm.

[26] http://db.eurocrim.org/db/en/doc/346.pdf.

[27] http://db.eurocrim.org/db/en/doc/330.pdf.

[28] http://eur-lex.europa.eu/legal-content/EN/TXT/PDF/?uri=CELEX:32011L0036&from=EN.

accommodation, medical treatment, psychological assistance, information, interpretation and translation services). Children must receive additional measures such as physical and psycho-social assistance, access to education, and, where appropriate, the option to appoint a guardian or a representative. During the investigation and criminal proceedings, victims must receive appropriate protection including access to legal counselling and representation, free of charge if necessary, and access to a witness protection programme, where appropriate.

Directive 2011/99/EU of the European Parliament and of the Council of 13 December 2011 on the European Protection Order[29]

As stated in this Directive, the provisions apply to protection measures which aim specifically to protect a person against a criminal act of another person which may, in any way, endanger that person's life or physical, psychological and sexual integrity. This Directive applies exclusively to protection measures adopted in criminal matters (Protection measures adopted in civil matters are covered by Regulation (EU) No 606/2013 of the European Parliament and of the Council of 12 June 2013 on mutual recognition of protection measures in civil matters).

Directive 2012/29/EU of the European Parliament and of the Council of 25 October 2012 Establishing Minimum Standards on the Rights, Support and Protection of Victims of Crime, and Replacing Council Framework Decision 2001/220/JHA[30]

This Directive deals with victims' needs in an individual manner, based on an individual assessment and a targeted and participatory approach towards the provision of information, support, protection and procedural rights. Special attention is given to special support and protection for victims of certain crimes, including victims of gender-based violence, predominantly women, in particular due to the high risk of secondary and repeat victimisation, of intimidation and of retaliation. The Directive also insists on a child-sensitive approach, whereby the best interests of a child victim must be the primary consideration throughout their involvement in criminal proceedings. Furthermore, the Directive is built on the key principle of the "role of the victim in the relevant criminal justice system".

Resolution of the Council on a Roadmap for Strengthening the Rights and Protection of Victims, in Particular in Criminal Proceedings, 9 and 10 June 2011[31]

The EU Council set out a Roadmap for strengthening the rights and protection of victims, in particular in criminal proceedings, establishing minimum standards on the rights, support and protection of victims of crime.

---

[29] http://ec.europa.eu/justice/criminal/files/directive_2011_99_on_epo_en.pdf.

[30] http://eur-lex.europa.eu/legal-content/EN/TXT/PDF/?uri=CELEX:32012L0029&from=en.

[31] http://victimsupporteurope.eu/activeapp/wp-content/uploads/2012/09/Resolution-of-the-Council-on-a-roadmap-for-strengthening-the-rights-and-protection-of-victims-in-particular-in-criminal-proceedings1.pdf.

6 Case Study

Research has not revealed any particular cases of CC/CT research in which issues relating to gender, religion, minorities, or social cohesion emerged. To illustrate the type of issues that may arise in some types of research, we can refer to a case in another field of research, namely genetics.

The *Arizona Board of Regents v. Havasupai Tribe* case[32] deals with the allegation of broadening research purposes and minority group interests and concerned genetic research of a small US tribe which focused on discovering a link between genes and diabetic risk. However, the research was later used for researching other disorders, including schizophrenia, which was a taboo issue for the tribe. The general consent form on which the research was based did not specifically refer to this. The case was eventually settled out of court thereby preventing a legal precedent on this issue, but it served to highlight the way in which research results can achieve a far-reaching impact that was not originally envisaged. The importance of group identity can be adversely affected by research results and those not directly involved can nevertheless feel the impact by association.

Research involving sensitive information related to ethnographic studies of certain cybercrime victims should be careful of the risks of function creep and of consent based on less than specific information. Similarly, researchers carrying out large-scale quantitative research involving Big Data analytics have a moral obligation to weigh the advantages of publishing information that may risk the marginalisation of the group or its members. Anti-discrimination law may also create a legal obligation in this respect. Care should also be taken in the presentation and interpretation of research results to avoid stigmatisation being caused by insensitive communication creating false impressions about the groups being studied.

7 Focus on Freedom of Speech

Engaging in research that addresses discrimination issues relating to specific gender, minority or religious groups could cause problems in 'closed' societies where ethnic or religious groups hold disproportionate amounts of power or where certain freedoms are limited. In such countries researchers could face censorship, which may affect the securing of funding and/or limit the publication of work.

Freedom of speech may also be limited in so-called 'open' societies. Numerous countries, regardless of their political, cultural or religious traditions, limit this freedom in case of targeted attacks involving discrimination, insult or defamation. Research of a particular groups might be perceived as such. On the whole, Europe aims to protect freedom of speech in cyber-space[33] and the European

[32] Information about the case is based on the case description on http://genetics.ncai.org/case-study/havasupai-Tribe.cfm.

[33] EUROPEAN COMMISSION (2013), EU Cybersecurity plan to protect open internet and online freedom and opportunity (Press Release). Available online at: http://europa.eu/rapid/press-release_IP-13-94_en.htm.

Commission, together with the High representative of the Union for Foreign Affairs and Security Policy, has published a cybersecurity strategy,[34] alongside a Commission proposed Directive.[35]

Any study of matters relating to the effects of CC/CT on victims and their rights could, in certain cases, clash with their legitimate expectation of privacy, which could generally curtail freedom of speech.

Most cases involving vulnerable victims would require anonymisation of the names of the people involved or even non-attention (understood as being voluntary, though induced, lack of interest) in respect of each single episode most harmful to victims' privacy; failing this, post-crime victimisation situations could come about. Generally, we speak about "secondary victimisation" (or "post-crime victimisation") when the victims of crime undergo a second victimisation by institutions or social workers, or otherwise on account of unwanted media exposure. In this regard, in 1985 the General Assembly of the United Nations formulated the "Declaration of the basic principles of justice for the victims of crimes and abuse of power"[36] in which the Member States, in close cooperation with representatives of the mass-media, are encouraged to elaborate and implement effectively guidelines for the media aimed at protecting victims and curtailing re-victimisation.

8 Focus on Academic Freedom

According to Encyclopaedia Britannica, academic freedom refers to "*the freedom of teachers and students to teach, study, and pursue knowledge and research without unreasonable interference or restriction from law, institutional regulations, or public pressure. Its basic elements include the freedom of teachers to inquire into any subject that evokes their intellectual concern; to present their findings to their students, colleagues, and others; to publish their data and conclusions without control or censorship; and to teach in the manner they consider professionally appropriate. For students, the basic elements include the freedom to study subjects that concern them and to form conclusions for themselves and express their opinions.*"[37]

[34] EUROPEAN COMMISSION AND THE HIGH REPRESENTATIVE OF THE UNION FOR FOREIGN AFFAIRS AND SECURITY POLICY (2013), Cybersecurity Strategy of the European Union: An Open, Safe and Secure Cyberspace. Available online at: http://eeas.europa.eu/policies/eu-cyber-security/cybsec_comm_en.pdf.

[35] EUROPEAN COMMISSION (2013), Proposal for a directive concerning measures to ensure a high common level of network and information security across the Union. Available online at: http://ec.europa.eu/information_society/newsroom/cf/dae/document.cfm?doc_id=1666.

[36] Resolution n° 40/34 of 29 September 1985 at the General Assembly of the United Nations (https://www.unodc.org/pdf/compendium/compendium_2006_part_03_02.pdf).

[37] ENCYCLOPAEDIA BRITANNICA. Available online at: http://www.britannica.com/EBchecked/topic/2591/academic-freedom.

In order to retain academic freedom, it is important for researchers to ensure they remain independent and neutral to avoid being pressured into either carrying out or abandoning certain types of research on specific groups of persons. Maintaining this stance also supports thoroughness and limits the possibility of bias induced by convictions, ideals or beliefs. Total independence from institutions such as universities and governments, as well as peers and the media is also important.

9 Country Studies

9.1 Italy

In Italy there is no specific legislation in respect of CC victims' rights. Laws governing victims' rights generally and CC laws will be used to analyse the issues.

In the Italian legal system, the term *victim* is not used, instead terms such as *the harmed person*[38] or *the harmed party*[39] are used which incorporate the concept of passivity and protection of interests damaged by the prosecutable act. A selection of laws includes this notion:

Law 547/1993 contains specific provisions contemplating and punishing traditional IT crimes, such as unauthorised access to IT systems, IT damage, computer fraud etc.

Legislative Decree 231/2001 introduces the principle of administrative liability for crimes committed by executives and employees of companies, including IT crimes.

Legislative Decree 196/2003 is the Italian Privacy Code governing personal data processing.

Law 48/2008 gives effect to the Council of Europe's Budapest Convention[40] which addresses cybercrime.

The Constitution of the Italian Republic provides for freedom of expression, press and religion in public places.

The Italian Penal Code is a source of Italian criminal law and addresses crimes against the person, property and morality and decency. It has been amended by **Law 547/1993** in respect of computer criminality.

Due to the difficulties already outlined, particular to victims of cybercrime, there are only a few legal cases in this area. One of these is the case of **Mr H**[41] which concerned a series of computer frauds committed against 28 users and online sales platforms. A characteristic of the case reflects the unfortunately typical problem of the absence of participation in the proceedings by the victims, resulting in no compensation for the affected persons.

---

[38] Articles 92 and 103 of the Criminal Procedure Code.

[39] Article 70, no. 2 of the Criminal Code.

[40] Council of Europe, Convention on Cybercrime, 23 November 2001, available at: http://www.refworld.org/docid/47fdfb202.html.

[41] Milan Law Court, with a single judge, Criminal Section VII, Decision n° 10397/2012 of 16 October 12.

Since 2007 there have been many instances of criminal proceedings brought against people accused of organising and carrying out computer fraud against Italian citizens. In certain cases, banks have been convicted with regards to cases of phishing, due to a failure to implement security measures to protect the rights of their account holders. However, information gathered shows that very few users chose to retain a lawyer to follow proceedings first hand.

Lastly, with regard to this type of crime, it is helpful to mention the ambiguous role, halfway between victim and accessory to the crime, played by what are known as financial managers. These people, who are contacted by perpetrators of computer fraud on which phishing is based on in order to launder money coming from the accounts of the original victims, have been alternatively considered by courts as further victims of fraud, rather than accomplices who are aware of the unlawful scheme.[42]

Over the course of the last few years, jurists and sociologists[43] have taken action frequently in the Italian landscape; the aim is to increase protection of and assistance to the victims of crime, not only with regard to a hoped-for, more active participation in criminal proceedings, but also to provide the protection they require with regards to the distribution of data and news pertaining to their condition and, consequently, to the phenomenon known as re-victimisation. In legislative terms, a constitutional bill put forward in 2006 has fostered discussion on whether it is advisable to amend article 111 of the Constitution in order to constitutionally recognise guarantees and rights for the victims of crime.[44]

9.2 France

This section provides a brief overview of some of the relevant national legislation and general issues in relation to victims' rights and non-discrimination in terms of gender, religion and minority in the context of research on CC/CT activities.

Although not legally defined in French law, the principles of equality and indivisibility of the nation are at the core of fundamental French laws, which prohibit distinction based on ethnicity, community religion etc. The French legal system also has a long history of taking account of victims in criminal proceedings, and recently introduced obligations to support, inform and protect victims at every stage.

---

[42] See the decision handed down by the Judge for Preliminary Investigations at Milan Law-Court, Dr. Luerti (http://robertoflor.blogspot.it/2009/06/phishing-misto-e-attivita-abusiva-di.html).

[43] See for example, the records of the Coordination of Democratic Jurists meeting at Turin on 9 June 2001 at the "The victim of Crime, this unknown entity" meeting (http://files.giuristidemocratici.it/giuristi/Zfiles/ggdd_20030723122357.pdf).

[44] Chamber of Deputies n° 1242, Constitutional Bill put forward by the deputy, Boato, Amendment to article 111 of the constitution regarding the guaranteeing the rights of victims of crime. Presented on 29 June 2006. (http://www.camera.it/_dati/leg15/lavori/schedela/apriTelecomando.asp?codice=15PDL0008750).

The **Preliminary Article** added to the **Criminal Procedure Code** by **Law n. 2000/516**[45] provides for the judiciary to ensure respect of victims' rights and to keep them fully informed throughout any criminal process.

The **Criminal Procedure Code**[46] provides the right for victims to receive full information about their specific rights, possible alternative actions and support and compensation services. Articles 2 & 3 of this Code also provides for the simultaneous exercise of a civil action by the victim, relating to any damage arising from the same crime. Several fundamental rights are, therefore, provided to the victim by the national law.[47]

Freedom of the Press Act[48] criminalises any defamation or insult committed against a person or group because of their origin, ethnicity or membership/non-membership of a nation, race or religion.

The **Constitution**[49] states that France "*shall ensure the equality of all citizens before the law, without distinction of origin, race or religion. It shall respect al beliefs*".[50] Equality of men and women is also addressed.

The **Act on the Fight Against Racism**[51] is the first law specifically designed to combat all forms of racism.

The **Act on Information Technology, Data Files and Civil Liberties**[52] prohibits the processing of personal data that either directly or indirectly reveals racial or ethnic origins, political, philosophical and religious opinions.

The **Act on Information Technology, Data Files and Civil Liberties**[53] addresses cybercrime.

The **Act for Building Confidence in the Digital Economy**[54] introduces a new section into the Criminal Code imposing a duty on internet service providers to participate in the fight against hate crime such as websites disseminating xenophobic, anti-Semitic or Islamophobic ideas. Article 6 was limited to the incitement of crimes against humanity and racial hatred, but the recent law

---

45 Of 15 June 2000.

46 Article 53-1 and Article 75.

47 More information about help and support services to victims in France are avaible at: https://e-justice.europa.eu/content_rights_of_victims_of_crime_in_criminal_proceedings-171-FR-maximizeMS-en.do?clang=en&idSubpage=4&member=1.

48 Loi du 29 juillet 1881 sur la liberté de la presse, JORF du 30 juillet 1881 p. 420.

49 Constitution du 4 octobre 1958, JORF n° 0238 du 5 octobre 1958, p. 9151 Available online at: http://www.conseil-constitutionnel.fr/conseil-constitutionnel/english/constitution/constitution-of-4-october-1958.25742.html.

50 Article 1.

51 Loi n° 72-546 du 1 juillet 1972 relative à la lutte contre le racisme, JORF n°0154 du 2 juillet 1972 p. 6803.

52 Loi n° 78-17 du 6 janvier 1978 relative à l'informatique, aux fichiers et aux libertés, JORF du 7 janvier 1978), p. 227–231.

53 Loi n° 78-17 du 6 janvier 1978 relative à l'informatique, aux fichiers et aux libertés, JORF du 7 janvier 1978, p. 227–231.

54 Loi n° 88-19 du 5 janvier 1988 relative à la fraude informatique, JORF du 6 janvier 1988 p. 231.

for true equality between women and men[55] has expanded the scope to hateful incitement against "sex, sexual orientation, sexual identity and handicap".

France is committed to the defence of freedom of expression, including on the internet, and has non-discrimination at the core of its concerns. The country was also one of the first to implement CC/CT legislation and to develop a system providing assistance and protection to victims. Although there are a high number of cybercrime victims, this may be as a result of having a high rate of internet connection, as France have pioneered computer law enforcement in Europe. The legislation also limits the freedom of speech on the internet to protect minority groups, however this could be a limiting factor for researchers studying such groups.

9.3 Spain

This section provides a general overview of victims' rights in Spain in relation to research activities. Current Spanish criminal legislation does not offer a clear definition of the concept of victim, however private individuals have the right to an effective role in the procedural criminal system and are considered private prosecution parties. General rights for victims, rather than specifically relating to CC/CT are found in several pieces of legislation.

Organic Law 19/1994 on the protection of witnesses and experts in criminal cases.

Law 35/1995 on aid and assistance to victims of violent crimes and crimes against sexual freedom.

Criminal Procedure Act and all its successive reforms.

Organic Act 1/2004 on integrated protection measures against gender violence.

Law 29/2011 on the protection of victims of terrorism.

Additionally, academic and governmental experts have proposed a **Victims' Statute** within the draft of the new Spanish Criminal Procedure Code.[56] Should this be approved it would incorporate a definition of 'victim' or procedural purposes, which would be understood as a natural person or legal entity harmed by the crime and the one who directly suffers loss or damage caused by the punishable acts. Article 69 of the draft law provides for a general prohibition on 'secondary victimisation' which is a frequent consequence of certain types of cybercrime. The provision of a specific regulation regarding this issue could make it easier to combat the likely rise in this phenomenon in the future. The new Article 76 establishes victims' rights automatically arising from the fact that the victim is injured or harmed by the crime. This would make it possible to file a suit more easily because the procedural position of 'victim' arises out of an objective fact, which can be easily verified in cases where proceedings are initially refused.

[55] Loi n° 2014-873 du 4 août 2014 pour l'égalité réelle entre les femmes et les hommes, JORF n°0179 du 5 août 2014 p. 12949.

[56] Anteproyecto de la Ley de Enjuiciamiento Criminal de 27 de julio 2011.

According to a report issued from the Interior Ministry in May 2014, Spain's security forces received 42,437 complaints in the previous 12 months for different typologies of cybercrimes. Of these cases, only 2,167 had been resolved. It means that around 95 % of cybercrimes, or offenses related to new technologies at that time, were going unpunished in Spain.[57] However, in 2013 Spanish authorities, in connection with the European Cybercrime Centre dismantled the largest and most complex cybercrime network dedicated to spreading *Police Ransomware*;[58] a type of malware that blocks the computer, accusing victims of having visited illegal websites and requesting payment to unblock it. There were more than 1200 reported cases in Spain, and the numbers could be much higher. This type of crime highlights the difficulties faced by cybercrime victims in terms of seeking redress.

Although legislation and self-regulatory initiatives to protect victim rights exist in Spain, and more effective rules are being proposed and debated, none of these laws or initiatives are specific to CC/CT; in the Spanish system the victim of any crime is entitled to initiate prosecution. A general catalogue of procedural and extra-procedural rights for victims of crimes, including CC/CT is needed, as well as official schemes offering assistance.

9.4 Country Overview

It can be seen from the examples provided, that research has produced no specific legislation, and therefore case law, relating to victims' rights and non-discrimination in relation to CC/CT research. The general principles which exist in national provisions and approaches afford protection, to a greater or lesser extent, in general terms, which can then be applied to CC/CT situations. It has been shown however, that there are conditions unique to CC/CT which may not be adequately accounted for by current measures. Different approaches and priorities adopted by various countries reveals an uneven landscape in terms of protection of rights. It might be said that this is an unsatisfactory situation for victims of certain cybercrime, who are frequently spread across several jurisdictions, resulting in inconsistent justice. This same feature applies to researchers in this field who need to be aware of the differences.

10 Conclusion

The analysis provided in this chapter addressed issues of social inclusion, discrimination on the basis of gender, religion or ethnic minority and the rights of victims, within the context of CC/CT research. Special considerations and examples of issues that may pose problems have been referred to, along with a variety of legislative instruments and standards that apply to certain aspects of

[57] Compare the news online at: http://elpais.com/elpais/2014/05/09/inenglish/1399628265_760093.html.

[58] Further information avaible online on the site of Europol at: https://www.europol.europa.eu/content/police-dismantle-prolific-ransomware-cybercriminal-network.

these areas. However, it appears from the research carried out, that no legislation or case law addresses the issues in this particular context.

This leads to the understanding that CC/CT researchers are free to engage in research in which minority groups are analysed in order to obtain scientific insight into the effects of CC/CT and how some societal actors might be disproportionately affected by either the crimes or measures to prevent them. Similarly, detailed research into the particular characteristics of actual or potential victims of CC/CT crime is to be encouraged for the same reasons, and to inform the development of further protective measures.

Nonetheless, there still exist potential dangers to violate victims' rights indirectly as a result of carrying out this type of research. There is also a risk, albeit small, that indirect discrimination could be a result of some research activities. It may therefore, be helpful to highlight some of these risks that researchers should be aware of.

10.1 Presentation of Research Results

Legal frameworks addressing discrimination are typically and intentionally open and flexible, focusing on principle rather than specific actions, which means context is an important factor in interpretation. This leaves open the possibility that the presentation of results of research in which particular groups are singled out might be perceived as discriminatory in some jurisdictions. Results published on the internet could trigger liability in several jurisdictions, however this risk seems negligible as long as the publication is not specifically targeted (e.g. in language), and particular efforts are made to anonymise or protect victims' personal data.

10.2 Potential Bias

Researchers should avoid the stigmatisation of groups or reinforcing prejudices[59] by the implied use of a white, heterosexual, Christian male as a reference point for studies. Although this is a moral obligation rather than a legal one, it is advisable to bear in mind the openness of the legal anti-discrimination norms. It is important that researchers question themselves as to which assumptions they, possibly unconsciously, making and they should be as transparent as possible in their reporting about all assumptions underlying their research.

10.3 Indirect Effects

There is some risk in CC/CT research, of indirect discrimination when seemingly neutral characteristics are used that correlate with minority groups or gender;

[59] For a discussion of stigmatisation, see for example Gross, S. R., & Livingston, D. (2002). Racial profiling under attack. Columbia Law Review, 1413-1438; van der Leun, J. P., & van der Woude, M. A. (2011). Ethnic profiling in the Netherlands?A reflection on expanding preventive powers, ethnic profiling and a changing social and political context. Policing and society, 21(4).

for example, social background, education level or zip code. Quantitative studies involving Big Data Analytics to discover people susceptible to being a victim of cybercrime, for example, could produce categories that may feed into policy decisions but which, in effect, serve as a proxy for ethnicity or gender. This would not constitute discrimination if sound research methods were applied preventing unintentional research biases, the research was objectively justified by a legitimate aim and a lack of alternatives was shown.

Another possible indirect effect of victim research studies is the 're-victimisation' of those taking part. This type of re-aggravation of the adverse effects of the crime is largely stigmatised by the EU and recent legislation protecting victims' rights. This is a difficult area and all necessary care not to harm the interests of those involved must be taken.

10.4 Neutrality of Research

The importance of independence and neutrality of researchers, in being free from pressure by the government, funding agency, media, or advocacy groups, has been pointed out. In some countries, researchers have limited possibilities of acquiring funding or publishing their work if the research goes against the grain of what the government or society at large deem appropriate. Researchers could face problems in more closed societies, in this regard.

11 Recommendations

Rather than the areas addressed here giving rise to particular legal issues in relation to carrying out CC/CT research, it can be said that the underlying ethical principles have a more direct effect. This is more likely to result in research that avoids stigmatisation or indirect discrimination of particular groups and fully respects the rights and interests of those who have been affected by these types of crime.

All those involved in developing, controlling and delivering research projects are advised to:

- take particular care in the presentation and interpretation of research results in order avoid stigmatisation, encroaching on victims' privacy and harming their reputation or 'right to be forgotten';
- as far as possible, use anonymised data for victims;
- limit the use of (sensitive) information that is capable of violating victims' rights to the amount necessary for research purposes;
- avoid bias in research design or analysis in terms of what is considered 'normal' behaviour, based on an implicit standard of white heterosexual Christian male as a reference point;
- foster inclusive formations of research groups, with an appropriate gender balance and where possible including researchers from different religious or ethnic groups, which can minimise the risk of biased research assumptions;

- Guarantee researchers' impartiality and extraneity, ensuring they are free from pressure from financing organisations, governments or social pressure groups.

Acknowledgement. The research leading to these results has received funding from the European Union Seventh Framework Programme (FP7-SEC-2013) as the COURAGE project under grant agreement no 607949.

Risks Related to Illegal Content in Cybercrime and Cyberterrorism Research

Alison Lyle[1(✉)], Benn Kemp[1], Albena Spasova[2], and Ulrich Gasper[3]

[1] Office of the Police and Crime Commissioner for West Yorkshire, Wakefield, UK
{alison.lyle,benn.kemp}@westyorkshire.pnn.police.uk
[2] International Cyber Investigation Training Academy, Sofia, Bulgaria
albaadvisors@gmail.com
[3] Cybercrime Research Institute, Cologne, Germany
gasper@cybercrime.de

Abstract. What follows here is an examination of the risks and issues related to illegal content within, and related to, the context of cybercrime and cyberterrorism research. Before any useful analysis can take place, it is necessary to create an understanding of the subject matter; therein lies the first challenge. The problem of establishing what the term 'illegal content' encompasses is addressed throughout. By outlining the particular relevance of illegal content with regard to research, we set out the key considerations which will assist in understanding what is required to successfully carry out valuable research and to understand the possible limitations. Some of these are related to the fact that the nature of much illegal content means that victim considerations are of utmost importance. Just as there is no specific definition of illegal content, there is no specific legislation addressing this type of criminal activity, therefore a wide range is presented and considered, which further assists in illustrating different perspectives. Countries too, have different perspectives and an in-depth examination of two of them reveal both similarities and differences. The general conclusion draws together the findings and the issues that have been addressed and provides a holistic view of the main points before key recommendations are presented.

Keywords: Illegal content · Anonymity · Pseudonymity · Risk Management · Data protection compliance · Research ethics

1 Definitions

The term illegal content is used across cybercrime and cyberterrorism (CC/CT) disciplines but it remains one which is not clearly defined within the European Union. In 2013 the European Data Protection Supervisor (EDPS) stated;

B. Akhgar and B. Brewster (eds.), *Combatting Cybercrime and Cyberterrorism*,
Advanced Sciences and Technologies for Security Applications,
DOI 10.1007/978-3-319-38930-1_7

"The EDPS is of the view that there is a need for a more pan-European harmonised definition of the notion of 'illegal content' for which the notice-and-action procedures would be applicable"[1]

In its consultation, the Commission had listed intellectual property rights infringements, consumer protection law breaches, hate incitement, child abuse content, terrorism related content, defamatory material and privacy-invading material among the examples of what could constitute 'illegal content'.[2] While this is a useful reference point, to a certain extent the definition is subjective and different member states will take different views depending on many variables. How certain acts are dealt with, whether they are dealt with and what sanctions are delivered can reveal different approaches. At this time, the definition is left to nations to use, adapt or make new legislation to tackle and define these. This leads to differences in the practical application of an illegal content definition across EU member states. Many crimes that might be considered as illegal content are dealt with by existing laws, as frequently it is the way they are carried out that categorises them as 'cybercrime' (CC). This involves the adaptation of rules of evidence, disclosure and laws controlling investigations rather than a requirement for a separate law.

2 Introduction

When researching issues relating to CC/CT, illegal content is a theme that will run throughout. An understanding of illegal content is key for researching and identifying what constitutes criminality in Member States across the EU. However, it is an area of prominence within CC/CT that currently lacks a clear definition, research agenda or legal framework. Many types of criminal behaviour can be incorporated into the broad understanding of illegal content, which have varying impacts on the victim. Without traditional boundaries, this category of crime represents a vast problem to be addressed in order to provide real solutions and protection for citizens.

Illegal content typically involves offensive, harmful or manipulative material aimed at individuals who suffer damage, usually psychological, or financial loss. It is important to understand the impact of illegal content crimes on those it affects. Whilst it might be considered that the internet and computer are used as tools to enable crimes to be committed more easily and may reduce the impact on the victim in absence of physical/personal contact, it might be the case that the invasive nature of these offences, committed in the personal 'space' of the

---

[1] European Data Protection Supervisor, "EDPS formal comments on DG MARKT's public consultation on procedures for notifying and acting on illegal content hosted by online MARKT's public consultation on procedures for notifying and acting on illegal content hosted by online intermediaries", European Union Data Protection Supervisor, Brussels 2012. Available at: https://secure.edps.europa.eu/EDPSWEB/webdav/site/mySite/shared/Documents/Consultation/Comments/2012/12-09-13_Comments_DG_MARKT_EN.pdf.

[2] Ibid.

victim, has a more profound effect. Illegal content on the internet is particularly effective due to the offenders being able to target specific groups or individuals more easily.

In order to address these problems, the traditional, developed research agenda can be used. Adapting this and developing new methodologies would move research in this area forward and potentially allow for more meaningful results and analyses. The way people live their lives, and the way others commit crime, has moved on and the research agenda must necessarily follow.

3 Relevance of Illegal Content with Regard to Research

An understanding of the relevance of researching illegal content should be fundamental to the development of any research methodology for counter terrorism and CC. In order to identify solutions required in this area, it is vital to understand the scope and nature of it. This requires the development of an effective research agenda, based on a common understanding of the issues involved. Examples of some of these are: purpose and intent of those engaged with this type of activity, types of material, methods of distribution and impact on victims or critical infrastructures. These are very wide and complex areas which need to be understood so that the various organisations, agencies and individuals involved in providing real solutions can do so in an effective way.

Issues such as confidentiality are relevant to this type of research. Victims of some forms of illegal content may be sensitive, embarrassed or fearful about participating in research, resulting in less information and potentially less reliable information. This reason is linked with the fact that due to the nature of much of the material and the way it is disseminated, discovering incidents is largely dependent on them being reported. This, in turn, is problematic in that what is offensive to one is not to another. Some content would cause harm to one and not to another and so the true scale may not be known, as much would go unreported. In addition, cultural and national differences need to be considered. For various reasons, being identified as a victim of a crime would discourage some people from reporting incidents or participating in research.

In respect of searching for illegal content online the researcher would face the difficult, if not impossible, task of identifying the context of the content; for example, blogs and social media 'conversations'. This may result in research only being possible using identified sources such as dedicated websites or secondary sources such as figures of crimes reported. The illegal content which cannot be defined or contextualised and has not been reported would not be included thereby potentially giving a skewed result.

There may also be a requirement for researchers finding distressing illegal content to take action either on ethical or legal grounds. It might be, depending on the circumstances, that the researcher has a legal duty to report material or disclosures encountered during their work. Even if no legal duty exists, an ethical dilemma might arise. The situation is potentially more difficult when researching illegal content online due to the context based nature of some material. For example, written messages on a social media website might be part of an 'innocent' conversation. Conversely, apparently innocent messages when read in context could have a damaging effect and a more sinister meaning. Researchers may also find themselves in the position of discovering disclosures which may result in illegal and dangerous acts[3], for example a communication revealing an intention to carry out a criminal act in the future.

Considering these ethical and legal issues is of particular relevance in respect of those carrying out the research and those participating. By its nature, much of what is considered to be illegal content is harmful, offensive and disturbing. Work carried out in this area must be sensitive to the potential effects on the researcher and the participant. It is of utmost importance that these considerations are paramount in order to protect both.

Other practical considerations in the area of illegal content include lack of access to some material, for example the Internet Watch Foundation (IWF)[4] who carry out research particularly in relation to child sexual abuse online, do not have the authority to pass payment barriers and are only able to conduct research on publicly available content. This provides limitations to research. An additional barrier is discovering and tracing material due to the nature of the cyber environment; this leads to a lack of data on the amount and availability of the content and the methods of distribution.

Much of illegal content is defined by the harmful effect is has, or is capable of having, on the victim. In turn, harm is a subjective concept which cannot be easily predicted or measured. In terms of research then, it is important to carry out quantitative studies in order to reveal the nature and amount of material as well as qualitative studies analysing the effect it has on those who receive or view it.

Combining quantitative and qualitative methods have proved useful when researching children, due to the qualitative study making the interpretation of the quantitative data more meaningful[5]. User-centric methods would include interviews, focus groups and surveys to identify real-world needs and experiences. This method could be adapted to 'server-centric' which would be carried out online using loggings and metrics to chart where people go. Online focus groups may be an effective way of adapting traditional research methods to suit the purpose[6]. This may be particularly useful with children and vulnerable

[3] Stern, S.R. 'Encountering Distressing Information in Online Research: A Consideration of Legal and Ethical Responsibilities' Chap. 11 in: Hughes, J (ed.) SAGE Internet Research Methods (SAGE 2012) Google eBook http://books.google.co.uk/books?id=A6mHAwAAQBAJ&dq=illegal+content+internet+research+methods&source=gbs_navlinks_s.

[4] Established in 1996 by the internet industry to provide a reporting point for illegal content online http://wwww.iwf.org.uk accessed 7 September 2014.

[5] Lobe, B; Livingstone, S; Olafsson, K and Simoes, J.A. (2008) Best Practice Research Guide: How to research children and online technologies in comparative perspective. London, EU Kids Online (Deliverable D4.2).

[6] Lobe, B; Livingstone, S and Haddon, L (eds.) (2007) Researching Children's Experiences Online across Countries: Issues and Problems in Methodology. London, UK Kids Online (Deliverable D4.1) http://www.lse.ac.uk/media@lse/research/EUKidsonline/EU%20Kids%20I%20(2006-9)EU%20Kids%20Online%20I%20Reports/D41_ISBN.pdf accessed 11 September 2014.

people who may feel more comfortable in a familiar environment in which their anonymity can be maintained.

Comparison studies across EU Member States may be problematic in relation to illegal content due to potential cultural and national differences in perceptions of harmful or offensive material. In addition, differing political agendas across EU Member States' governments may mean that some content is deemed illegal in one country and acceptable or merely controlled in another; for example, online gambling. This highlights the difficulty with creating a universal definition of illegal content and the need for a wider understanding of different approaches to enable a collaborative effort to tackle it.

4 Most Relevant Aspects of Illegal Content as a Possible Limitation of Research

- There is a lack of definition and clear understanding of what constitutes illegal content.
- Cultural and national differences across and within Member States mean different approaches to what is deemed offensive or harmful, and therefore illegal, content. EU legislation allows for this freedom. This adds to the difficulty of creating a common understanding of the subject which in turn leads to difficulties in researching.
- No clear research agenda means addressing the specific issues related to this area is problematic.
- The area affected by illegal content is vast, transcending traditional boundaries and incorporating many topics and issues.
- The lack of common terminology causes difficulties in identifying areas to be researched, and would make literature reviews problematic.
- Complex and inadequate legal framework across the EU and between Member States makes a researcher's position precarious when accessing material and working with vulnerable participants.
- Victims of some crimes might belong to a particular section of society which may be reluctant to participate in research due to shame or stigma attached to being a victim of some types of crime (e.g. victim of sexual abuse).
- Potential conflict of legal and ethical duties of the researcher when discovering certain types of illegal content.
- Difficulties due to the nature of some material being context based makes identification of material problematic.
- Potential practical problems of using traditional methodologies due to the large area involved. Online methodologies also involve additional considerations such as handling and protecting data, which are covered by legislation in all Member States.
- Accessing some material would not be possible, for example passing payment barriers on illegal websites.

5 Inventory of European Union Standards

Art. 288 (ex. Art. 249 TEC) Treaty on the Functioning of the European Union – 'A Directive shall be binding, as to the result to be achieved, upon each Member State to which it is addressed, but shall leave to the national authorities the choice of form and methods'.

Directive 2011/92/EU – combating sexual abuse and sexual exploitation of children and child pornography (and replacing Council Framework Decision 2004/68/JHA).

'E-Privacy' Directive 2002/58/EC – amended by Directive 2009/136/EC.

Regulation (EC) No 460/2004 of 10 March 2004 – establishes the European Network and Information Security Agency.

Council Framework Decision 2008/913/JHA 28 November 2008 – on combating certain forms and expressions of racism and xenophobia by means of criminal law.

Council Framework Decision 2002/475/JHA of 13 June 2002 – on combating terrorism.

Council Decision 2002/187/JHA of 28 February 2002 – setting up Eurojust with a view to reinforcing the fight against serious crime.

Communication from the Commission concerning terrorist recruitment: addressing the factors contributing to violent radicalisation, COM(2005) 0313 final.

Digital Agenda for Europe COM(2010) 245 final – one of seven initiatives of the Europe 2020 Strategy.

"The EU Internal Security Strategy in Action: Five steps towards a more secure Europe" COM[2010] 673 – created EU Radicalisation Awareness Network (RAN) to promote actions to empower communities and key groups engaged in the prevention of violent radicalisation and recruitment.

Proposal for a Directive concerning measures to ensure a high common level of network and information security across the Union – COM(2013) 48 final 07/02/13.[7]

Communication on Preventing Radicalisation to Terrorism and Violent Extremism: Strengthening the EU's Response COM(2013) 941 final – refers to terrorists and extremists capitalising on technological advances and using social networking sites, online video channels and radical chat rooms.

Cybersecurity Strategy of the European Union: An Open, Safe and Secure Cyberspace – JOIN(2013) 1 final 07/02/2013 – the first comprehensive policy document that the EU has produced in this area.

Charter of Fundamental Rights – became primary EU law under the Lisbon Treaty 2009.

---

[7] 'The Security Directive' aims to implement cybersecurity strategy across EU. At June 2015 the main principles have been agreed in a fourth trilogue meeting.

6 Inventory of Council of Europe Standards

European Convention on Human Rights

Convention on Cybercrime (the Budapest Convention) ETS 185, 2001.

Additional Protocol to the Convention on Cybercrime ETS 289, 2003 – concerning the criminalisation of acts of a racist and xenophobic nature committed through computer systems.

Convention on the Protection of Children against Sexual Exploitation and Sexual Abuse (ETS 201).

Convention on the Prevention of Terrorism (ETS 196).

Additional Protocol to the Convention on Cybercrime, concerning the criminalisation of acts of a racist and xenophobic nature committed through computer systems (CETS No.: 189).

7 Inventory of International Standards

United Nations Convention on Rights of the Child – Adopted and opened for signature, ratification and accession by General Assembly Resolution 44/25 of 20 November 1989. Entry into force 2 September 1990.

ITU – International Telecommunication Union; launched the Global Cybersecurity Agenda in 2007, a framework for cooperation and response to cyber threats. http://impact-alliance.org/download/pdf/resource-centre/brochure/ITU-GCA-brochure.pdf.

IMPACT - International Multilateral Partnership Against Cyber Threats, a key partner of the ITU http://impact-alliance.org/aboutus/mission-&-vision.html.

8 Illegal Content v. Freedom of Speech

The universal right to freedom of opinion and expression has a rich history in societal, political, ethical and legal contexts. In particular, it is protected at European level within the core legal documents and the underlying principles of both the Council of Europe and the European Union. It is considered fundamental and although conditional, is only limited in strict circumstances. The Committee of Ministers of the Council of Europe has stated that the right to freedom of opinion and expression is a universal right which needs to be protected everywhere and for everyone. It is emphasised that this applies equally online as well as offline.[8]

[8] CM/Rec(2014)6 on a Guide to human rights for Internet users. Available at: https://rm.coe.int/CoERMPublicCommonSearchServices/DisplayDCTMContent?documentId=09000016804d5b31.

Whilst illegal content is lacking a clear definition across EU member states, all national laws provide for circumstances where online content is clearly illegal, such as child sexual abuse images. Authorities will take direct action to monitor and tackle this but the challenge remains to identify where the line is drawn between freedom of speech and offensive material.

Expectations in relation to freedom of speech can differ between various societies across the world; some regimes are particularly restrictive whilst others promote and protect the right. The internet has gone some way to opening up opportunities for all people to exercise their right to freedom of speech, however the indiscriminate blocking of content by some States, in the name of preventing dissemination of perceived illegal content, is capable of being seen as unfairly curtailing this freedom. Achieving a fair balance between protecting citizens from harmful criminal acts and respecting fundamental freedoms is a challenge faced across the globe.

9 Illegal Content v. Academic Freedom

Academic freedom is the belief that the freedom of inquiry by faculty members is essential to the mission of the academy as well as the principles of academia generally, and that scholars should have freedom to teach or communicate ideas or facts (including those that are inconvenient to external political groups or to authorities) without being targeted for repression, job loss, or imprisonment.

There is an expectation that this freedom is exercised in compliance with the law for example within the UK the Education Act[9] has explicit reference that this freedom shall be within the confines of the law.

In relation to illegal content, academic freedom is necessarily restricted where researchers may need to access illegal material for the purposes of their studies. It is neither desirable nor practical to allow certain people to have access to such content, particularly if it is of a disturbing or harmful nature. There are legal as well as ethical implications arising from this suggestion.

Just as with most other fundamental rights, the right to academic freedom is sometimes restricted in specific circumstances. One of these restrictions relates to the overriding, competing interests of others and this may be engaged in the case of victim rights in relation to illegal content.

10 Country Studies

10.1 Estonia

Estonia is a very technologically advanced country. One of the strategies when the country gained independence was to invest heavily in new technologies, which has resulted in the internet and electronic systems being an integral part of the

[9] Education Act 2011 (c21).

citizens' lives. ICT education was built in to the school system[10] in the 1990's and still continues, creating an 'e-population'. Digital infrastructures have been created including banking systems, ID cards and government services. A whole range of systems and solutions have been developed, and continue to develop.

A recent survey[11] shows that Estonians are less likely to be concerned about security of online information by public authorities than they were a year ago. Only 30 % of respondents from Estonia expressed a degree of concern about encountering illegal content on the internet, which is the third lowest of all EU countries. The apparent trust and confidence may be reflected in the figures representing action taken by Estonians to ensure their children are safe online; 63 % answered that this is not applicable while 1 % said they would not know how, which may be because of other measures being taken in this respect.

There are currently 14 laws and regulations[12] in Estonia which control the security of various aspects of the information network; there are also 12 strategies, frameworks and action plans which provide guidance and promote good practice in this area. These include:

Electronic Communications Act 2005[13] – assists in the protection of users of electronic communications services.

Information Society Services Act 2004 – outlines liability for information society service providers.

Personal Data Protection Act 2008[14] – re. the processing of personal data of natural persons of fundamental rights and freedoms in accordance with the protection of the public interest.

Penal Code 2008 – criminalises actions including those carried out using computers. In response to the European Council Framework Decision 2008/913/JHA[15] the Estonian government took measures to amend the penal code in respect of hate speech. It is a criminal offence in Estonia to incite hatred, violence or discrimination on grounds of sexual orientation[16].

The Penal Code was amended on 1.5.2015. Chapter 10, Division 1 'Offences against Equality', s151 'Incitement of Hatred'. This legislation also crimi-

---

[10] The Tiger Leap programme was introduced in 1996, establishing ICT in all schools by 2000 and promoting ICT education. Still active today.

[11] Conducted by TNS Opinion and Social at the request of the European Commission 'Special Eurobarometer 423 on Cybersecurity' Published February 2015. Available from: http://ec.europa.eu/public_opinion/archives/ebs/ebs_423_en.pdf.

[12] www.riso.ee/et/oigusaktid accessed 9 September 2014.

[13] To be amended on 1.1.2016.

[14] Amended on 1.1.2015.

[15] Council Framework Decision of 28 November 2008 on combating certain forms and expressions of racism and xenophobia by means of criminal law. http://eur-lex.europa.eu/LexUriServ/LexUriServ.do?uri=OJ:L:2008:328:0055:0058:EN:PDF accessed 30 August 2014.

[16] Factsheet on Hate Speech and Hate Crimes against LGBT Persons, FRA (European Union Agency for Fundamental Rights) http://fra.europa.eu/sites/default/files/fra_uploads/1226-Factsheet-homophobia-hate-speech-crime_EN.pdf accessed 10 September 2014.

nalises unequal treatment of persons on the same basis. Division 2 of this Act addresses 'Violation of Fundamental Freedoms', specifically referred to are freedom of religion, association, confidentiality of messages, illegal disclosure of personal/sensitive data and illegal use of another's identity.

Digital Signature Act 2000 (amended 2004) – provides for the necessary conditions for the use of digital signatures and certification and time-stamping serves oversight procedures.

Security Authorities Act 2001 (amended 2014) – sets out the functions and powers of security agencies national security and the constitutional order and security authorities for the oversight regime.

Information Systems data exchange layer 2008 (amended 2011) – the crossroad between the agencies and individuals to provide a safe internet-based data exchange and state information system allowing secure access to technical infrastructure and organisational environment.

Classifications System 2008 – regulation establishing the classifications system, for the management and use of classifiers, identification codes for data.

S25 of the Child Protection Act 2014, which comes into force in 2016, prohibits the dissemination of objects with pornographic content and promoting violence or cruelty, in respect of children. Though the section refers to 'printed matter, films, audio and video recordings" also referred to are 'objects'.

A review of the legislation in the wake of the 2007 attacks highlighted weaknesses and shortcomings in respect of dealing with CC. Improvements such as clearer legislation, increasing punitive measures and widening areas covering by existing laws to include new offences specific to ICT were made.

Although Estonia has great awareness in protecting the infrastructures, and has various strategies and mechanisms in place for keeping information secure, the approach to illegal content is somewhat less restrictive. In a recent report, published by Freedom House,[17] Estonia is referred to as being one of the 'lightest in the world' when it comes to restrictions on internet content. Most forms of illegal content other than those referred to above, fall under privacy laws or are dealt with by civil actions. The emphasis is on website providers to monitor content. In 2013, the European Court of Human Rights upheld an Estonian Supreme Court Decision which stated that content hosts may be held liable for comments made by third parties on their website.[18]

The overall picture in Estonia appears to be one where illegal content is largely controlled by the internet users and the service providers. Freedom of speech is an important right which might be seen as prevailing over control of illegal content, both by definition and by action.

[17] Freedom on the Net 2013, Estonia. http://www.freedomhouse.org/sites/default/files/resources/FOTN%202013_Estonia.pdf accessed 10 September 2014.

[18] Delfi AS v. Estonia (App no 64569/09) [2014].

10.2 United Kingdom

The question of defining illegal content in the UK was addressed in June 2014 by the House of Lords Communications Committee.[19] In Chapter 2 of the report some of the content is referred to: cyber bullying, revenge porn, trolling and virtual mobbing. It is noted that these definitions are not official ones. Other specific crimes are also listed: "*Harassment, malicious communications, stalking, threatening violence, incitement are all crimes and have been for a long time.*"[20] The Committee express the opinion that this criminal behaviour is the same behaviour as existed before the internet, but is now being carried out in a different environment; "*It's not about the medium, it is about the offence.*"[21]

The national legislation and policies which could apply in the area of CC/CT research into illegal content are as follows:

Malicious Communications Act 1988 – s1 deals with sending to another any article which is indecent or grossly offensive, or which conveys a threat, or which is false, provided there is an intent to cause distress or anxiety to the recipient. The offence includes electronic communications.

Protection from Harassment Act 1997 – can be used in conjunction with the **Crime and Disorder Act 1998** for offences which are racially or religiously aggravated.

Protection of Freedoms Act 2012 – amends the Protection from Harassment Act to include stalking and provides for racially/religiously aggravated stalking.

Domestic Violence, Crime and Victims Act 2004 – provides courts with the power to issue a restraining order, even where the offender is acquitted in cases of harassment.

Computer Misuse Act 1990 – amended by Police and Justice Act 2006 re. making, supplying or obtaining of articles. Criminalises any unauthorised access to computer program or data with intent to obtain information. Amended in 2015 by the Serious Crime Act 2015 in respect of serious harm/damage and extending jurisdictions to other countries with a significant link to domestic country.

Sexual Offences Act 2003 – includes a defence for those working with material (s46). Updated by the Sexual Offences Act 2003 (Notification Requirements) (England and Wales) Regulations 2012.

Protection of Children Act 1978 – amended by the Criminal Justice and Immigration Act 2008 re. definition of indecent photographs. S1(b) refers to data stored by electronic means. Amended by the Criminal Justice and Public Order Act 1994 re. offences of taking or distributing indecent photographs. Amended by the Criminal Justice Act 1988 making simple possession of indecent photographs of children an offence.

---

[19] Parliament UK (2014) 'Social Media and Criminal Offences' Communications Committee, First Report. Available at: http://www.publications.parliament.uk/pa/ld201415/ldselect/ldcomuni/37/3702.htm.

[20] Ibid Section 2, para. 13.

[21] Ibid Section 2, para. 12.

Communications Act 2003 – s127 refers to communications of a menacing character.

Defamation Act 2013 – amends the Defamation Act 1996. A statement is not defamatory unless it causes 'serious harm'. The operator of a website is not liable for statements posted if it was not the operator who posted it (this defence is defeated if it cannot be known who did post it, or the operator failed to respond to a complaint).

Electronic Commerce Directive (Hatred against Persons on Religious Grounds or the Grounds of Sexual Orientation) Regulations 2010.

The Economic and Social Research Council[22] has published a guide for ethical considerations during research[23] which emphasises the importance of protecting all those involved in research, including research subjects and researchers as well as institutions and funders. This is particularly important when researching sensitive issues which are an inherent part of illegal content. In a recent article[24] it was recognised that both organisations and researchers in this area should provide a strategy to deal with the potential harm to the persons undertaking the research in order to deal with this. Additionally, the Research Council UK published policy and guidelines for good research conduct[25] which establishes standards of practice and addresses unacceptable practices.

The relevant legislation in the UK which may apply to researchers of illegal content includes the Computer Misuse Act 1990[26] which criminalises unauthorised access to computer materials, and the Police and Justice Act 2006 which criminalises the use of tools in the offences outlined in the 1990 Act and amends it. The Crown Prosecution Service guidance[27] refers to the mens rea of unauthorised access to computer materials which must include knowledge of unauthorised access and an intention to obtain information or data from the computer. The question of unauthorised access is decided on a case by case basis. It must also be remembered that the Crown Prosecution Service would always consider whether it is in the public interest to pursue a prosecution; it is highly probable that the actions of a researcher would fall outside of this.

Some researchers working with illegal content may have to access material that contains indecent images of children. The legislation which applies to this in

[22] https://www.esrc.ac.uk/about-esrc/what-we-do/ accessed 8 September 2014 '... the UK's largest organisation for funding research on economic and social issues.'

[23] ESRC Framework for Research Ethics (FRE) 2010. Updated September 2012.

[24] Jan Coles, Jill Astbury, Elizabeth Dartnall, and Shazneen Limjerwala, 'A qualitative exploration of researcher trauma and researchers' responses to investigating sexual violence.' (2014) 20 Violence against women 95–117.

[25] RCUK (2013) 'Policy and Guidelines on Governance of Good Research Conduct'. Available at: http://www.rcuk.ac.uk/RCUK-prod/assets/documents/reviews/grc/RCUKPolicyandGuidelinesonGovernanceofGoodResearchPracticeFebruary2013.pdf.

[26] As amended by the Serious Crime Act 2015.

[27] https://www.cps.gov.uk/legal/a_to_c/computer_misuse_act_1990 accessed 11 September 2014.

the UK is the Sexual Offences Act 2003 which makes possession or publication of 'indecent images' of children less than 18 years of age, a criminal offence. If research is planned that will include accessing this type of material, it would be advisable for the institution to have a policy in place[28] which outlines the use to which their computers will be put and what type of research will be carried out. This should be made under the guidance of senior police authorities and a declaration made by the researcher saying that they are aware of the legislation and their research is legitimate.

In addition, in the UK, there exists a 'Memorandum of Understanding Between Crown Prosecution Service (CPS) and the Association of Chief Police Officers (ACPO) Concerning Section 46 Sexual Offences Act 2003'.[29] This document refers to the Protection of Children Act 1978 (in respect of 'making' indecent photographs or pseudo-photographs of a child) and the Sexual Offences Act 2003. It serves to protect those who access or particularly copy or download (which constitutes 'making') images for legitimate reasons such as reporting or investigating crime. It provides guidance to organisations and those whose work involves them in the discovery or reporting of indecent images of children in electronic communications media.[30] Individuals in smaller organisations should show that their actions are justified in pursuit of the purposes set out.

Some areas of illegal content may require research to be carried out involving children or vulnerable people and may require the researcher to register with (in the UK) the Independent Safeguarding Authority[31] to ensure their suitability for working with these groups.[32]

Other practical considerations in the area of illegal content include lack of access to some material, for example the Internet Watch Foundation (IWF)[33] who carry out research particularly in relation to child sexual abuse online, do not have the authority to pass payment barriers and are only able to conduct research on publicly available content. This provides limitations to research. An additional barrier is discovering and tracing material due to the nature of the cyber environment; this leads to a lack of data on the amount and availability of the content and the methods of distribution.

---

[28] JISC legal information 1 February 2007 'Cybercrime Essentials' http://www.jisclegal.ac.uk/Portals/12/Documents/PDFs/crimeEssentials.pdf accessed 11 September 2014.

[29] http://www.cps.gov.uk/publications/docs/mousexoffences.pdf accessed 11 September 2014.

[30] Ibid page 3.

[31] Set up as a result of the Bichard Inquiry which led to the Safeguarding Vulnerable Groups Act 2006. http://www.criminalrecordchecks.co.uk/crb/isa-independent-safeguarding-authority.

[32] ESRC Framework for Research Ethics (FRE) 2010, updated September 2012 http://www.esrc.ac.uk/_images/framework-for-research-ethics-09-12_tcm8-4586.pdf accessed 7 September 2014.

[33] Established in 1996 by the internet industry to provide a reporting point for illegal content online http://wwww.iwf.org.uk accessed 7 September 2014.

The United Kingdom carries out extensive blocking and filtering of illegal content particularly in relation to child sexual abuse and extremist and terrorist material. Many thousands of URLs and search terms have been prevented from being used. Whilst this has the desired effect of blocking access to such material, it could also serve to hamper the extent and type of research which can be carried out. If the research were aimed at measuring the amount and type of material posted, this would have to rely on data produced by the blocking companies, which presents additional difficulties.

11 General Conclusion

Whilst some offences are categorised by the way they are carried out and can be thought of as cyber-defined crimes, most of what falls within the understanding of 'illegal content' can be thought of as cyber-enabled crimes, where the 'cyber' element refers to the tool used and the environment in which it is carried out. This perception is endorsed by the House of Lords' Committee on Communications.[34]

At European level, there is recognition of the need for a combined, multi-layered approach to tackling cybercrime. The co-operation of all stakeholders is required to combat the damage and disruption caused; this is particularly relevant in the area of illegal content with its various issues and problems. It impacts on the fundamental human rights which are of central importance to all citizens and are protected by national and European law. However, an additional problem that has been identified is the difference in levels of seriousness of crimes that could fall within the definition of illegal content. In addition to compounding the problem of common understanding, it also requires implementation of a balancing act when enforcing protective measures. As well as protecting the human rights of the victims of these crimes, it is of equal importance to allow the freedom of others to express themselves and carry out their work, unimpeded by inappropriate sanctions.

It is essential that relevant and effective research is carried out in respect of illegal content so that all those involved can have an informed understanding of what is incorporated and positive strategies can be developed to combat it. The problems related to identifying what is required and the practical difficulties encountered as part of the research process, serve as barriers to understanding; barriers which are delaying the response to a rapidly growing (in size and seriousness) area of criminality.

A more effective way of understanding illegal content might be to adopt a different viewpoint; rather than defining 'types' of illegal content as the starting point, a better approach might be to define the effect it has on the victims. Rather than categorising similarities in type of material, the yardstick could be the outcome, i.e., the damage caused. After all, crime is defined and measured

[34] Parliament UK (2014) 'Social Media and Criminal Offences' Communications Committee, First Report. Available at: http://www.publications.parliament.uk/pa/ld201415/ldselect/ldcomuni/37/3702.htm.

by degrees of harm caused to another. Sentences are determined according to the degree of actual or potential harm caused by an action/intent. Adopting this alternative approach (from the other end, as it were), would also reveal what material caused the damage and the potential source of it.

In this transitional phase in the development of society, where 'true' reality and 'virtual' reality is beginning to merge, it seems appropriate to view CC/CT as a single area, defined by the space they share. However, this may prove unhelpful in identifying a common platform on which to create an understanding of a very complex subject. It is arguable that CT rather than being part of 'illegal content', is a separate field with its own emerging research agenda[35] and may favour different methodologies. It might be unhelpful to try to fit these types of offences into the category of illegal content.

12 Recommendations

- To undertake appropriate risk assessments concerning the planned research and prepare clear safety plans and processes to ensure the maximum physical, psychological and emotional safety of all those involved;
- The have an awareness of and sensitivity to, cultural differences both within and between member states;
- Develop methods of research which are sensitive to the needs of those involved; the use of anonymity/pseudonymity would provide reassurance and encourage participation;
- Establish a clear set of guidelines to create a common understanding of what constitutes illegal content;
- Establish a European legal and ethical guide for researchers, covering all aspects of potential work including legal and ethical issues and data protection;
- A solution put forward by the European Commission[36] is the development of public-private partnerships at EU level. Such cooperation and unity between policy makers and service providers would create a more effective defence and response to illegal content achieving its various aims.

Acknowledgement. The research leading to these results has received funding from the European Union Seventh Framework Programme (FP7-SEC-2013) as the COURAGE project under grant agreement no 607949.

[35] Freilich, J.D.; Chermak, S.M. and Gruenewald, J (2014) *The Future of Terrorism Research: a review essay* International Journal of Comparative and Applied Criminal Justice.

[36] 'Protecting Europe from large scale cyber-attacks and disruptions: enhancing preparedness, security and resilience' COM (2009) 149 final. Brussels 30.3.2009. http://eur-lex.europa.eu/LexUriServ/LexUriServ.do?uri+com:2009:0149:FIN:EN:PDF accessed 5 September 2014.

Part III:
Technologies, Scenarios and Best Practices

Cybercrime Economic Costs: No Measure No Solution

Jart Armin[1(✉)], Bryn Thompson[1], and Piotr Kijewski[2]

[1] CyberDefcon, Hove, United Kingdom
jart@cyberdefcon.com
[2] NASK Research and Academic Computer Network, Warszawa, Poland

Abstract. Governments need reliable data on crime in order to both devise adequate policies, and allocate the correct revenues so that the measures are cost-effective, i.e., the money spent in prevention, detection, and handling of security incidents is balanced with a decrease in losses from offenses. The availability of multiple contrasting figures on cyber-attacks checks the accurate assessment of the cost-effectiveness of current and future policies for cyber space. What factors contribute to the costing equation is not clearly understood with wide variation in methodologies used. The most relevant literature in this field is reviewed and analysed against quantitative insights provided by the CyberROAD survey to stakeholders. Research gaps are highlighted to determine the issues that need addressing to provide a solid ground for future legislative and regulatory actions at national and international levels.

Keywords: Cybercrime · Economic costs · Measurement · Methodology · Security · Cyber security · National Security · Cyber threats · Research gap · CyberROAD · DDOS · Botnet · Trust · Taxonomy · Metrics · Standards · Benchmarking · Data · Definitions · Government · Budget · ENISA

1 Introduction

In a response to the 2015 CyberROAD survey question to stakeholders: "Have you experienced a cybercriminal action in the last 5 years?" 78 % of the respondents responded they had, either in a personal capacity (31 %) or through work (47 %). When asked "To make the Internet a safer place and to fight cybercrime, what are the topics we should research into?" most respondents rated "Better metrics and statistics on cybercrime" as their second choice (out of six) in order of importance.

Cybercrime (CC) has climbed to the top tier in the National Security Strategy of many EU states e.g. France, the Netherlands and the UK, becoming the #1 threat above organized crime and fraud generally. However as indicated within a recent 2013 study for the European Parliament - Directorate General

B. Akhgar and B. Brewster (eds.), *Combatting Cybercrime and Cyberterrorism*,
Advanced Sciences and Technologies for Security Applications,
DOI 10.1007/978-3-319-38930-1_8

for Internal Policies; "The Economic, Financial & Social Impacts of Organized Crime in the EU", "estimates of cybercrime costs are highly contested". It concludes by saying "So is cybercrime a threat, and to whom? It is a threat to all of us. The question is how much of a threat, and how can we better understand how much of a threat it is" [1].

Using property crime, for example, as a comparison, in most countries the metrics are mostly readily available. In the US, the FBIs "Uniform Crime Report" [2] details how many offenses were committed nationally in 2011 (9,063,173) and of what type (burglary 24 %, larceny 68 % and motor vehicle theft 7.9 %). It is not too difficult from this point on to provide an accurate estimate of the overall cost of property crime to the US economy in 2011 (14bn). "However, when inquiring about the direct costs of CC to any economy, individual industries, or companies and you get no straight answers" [3].

Worryingly, it seems that awareness to the extent of the problem has advanced very little over the years. At the turn of the millennium CC was recognised as "the organized crime of the 21st century" [4]. An article published in Bloomberg Business in 2006, announced that in the previous year, for the first time, "proceeds from CC were greater than proceeds from the sale of illegal drugs, according to an adviser to the U.S. Treasury Dept." [5]. In truth, we are no closer now in knowing how accurate an assessment that was, despite the vast sums spent in the meanwhile. The 2006 Bloomberg article and the problems it summarises could have been written today.

Certainly, there is no lack of reporting on the cost of CC; these make the headlines on a regular basis. But how well do these stand up on closer inspection? Without fundamentally accurate data, how do we know where the research money should be spent? How can policy makers plan for the future? How can boards budget correctly? How can risk be evaluated when data is patchy and unverifiable?

As part of the CyberROAD project this area was viewed from its core foundations. The project established a perspective of where the state of the art is now and needs to be to meet the challenges of the future.

2 The CyberROAD Cybercrime Survey

The CyberROAD project, described earlier in Chap. 4, designed a broad-based survey in order to gain an understanding of the impact of CC on stakeholders which could be weighed against current research results. It was decided to follow the Delphi approach consisting of an initial poll followed by 2 further questionnaires where participants of the first round are invited to complete at least one, or possibly two, subsequent polls. Answers from the first survey are used to generate more specific questions in the fol-lowing rounds. A principal area of the CyberROAD surveys centers on; "The cost of cybercrime" in relation to everyday life and business.

Purpose. The purpose of the CyberROAD survey is to explore and establish the needs of stakeholders and to find out what they see as the potential threats both now and into the future. As perceived threats may be different from real threats, it is important to try to correlate stakeholders experiences of CC with the situation as reflected in current reports and analyses. A mismatch between the two can be costly in terms of money spent on research and to stakeholders' understanding of what should or could be done to alleviate risk, i.e., are the right threats being targeted at present?, Can a blanket approach to security be taken or would a more flexible system be of more benefit?

Methodology. Survey 1 was prepared using specialist online software and designed following the Delphi method. The questions for this survey were of a generic nature as Surveys 2 and 3 would explore resultant themes at a deeper level. To exploit the CyberROAD Cybercrime Survey a number of distribution methods were employed by project partners. These included the project website, a dedicated website, announcements via social media, and prompting by email to interested parties. The surveys were split into two versions: one for English speakers worldwide and the other translated into Polish and aimed at Polish users.

Macro to Micro (World, Europe, Poland Case Specific). For the purposes of the CyberROAD project it was decided that the greatest value would be obtained from a comparative study using participants worldwide but with a bias towards European citizens. Using the Delphi method for the surveys made it possible to draw down and to probe further using selective criteria, if required. For this project, it made sense collate at a macro level i.e., world (with a European bias), and at a micro level i.e., a specific country; Poland. Poland was selected because it is one of the larger EU countries and is also represented by a national CERT team (CERT Polska) in the CyberROAD consortium. The participation of a national CERT allowed for convenient access to various statistics on the threats affecting Poland and good potential outreach to other entities in the country as well as the general public which is especially important when disseminating surveys.

2.1 An Overview of Survey Findings

Survey respondents see CC as a problem rooted primarily in economic and technological interests. The vast majority of all participants believe the main driver of CC is the opportunity for easy and minimal-risk money.

Most respondents consider "better education of users of the Internet" as the single most important topic that should be researched in order to make the Internet a safer place (75 % of respondents). "Improved technology for our networks and operating systems" scored the next highest in the 'Very Important' category (only 58 % viewed this as Very Important), while "better laws and regulations" were viewed as 'Very Important' by only 40 %. Most respondents, however, rated

"Better metrics and statistics on cybercrime" as their 2nd choice after selecting their top choice of topic for more research. Indeed, the above responses seem to correlate with the response to another question, concerning training within their organization: 59 % of respondents were not trained in cyber security issues at all or only if there was a problem (note: we included "don't know" responses in this category as well). Even though many respondents considered CC to be a concern and many had been victims either personally or as part of their organization (as many as 78 %) most respondents declared that the main consequence of the CC action was inconvenience (50 % of respondents). Nevertheless, many claimed enormous losses to their country or worldwide economy as a result of CC in general (although in contrast most respondents said they had no idea what the losses were). Perhaps this seemingly contradictory response (large losses vs the primary loss being inconvenience) is due in part to the term "cybercrime" being often understood in very different ways, as other responses in the survey indicated.

Another very visible problem is the relatively low reporting rate of CC to the Police (44 % of CC cases not reported) and/or national CERTs (72 % of CC cases not reported). This is followed up by a low successful prosecution rate: only 8 % of the cases were successfully prosecuted. Information sharing in general was found to be a problem (only 43 % respondents said they or their organization shared information on cyber-attacks) - an issue that also hinders effective measurement of CC.

The responses to the Polish survey (the same survey but translated into Polish) were in many aspects similar, but in general tended to show slightly worse results in regard to user awareness and experiences with CC. In part, this is possibly because the responder base was nearly the opposite of the English speaking one (con-sumer group vs a more specialist group).

Overall, however, the initial findings appear to confirm that there is a tangible need for better definitions, metrics and statistics for CC together with more training. Initial analyses tend to support the view that current definitions on CC are confusing to stakeholders whose experiences do not align with the information readily available. This mismatch of messages is a stumbling block in cyber-crime prevention which could be alleviated with better quantification.

3 Review of the State-of-the-Art of the Metrics and Economics of CC

Within the 5 years 2011 to Jan 2015 there were 3,920 web searchable scholarly articles, papers and books relating to the "economics or costs of cybercrime"[1]. Added to this is the wide spectrum of commercial sources collecting, collating and disseminating related information and data, some of which is not publicly accessible.

---

[1] Google search on 13/02/15.

For the CyberROAD project a comparative analysis of five major reports on the theme of the "cost of cybercrime" was carried out[2]. The reports were selected as representative of their genre in presenting a breakdown on the "cost of cybercrime", offer recommendations and advice on how costing and metrics can be improved or convey specific quantitative data. The studies come from academia, consumer groups, technology providers and policy advisors and align to the criteria of the CyberROAD Triad approach through a combination of evidenced-based practices (Fig. 1). A short overview of this research is presented here together with the outcomes summarized in the form of research gaps.

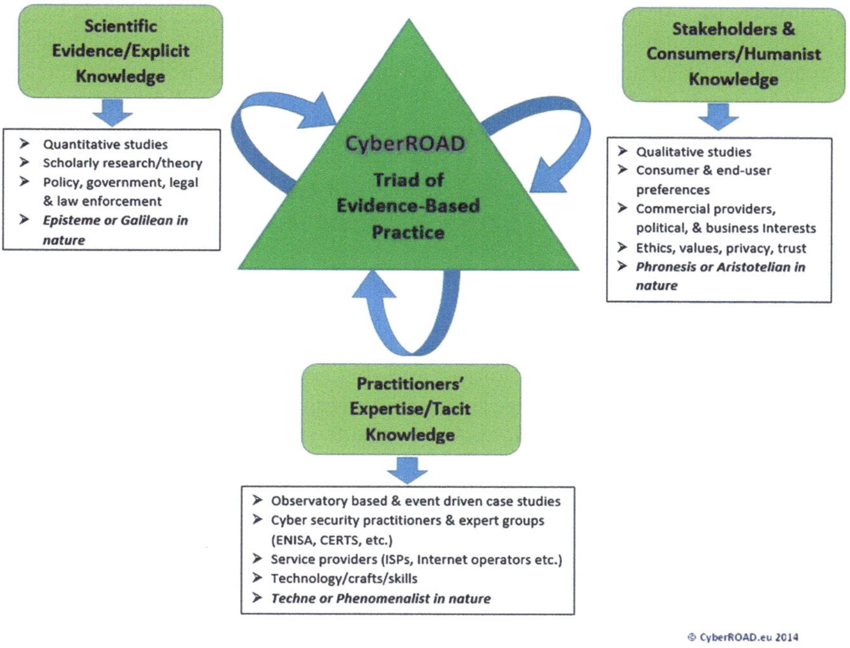

Fig. 1. CyberROAD Triad of evidence-based practice

An important consideration is the source of data used in CC costing equations. The degree to which data is designated as open or publicly accessible is sometimes questionable. For instance, the intended motive/aims of the data provider, whether altruistic in nature or commercially interested, is difficult to quantify. It follows that any related data may be regarded with suspicion and its validity questioned; whose data can be trusted, how can a "trusted" environment be measured? Methodologies used to collect and collate information can be unique to the entity, unclear or not fully disclosed. Data may be

[2] http://www.ares-conference.eu/conference/.

incomplete without standard modus operandi, guidelines on best practices for data collection or benchmarks for data measurement.

Additionally, a rapidly changing digital era brings new challenges into play as big data becomes integral to the everyday experience. For example, what value can be attached to privacy? This topic is considered in brief in the context of an overview of a recent report on privacy.

Anderson et al. Study, 2012. The "first systematic study of the costs of cybercrime" [6] concludes that available statistics are "insufficient and fragmented" [6, p. 12] despite more than 100 different sources of data on CC having been counted in early 2012. The unequivocal message is that a lack of cohesion between different sources clouds the issue, leads to inconsistency of data and engenders mis-trust of the numbers. As a consequence, policy makers, who depend upon reliable figures, are left with little to go on, while the problem's true extent is obscured by the absence of easy-to-understand metrics. This report supports the widely held opinion that despite eye-catching headlines suggesting otherwise, it remains the case that few straightforward numbers exist on CC and its true cost politically, economically, socially and ethically.

This "Cost of Cybercrime" study details a simplified framework for standardizing measurements, arrived at by decomposing an earlier, and much criticized [7], report from Detica [9], where "difficult to assess" categories were used. Anderson et al. suggest that "cost to society" can be calculated through the application of "sum of direct losses, indirect losses, and defense costs", to "known data" on CC and supporting infrastructures. It is important, too, that the definition of CC has the flexibility to accommodate fluidity between traditional, transitional and modern crimes as cyberspace continues to evolve. Using this method of calculation, the report claims that "new computer crimes" actually cost only "tens of pence/cents" per person and not the vast sums as reported elsewhere.

The report highlights the subjective or 'obvious' agenda when organizations (such as vendors, police agencies or music industry lawyers) are the authors of studies of CC [6, p. 12]. Important areas for further research are highlighted as; what data can be trusted and from where should it be sourced, what are the determining metrics to be used, the need for benchmarks, why does CC have high indirect costs and low indirect costs, [6, p. 26]. Additionally, Anderson et al. conclude that less should be spent on "anticipation of computer crime (on antivirus, firewalls etc.)", and more on "catching and punishing the perpetrators" [6, p. 12].

Ponemon Institude Study, 2014. The Ponemon Institute, an independent U.S.-based research group used by major corporations, U.S. federal and state departments, consumer groups, has been conducting "The Cost of Cyber Crime Study" [29] since 2009. The 2014 Ponemon Institute report is based on the findings from surveys conducted with 257 organizations using a cross-section of industry sectors in 7 countries U.S.A, U.K., Germany, Australia, Japan, France and the Russian Federation. The research is field-based via interviews with senior-level personnel "...about their organizations" "actual CC incidents..." from large sized entities with more than 1,000 direct connections to the network or its systems (enterprise seats).

Criteria such as the "costs to detect, recover, investigate and manage the incident response" along with costs that "result in after-the-fact activities and efforts to contain additional costs from business disruption and the loss of customers", excluding the cost of "expenditures and investments made to sustain an organization's security posture or compliance with standards, policies and regulations", are used to compute the total cost of CC incurred by an organization. Some categories used in this report are indicative of those branded as "difficult to assess", such as lost opportunity cost, by the 'Anderson et al' report with the claim that resulting costs are difficult to substantiate. Another immediate problem is revealed in that comparison of two counterpart studies with the same title may be untenable where there is no commonality of approach in methodology or criteria. Results will therefore be, unsurprisingly, disparate.

McAfee Annual Cybercrime Reports. The McAfee report of June 2014 "Net Losses: Estimating the Global Cost of Cybercrime" [8] highlights some of the current pitfalls in the capacity to collate accurate data. The problems accentuated include "Estimating global loss from incomplete data" (p4), and "International agreement on a standard definition of CC would improve the ability to collect consistent data." The lack of effort made by most countries in collecting data on CC losses, along with widespread inconsistencies and poor quality of the data that is gathered, is a re-occurring theme in this report. The three example methods used to "extrapolate a global loss figure" highlight this very problem. Method 1 uses the loss by high-income countries to deduce a global total, method 2 totals the amount for all countries where open source data is available, and method 3 "aggregate(s) costs as a share of regional incomes."

This report acknowledges the inadequacies of the methods employed which, due to the lack of reliable data, could either be an "overestimate" or "underestimate" of the true cost of CC worldwide.

A focus of this report, and a major research gap, is the lack of reliable data and the issue of corporate entity participation in this field. Is it possible to assess whether information delivered from the private sector is necessarily biased towards its own agenda? Many different types of organizations currently provide critical services and share data to help protect against cyber-attacks. How can these be more effectively used, and trusted, to provide the types of figures that are missing. What can be done to improve the availability of data in countries around the world? Who can be trusted to provide this service in other countries? Should this be a role for a new, independent entity?

East West Institute Study, 2013. One of the few global studies into the need for improved methods of measurement was undertaken in 2013 by the East West Institute (EWI) [11], an international, non-partisan, not-for-profit policy organization that focuses on confronting critical challenges. "Measuring the Cybercrime Problem" [12] examines how trusted metrics and performance benchmarks can be established, and a trusted centralized data collection entity created, both research gaps previously identified in this review. The EWI study "presents a bold solution to this problem that involves private sector leadership aimed at promoting trust and cooperation". The report concludes with three recommendations and calls for "volunteers from all sectors–ICT, energy, financial services, transportation, retail, medical and others" to carry these out.

In this study the relevant capabilities of existing information sharing entities are benchmarked against "Target Criteria" based on three key areas: Governance-Related, Breadth-Related and Information-Related with the resulting "Gap Analysis" depicted in a table format.

Commercial entities are excluded from the analysis on the grounds that, "they are seen as likely to try to influence market conditions, whether or not this perception is justified." The resulting "Gap Analysis" reveals that not a single entity reached all the Target Criteria, one achieved 5 out of 7, and 5 scored 4 out of 7, giving justification to EWI's call for the creation of a trusted entity for data measurement, as one could not "be found".

Widening the sample set to include corporate entities willing to be tested against set standards or benchmarked criteria as a means to verify the quality of their data would expand the potential data pool. The application of tools for data suitability assessment is an area for further research.

Neustar UK Annual DDoS Report, 2014. In May 2014 Neustar published its second annual "UK DDoS Attacks and Impact Report" [13]. Neustar began as an operating unit managing large datasets under Lockheed Martin, a global aerospace, defense, security and advanced technological company. Today, Neustar handles billions of DNS queries and millions of text messages and phone calls. The report is based on findings from Neustar's survey of 331 UK companies across a variety of industries including financial services, technology, retail, government/public sector, health care, energy/utility, telecommunications, e-commerce, Internet services and media.

The scope of the inaugural 2012 survey was further developed with additional questions for the latest report. Each question targets specific information and data builds into a year-on-year profile of DDoS patterns and related changes. Examples questions include: What are the sizes and velocities of DDoS attacks? How long are DDoS attacks lasting? Are DDoS attacks a bigger or smaller threat to your business versus a year ago? and, how often were you attacked?

This seems a simple yet effective way of gathering quantifiable information and a good example of how the data can be displayed in an easy-to-understand format.

Even though this report appears to provide a model template for measurement and metrics there are still a number of issues when tested by the EWI method of analysis. Straightaway, it seems that Neustar would not qualify as a "trusted" data provider using the EWI suitability method due to its for-profit status. So, to what extent can this data be trusted? In the absence of benchmarks or standards, this is an unknown entity. Further research is required in this area to establish the criteria for cross-industry best practices and benchmarks.

Private, public and non-profits may each have a role to play in improving measurement and metrics. Used in this way, metrics can point to security vulnerabilities and provide a valuable source for gap analysis research. The Neustar report specifically highlights the vulnerability of the DNS/NTP servers to amplification attacks, when there are server misconfigurations. As a vulnerability, this has been highlighted by several other sources[3]. Any data, no matter what the source, should be viewed as a potential valuable asset, and put to the test. Currently, the problem is necessarily "bad data" as a lack of testing of its worthiness.

3.1 CyberROAD Review of the Economic State-of-the Art

To complete this review of the current economic state-of-the-art an analysis was made of the most relevant, and readily accessible data, fundamental to a study on CC metrics and measurement. A surprisingly large amount of information can be gathered from just a few sources which, taken at nominal value, yield a set of straightforward figures on some of the most contentious issues in CC. In summary:

1. Costs of CC.
   - The annual cost to the global economy from CC is more than 300 billion Euros [14]
   - Cost of CC for the EU 0.4 % of its GDP[4] = 13 billion/annum [15]
   - Sample EU countries estimates for the cost of CC[5]:
     - * Poland: 377 million/annum
     - * Germany: 2.6 billion/annum
     - * UK: 2 billion/annum
   - Cybercriminal revenues (estimate of the CC market itself) 15 billion/annum[6] [16]
   - Market for security products and services 50 billion/annum [17]
2. Examples of CC Metrics.
   - 3 Billion Users of the Internet (~39 % world population) [18]
   - Over 200 billion emails processed/day [19]
   - 917.9 million Websites (variable) – 39 million/month added (4 %) [18]
   - IP addresses - IPv4 = 4,294,967,296 (2) - IPv6 = 128-bits (2) [20]
   - 2.3 billion mobile-cellular subscriptions worldwide [21]
   - 1.4 million Browser user agents - bots

[3] http://www.pcworld.com/article/2013109/report-open-dns-resolvers-increasingly-abused-to-amplify-ddos-attacks.html.

[4] Estimate of average - range is up to 0.9 % of GDP - high-income countries incur higher losses.

[5] Based on share to EU GDP. Figures on GDP are available on the IMF website.

[6] CyberDefcon estimate which if only allowing for inflation & not increase is revenues.

3. Technical and Quantitative Metrics of CC Activity Indicators.
   - 85 % of processed emails are spam [23]
   - 7 % of all URLs malicious [24]
   - Public Block List count: 1,018,203,532 IP addresses [25]
   - 350 million+ in total identifiable malware [26]
   - 1 million+ measurable cyber-attacks (variable) [27]
   - 330 active Real-time Blackhole Lists (RBL & DNSBL) [28]
   - 7.9 million is the average annualized cost of data breaches [29]
   - 10.4 % net increase cost of data breaches over the past year [29]
   - 250,000 – 500,000 malicious binaries/day [30]
   - ˜280 million malicious binaries collected [30]
   - 6/10 million unique IP's sink holed/day [30]
   - 900,000 malicious domains/day [30]

3.2 Overview of Current Estimates

The above examples demonstrate that a variety of data types on CC metrics are available. This provides a good point from which to start. The next step requires evaluating the preferred statistics to be included in an innovative framework which will form the foundation for further study.

A result may be that several costing models are necessary as a single methodology that works across the board may not be achievable. To accommodate difficult to assess areas, such as loss of reputation or the value of privacy, a deal of flexibility will be needed for such a framework to be fit-for-purpose whether costing is to be applied to budgets, insurance or any other function.

The development of a working model is an essential research area if the impact of CC is to be fully understood and appreciated.

The Economics of Privacy (Acquisti et al. 2015). 'The Economics of Privacy' study (Acquisti et al. 2015) [31] provides an updated survey on the economics of privacy. The main focus is not on the abuse of personal data stored on computers, nor on data breaches, but on the value that can be attached to private data.

As soon as people consent to the use of their data for marketing purposes, then the value of the data can be associated to the gain that the user may acquire in terms of discounts or other privileges in their purchasing activities. On the other hand, when personal data is stolen or misused, than the task of assigning a cost based on worth is still an open problem.

This study clearly points out the three factors affecting the value of private data stored and shared over the Internet: individual responsibility, market competition, and government regulation. Individual responsibility requires awareness of the benefits and risks that sharing data brings in itself. Market competition exists to the extent to which to a value can be attached to this data. Finally, governments can regulate this market as it happens in other sectors.

At present, this topic is addressed in different ways in the EU and the US. While EU is steering towards government regulation on the management of private data, the US is drawing a framework that would allow different sectors to self-regulate this market. It turns out that no clear figure currently exists on the value of data breaches when related to individual data.

4 Review of the State-of-the-Art of Stakeholder Impacts

Throughout the EU independent initiatives in the form of projects and surveys provide valuable insights and perspectives on the impact of cybercrime, an important, but sometimes overlooked, appraisal of real-life scenarios. Groups, associations and organisations with an interest in, and knowledge of, CC prevention can tap into resources and reach specific stakeholders that may, otherwise, be unavailable. If data from this valuable resource is excluded, it is probable that an unrepresentative set of metrics will result and inappropriate solutions will ensue.

An overview follows of a sample collaborative project from the ICSPA 'International Cyber Security Protection Alliance' which is supported by EC3 at Europol, ENISA, the City of London Police and a number of industry players such as Atos, McAfee, CGI Canada, Trend Micro, Cassidian and Visa. CyberROAD classifies this as a macro project due to the size and number or participants.

Additionally, at a micro level, there is an overview of stakeholders' needs in the retail industry is provided from an assessment of The British Retail Consortium (BRC) together with stakeholder views from the perspective of the Federation of Small Businesses (UK).

Understanding, and measuring, the impact on all types of stakeholders is a necessary step in the assessment of the cost-effectiveness of solutions for the future, across all sectors. For example, money currently spent on anti-virus solutions may be more appropriately spent in providing other types of defenses for stakeholders facing specific types of cyber-attacks, such as DDOS.

Project 2020 - ICSPA International Cyber Security Protection Alliance (ICSPA, 2012) – a Macro View. The aim of this ongoing project, which began in 2012, is to provide an assessment of future challenges and opportunities, as a means of preparation for governments, businesses and citizens. An early output is the report 'Project 2020 Scenarios for the Future of Cybercrime' (ICSPA, 2012).

The methodology provides a number of scenarios from the perspective of an ordinary Internet user, a manufacturer, a communications service provider and a government. An analysis of the threat landscape in 2012 comes from evidences provided by ICSPA members across a range of Internet security companies via collaboration with Trend Micro.

A number of key uncertainties for the future were identified and summarized by Project 2020 as 'Implications for Cybersecurity Stakeholders'. Incidentally, these are major considerations for many types of data analysis including cybercrime measurement.

- Who owns the data in networked systems, and for how long?
- Who will distinguish between data misuse and legitimate use, and will we achieve consistency?
- What data will the authorities be able to access and use for the purposes of preventing and disrupting criminal activity?
- Who covers (and recovers) the losses, both financial and in terms of data recovery?
- Who secures the joins between services, applications and networks?
- Do we want local or global governance and security solutions?

A Stakeholder's View (Macro) – Survey Results from the British Retail Consortium. The British Retail Consortium (BRC), a leading trade association representing the retail industry, conducts an annual survey of retail businesses in British. 'The BRC Retail Crime Survey 2014' (The British Retail Consortium, 2014) details incidences of crime affecting retail businesses. In 2013–14 the number of cyber-enabled attacks increased with retailers reporting that they posed a significant threat to their business.

4.1 Major Outcomes from the Survey

- Businesses are increasingly the victims of crime committed online, such as cyber-enabled fraud. In 2013-14, fraud increased by 12 % and accounted for 37 % of the total cost of crime. The majority of fraud is committed online.
- An estimated 59 % of fraud is committed by organized groups and can often operate across several geographic areas.
- Credit and debit card fraud accounted for 81 % of fraud by volume.
- Theft of data and hacking were considered to pose the most critical threats.

Impacts on Stakeholders. Loss of staff time and distraction from business purpose together with reputational damage to the brand were cited as having the most significant impacts. Another highly ranked consequence, which the report highlights as an overlooked area, is that of damage to employee morale.

Challenges that Need to Be Met. Retailers cite a number of failings in the way that cyber-enabled crimes are re-ported and a lack of subsequent prosecutions. These issues are summed up as follows:

- The capacity of law enforcement to respond effectively to cyber-enabled crime. Only a tiny proportion of fraud cases result in any action being taken.
- An apparent inability of law enforcement to respond to offending that crosses police force borders.
- A lack of intelligence sharing from the National Crime Agency about emerging cyber threats.
- No confidence in the police response. This was cited as a major reason for failing to report incidents of fraud (cyber-enabled or otherwise.)

The challenges relating to CC reporting cited by British retailers are consistent with the findings of the CyberROAD survey which suggests this to be a common problem throughout EU countries. A quote from the United Nations on Drugs and Crime (UNODC) report (UNODC, 2013), Annex 2 entitled, 'Measuring Cyber-crime' (pp. 259–266) (UNODC, 2013), sums up the problem of underreporting "...for CC events, the difference between victimization and police-recorded crime can be many orders of magnitude."

In CyberROAD Survey #1 CC, 36.6 % of respondents said they had not reported being a victim of CC to the police. 27.6 % did report an incident but the police took no further action, while only 7.2 % of respondents stated that the police had achieved a successful prosecution.

A Stakeholders' View (Micro) – from the Federation of Small Businesses (UK). The Federation of Small Businesses (FSB) report 'Cyber security and fraud: The impact on small businesses'[7] (FSB, 2013) details 2,667 responses focusing on the specific interests of small and micro businesses.

The report recognizes that online crime, and fraud in general, whether real or perceived, presents a number of distinct problems for small businesses and, as a consequence, the costs involved can be a barrier to growth in the e-commerce market.

When asked the question: "How much money has your business lost as a result of fraud and/or online crime over the past 12 months?" 41 % reported being a victim with an average of 3,926 (EUR 5502) lost. The most prevalent CC's experienced were 'virus infections' (20 %), 'hacking or electronic intrusions' (8 %) or 'system security breach/loss of availability' (5 %). 73 % of respondents were concerned that they may be unaware that their computer systems had been compromised.

Preventing cyber-enabled fraud was reported as being a significant cost to the business. Bring Your Own Devices (BYOD) brings additional risks to small businesses through possible malware infection to company data and systems, loss of data and unauthorized access. Managing this risk with extra security measures including encryption, mobile security solutions and Network Access Control (NAC) adds to small business overheads.

Small businesses expressed concerns about compliance being weighted towards larger organisations. Although recognizing that standards are designed to improve data security through the adoption of good business practice, paying for an assessor or completing lengthy self-assessment forms adds additional pressures for SME's.

Challenges that Need to Be Met. Small businesses in the UK expressed the need for customized and realistic practices to enable SME's to meet the growing challenges from cyber-enabled fraud. Improvements are required in:

---

[7] http://www.fsb.org.uk/.

- Customized security guidance for small and micro businesses
- Improving law enforcement responses to online crime
- Improved cooperation from banks and payment providers in the cyber security area
- More information sharing within the private sector
- More efficient reporting methods for all crime including online crime and fraud
- Simplified and streamlined standards and benchmarks aimed at SME's

A number of research gap consistently emerge from the challenges outlined with in-formation sharing a key feature. It is clear that a one-size-fits-all policy does not meet the needs of SMEs where simplified and streamlined guidelines would be more suited. Currently, SMEs growth potential is limited by procedures that are not fit-for-purpose with it ensuing that accurate measurement of cybercriminal activity is hampered.

The views expressed by SME's in the UK align with responses to the question posed on the subject of the importance of information sharing on cyber issues in the CyberROAD Survey #3 Social, Economic and Political issues (Fig. 2). For English speakers the major concern is not knowing where information can be shared while Polish speakers express a lack of trust in sharing their information.

The responses show that end-users whether they are consumers, business owners or employees understand the value of information sharing but issues of trust, lack of knowledge on where or how to report problems and associated cost are major challenges at the present time. Accommodating the need for improved information sharing and metrics may well determine the success of the eco-systems of the future.

5 Further Findings from the CyberROAD Survey of Stakeholders

A selection of responses from both English and Polish speaking participants serves as a comparative example of the experiences of stakeholders in the EU.

CyberROAD Survey #1 asked: If you have been a victim of CC in the last 5 years, what was the effect of the action? The results solicit further inquiry into actual impacts of cyber intrusions as nearly half of all respondents, both English and Polish speaking, stated that the greatest effect was 'inconvenience' and 'not loss of money' (Fig. 3). If loss of money is not always the greatest issue, in what way does this impact on cost assessments?

To explore this area in greater depth the following question was asked in Survey #3 Social, Economic and Political issues: (For previous victims of CC only.) Survey 1 participants describe the two greatest effects of CC as: "down time" and "inconvenience". How much time would you estimate you lost when you became a victim of CC? (Fig. 4)

Polish speaking respondents report more time lost than their English speaking counterparts. 'Time lost' is not necessarily equated to 'loss of money' and, as such, is a difficult to assess area. Time lost has an obvious economic impact but is only accurately assessed on a case-by-case basis.

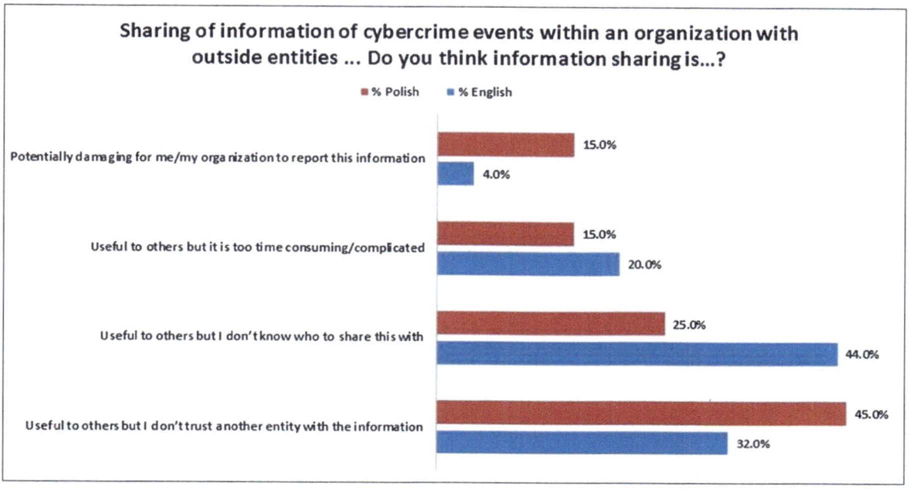

Fig. 2. Information sharing CyberROAD Survey #3 Social, Economic and Political Issues

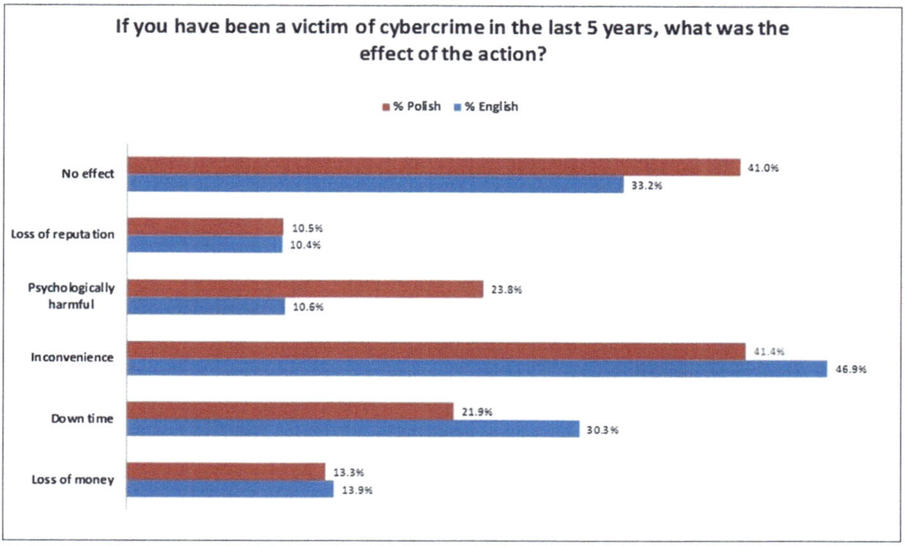

Fig. 3. Impacts of CC: English speaking v Polish speaking

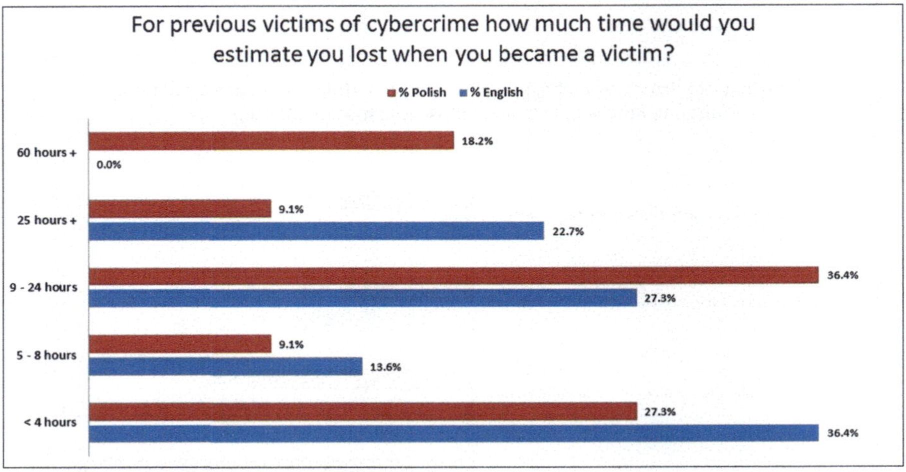

Fig. 4. Time lost as a victim of CC

6 Gap Analysis

The lack of quantifiable data is a theme consistently found in the bodies of work analysed in this study. The absence of standards and benchmarks in this area, together with confused definitions of CC, allows a variety of different methodologies to be adopted which makes like-for-like comparison of the metrics problematic. These issues are intertwined with questions of trust in the data as stakeholders' express doubts and ambiguity about information sharing on cyber-attacks and which further suggests that CC goes largely unreported.

From this summary of identified common problem areas it seems possible to thematically group together the gaps in research into five key areas with further sub-sets, as follows:

1. Definitions/Taxonomy
   (a) The definition of CC is currently open to interpretation; a taxonomy agreed at international level is needed to avoid confused/distorted data analysis.
2. Metrics/measurement
   (a) How can difficult to assess areas be valued, such as loss of reputation, privacy, etc.? Flexibility is needed to accommodate changes in a fast-paced new digital era.
   (b) What is a trusted data source?
   (c) Data is often incomplete and, at worse, inaccessible. Is open data achievable?
   (d) Standard formats for data collection and analysis can help improve quantification.

3. Trusted Data
   (a) Should there be a separate entity for data collection?
   (b) Who owns the data/for how long?
   (c) Can obvious agendas be obviated?
   (d) There is currently a lack of information sharing between entities
4. Standards/Benchmarks
   (a) The lack of standards/benchmarks limits consistency of data analysis
   (b) One size fits all policies may hamper SME stakeholders
   (c) Should governance be local or global?
5. Threats/cyber attacks
   (a) Low reporting rate to police due to lack of trust
   (b) Lack of knowledge/ambiguity about who to report CC to
   (c) Low prosecution rates
   (d) Cross border offending obstructs police action
   (e) Lack of information sharing

Fast-paced digital era encourages innovative new threats

Trust emerges as a central issue to each of the other identified research gaps. This is depicted here as a central pivot for research topics (Fig. 5):

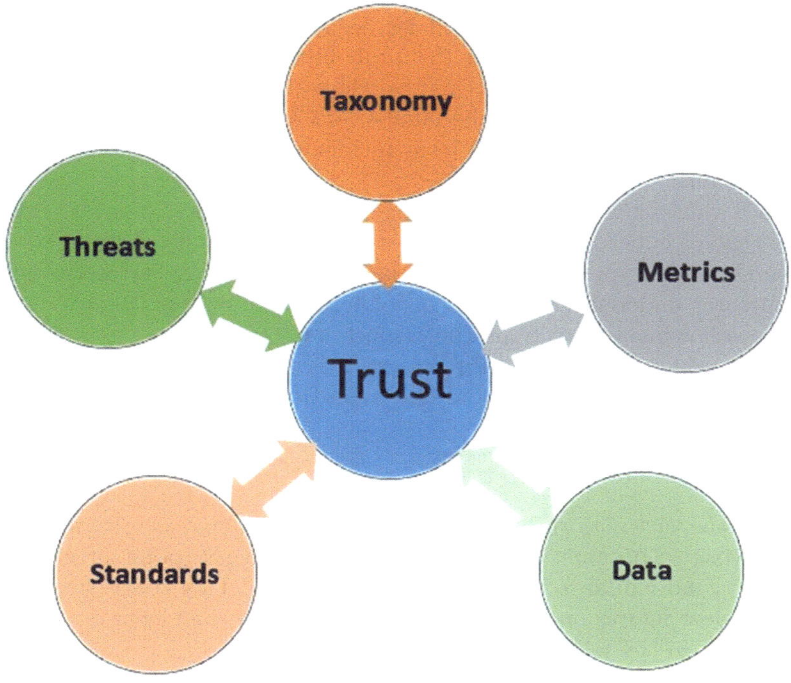

Fig. 5. The Pivot of Trust

The groups surrounding the "Pivot of Trust" provide a basic framework from which to elicit research gap scenarios. Each set is worthily of study in its own right but together satisfy many of the outstanding issues. The subject matter within each study area may be disproportionate in terms of range and depth but, in terms of improved Trust, each is of equal value.

As a scientific discipline, CC is still in its infancy. Value can, therefore, be gained from the evolutionary experiences of other sciences. For example, research without some form of taxonomy/definition would be chaotic in any circumstance.

Accuracy of data is fundamental to other scientific research areas and is dependent upon tried and tested methods of measurement. In some areas data that is unreliable or untrustworthy could be life threatening. With the advent of the Internet of Things, this could become a critical issue. How and what to measure is essential to know if, for example an accurate risk assessment is to be carried.

As the digital era evolves trust as a perception as well as a reality is important. Consistent and fair analysis can help change perceptions which can be achieved through the introduction of industry standards which provide the cornerstones to improved safety and reliability and trust in a variety of circumstances. Currently, cybersecurity and trust are not words that harmonize well.

The notion of Trust is central in the security domain, as all the relationships among people, associations, companies, etc. are based on trust. Moreover, when decisions are to be taken on the policies needed to prevent security incidents, reliable information is needed on the probability of the events, on the data that can be targeted by attacks, and on the value of data loss and recovery. Consequently, sound metrics on the number of CC events, their effects, and the damage that are actually caused from incidents is necessary for defense and recovery actions.

On face value, it might seem that the most importance area for additional study is that of cyber threats but it is essential to know if the money is being spent on the right type of research. To know this with any certainty there has to be a greater under-standing of co-dependent disciplines.

In the following sub-sections the importance of measuring economic costs on the state of CC in 2020 is enumerated from current scenarios and weighed against some of the in findings from the CyberROAD Cybercrime Survey 1.

6.1 Current Scenario

At present, the vast majority of governments address cybersecurity more within the framework of national defense rather than from the point of view of the protection of individual, social, end economic assets. This study suggests that the lack of clear figures on the real impact of computer incidents serves to limit understanding in the following areas:

- The extension of the threat (i.e., number of computers, individual, enterprises, etc. that have been victims of attacks)

- The total loss that was caused by attacks, both in terms of tangible and intangible assets In such a scenario, it is quite difficult if not impossible, to take decisions on:
- The policies to set up in terms of education, training, awareness, as well as in terms of software and system verification and certification
- The money to spend to implement the above policies, are today quite limited as the real impact in terms of saving is not well defined.

In fact, laws and regulations need to be grounded on reliable data in order to clearly show how the money spent in prevention and monitoring actually decrease the likelihood of more serious consequences.

It turns out that the current scenario poses a serious threat as the lack of coordinated and focused actions from the legislative and government bodies paves the way for various forms of criminal activities that, if not properly tracked and recorded, cannot provide evidence of the existence of a real threat.

6.2 Future Scenario

A desirable future scenario is one in which governments can rely on solid methodologies to collect reliable figures about the real impact of CC on companies, individuals and the public sector in order to take decisions, and allocate budget that is proportionate to the real threat.

In this scenario:

- Individuals, companies and the like have a high level of awareness on the possible uses of their data by public and private bodies, thus assigning a value to their data
- The market is mature enough so that a value can be assigned to each piece of information
- It is mandatory to disclose cyber-attacks and data breaches to a central authority, associating the costs incurred in terms of lost assets, lost business, repair/refactoring of software, and of business procedures.

The above obligation implies that novel techniques are in place that allow assessing the influence of the attack and data breach.

On the basis of past data, and of the actual market values, cost estimates are possible. Consequently, it is possible to devise policies that are cost-effective in containing the vulnerability of software and systems, handling security incidents, and preventing their rapid diffusion.

7 Conclusion

Reliable data is a fundamental on which revenues and budgets rely from the top at government level down to board level and individual stakeholders. To understand a problem, to know what is and how to tackle it, is a task that presents greater challenges when size and extent of that problem remains very much shrouded in mystery. The CyberROAD project is working towards a roadmap for CC and CT to reveal the research gaps that can help policy makers make more informed decision on where money should be directed to return the best possible outcomes.

CC as a subject of study is still in its infancy and much can be learned from the evolutionary development of other recently established sciences. To begin, a clear taxonomy is an essential element from which a framework for further study can be developed. Our investigation of current and future scenarios via focused surveys and comparison of the cost of CC reports reveals a number of research gaps that require attention if the scenarios outlined are to be achieved by 2020. Fundamental to the issue is the ability to quantify what we have and where we want to go. Currently, there is a mis-match between the experiences of stakeholders and the information to hand which can be improved with quantification of the issues and a reliable model for costing. Central to this information is the issue of trust, as without it there will be no confidence in the way forward with more time and money being wasted. Indeed, it is not an exaggeration to say that without quantification and measurement there will be no solution to the problem of CC by 2020 or beyond.

Acknowledgement. The research leading to these results has received funding from the European Union Seventh Framework Programme (FP7-SEC-2013) as the CyberROAD project (Development of the Cybercrime and Cyber-terrorism Research Roadmap) under grant agreement no 607642. Davide Ariu & Giorgio Giacinto of UNICA contributed to an early version of the document.

References

1. Levi, M., Innes, M., Reuter, P., Gundur, R.V.: The economic, financial & social impacts of organised crime in the European Union. Publication Office of the European Parliament (2013)
2. FBI: Uniform Crime Reports (2011). http://www.fbi.gov/about-us/cjis/ucr/ucr. Accessed Oct 2014
3. The Economist: Whats in a number? Estimating the cost of cybercrime. http://www.economistinsights.com/technology-innovation/analysis/measuring-cost-cybercrime/custom. Accessed Mar 2015
4. Center for Strategic Studies (CSIS): Cyber Threats and Information Security. Publisher CSIS report (2001)
5. Horn, P.: It's Time to Arrest Cyber Crime. Bloomberg Business (2006)
6. Anderson, R., Barton, C., Bohme, R., Clayton, R., Van Eeten, M., Levi, M., Moore, T., Savage, S.: Measuring the Cost of Cybercrime (2013). http://weis2012.econinfosec.org/papers/Anderson_WEIS2012.pdf
7. Anderson, R.: Debunking cybercrime myths, University of Cambridge Computer Laboratory (2012). https://www.lightbluetouchpaper.org/2012/06/18/debunking-cybercrime-myths/. Accessed Mar 2015
8. Detica Ltd.: The Cost of Cyber Crime (2012). https://www.gov.uk/government/news/report-released-into-the-cost-of-cyber-crime

9. Ponemon Institute: Ponemon Institute Research Finding (2014). http://www.ponemon.org/. Accessed Mar 2015
10. McAfee and CSIS: Economic Impact Cybercrime 2 (2014)
11. East West Institute. http://www.ewi.info/
12. Rauscher, K.F., Cox, E.N.: East West Institute - Measuring the CyberSecurity Problem (2013). http://www.ewi.info/. Accessed Mar 2015
13. Neustar: UK Annual DDOS Report (2014). https://www.neustar.biz/ddos-attacks-report. Accessed Mar 2015
14. McAfee and CSIS: Stopping Cybercrime can positively impact world economies, 6 June 2014. http://www.mcafee.com/uk/about/news/2014/q2/20140609-01.aspx. Accessed 13 Oct 2014
15. McAfee and CSIS: Economic Impact Cybercrime 2 (2014). http://www.mcafee.com/us/resources/reports/rp-economic-impact-cybercrime2.pdf. Accessed 13 Oct 2014
16. Group-IB: Group-IB (2011). http://www.group-ib.com/. Accessed Oct 2014
17. IDC: Security Products and Services. http://www.idc.com/prodserv/maps/securityproducts.jsp. Accessed Oct 2014
18. Internet Live Stats. http://www.internetlivestats.com/internet-users/. Accessed Mar 2015
19. The Radicati Group, Inc: Email Statistics report, 2015-2019 Executive Summary (2015). http://www.radicati.com/?p=10644
20. RIPE Network Co-ordination Centre. https://www.ripe.net/internet-coordination/press-centre/understanding-ip-addressing. Accessed 20 March 2015
21. International Telecommunications Union: The World in 2014: ICT Facts and Figures (2014). http://www.itu.int/en/ITU-D/Statistics/Pages/facts/default.aspx. Accessed Mar 2015
22. Bots vs Browsers. http://www.botsvsbrowsers.com/. Accessed Mar 2015
23. Barracuda Central: Spam Data. www.barracudacentral.org/data/spam. Accessed Apr 2015
24. Barracuda Central: Web Data. http://www.barracudacentral.org/data/web. Accessed Apr 2015
25. Spamhaus: Spamhaus Block List. www.spamhaus.org. Accessed Oct 2014
26. AV-TEST: Malware. http://www.av-test.org/en/statistics/malware/. Accessed Apr 2015
27. Akaimai: Real-time Web Monitor. http://www.akamai.com/html/technology/dataviz1.html. Accessed Oct 2014
28. Squid Blacklist. www.squidblacklist.org/downloads.html. Accessed Oct 2014
29. Ponemon Institute: 2014 Global Report on the Cost of Cyber Crime. http://www.ponemon.org/. Accessed Oct 2014
30. Shadowserver: Malware. https://www.shadowserver.org/wiki/. Accessed Oct 2014
31. Acquisti, A., Taylor, C., Wagman, L.: The economics of privacy. J. Econ. Lit. (2015, in press)

Towards the Development of a Research Agenda for Cybercrime and Cyberterrorism – Identifying the Technical Challenges and Missing Solutions

Borka Jerman-Blažič and Tomaž Klobučar(✉)

Jožef Stefan Institute, Ljubljana, Slovenia
{borka,tomaz}@e5.ijs.si

Abstract. Cybercrime and cyberterrorism research faces a number of challenges, such as the rate of change in technology, field complexity and interdisciplinarity. This chapter aims at identifying the major technical challenges that require solutions to be developed for the successful prevention and fight against such contemporary problems. The following solutions have been elicited as a leading contribution towards the design of a cybersecurity research agenda. The identified and selected solutions include technologies and techniques for computer fraud prevention, investigation and detection methods and tools, and crime prevention methods that address human elements.

Keywords: Cybercrime · Cyberterrorism · Research agenda · Technical challenges · Fraud prevention · Data sharing · Big data · Human elements

1 Introduction

Cybercrime (CC) is one of the fastest growing forms of crime, with more than one million people worldwide becoming its victims each day. Cybercriminals and CC network attacks are increasingly present in the everyday life of civilians, organizations, enterprises and government institutions. The longer we live in a digital world, the more opportunities will be present for cyber criminals or terrorists to exploit the vulnerability of networks, organizations and human lives. In discussing CC, the appearance and the relation to cyberterrorism (CT) should be mentioned here, as the dividing line and differentiation in the research approach are not very clear and sharp. Some authors have suggested that the key feature that makes the difference between CC and CT is the motivation of the actors, as crime is considered to be driven more by "personal gain or revenge" while terrorism is driven by dominance of "political" reasons to cause damage to an organization or a political system. Addressing a particular problem and developing prevention methods and technologies for specific CC/CT attacks is usually considered an unsustainable, non-scalable and inadequate approach, as this approach does not provide protection for all facets of cyberspace. In addition, the fight against CC/CT by the relevant authorities, e.g. law enforcement

B. Akhgar and B. Brewster (eds.), *Combatting Cybercrime and Cyberterrorism*,
Advanced Sciences and Technologies for Security Applications,
DOI 10.1007/978-3-319-38930-1_9

agencies, cannot assure the envisaged security and safety without cooperation with the private as well as the public sector. These large parts of the society acting in the digital world need to adopt a different approach for the security architecture (e.g. trusted computing, ubiquitously embedded security automation technologies, information sharing). Building security as a robust and solid foundation for citizens and economic entities to conduct transactions in the digital world is a must. This finding reflects this chapter of the book intended to identify the major technical challenges that require solutions to be developed for a successful prevention and fight against CC and CT. The selection of missing solutions includes technologies and techniques for computer fraud prevention, investigation and detection methods and tools, and, crime prevention methods addressing human elements.

2 Understanding Cybercrime and Cyberterrorism

2.1 Rate of Change in Technology

Today, information and communication technologies (ICTs) are omnipresent and the trend towards digitization is growing. The demand for Internet and computer connectivity has led to the integration of computer technology into products that have usually functioned without it, such as cars and buildings. Electricity supply, transportation infrastructure, military services and logistics – virtually all modern services depend on the use of ICTs. Today almost everyone in the world is connected either to the Internet or to some other phone network [1]. The estimated number of Internet users is close to five billion [2] and the expectations are that this will increase steadily over time. No society, no country, no individual will be unaffected in the close future [3]. The second contributor to the extreme speed of change in technology is the explosion of data. The amount of data being produced is rapidly growing and will grow ten times over the next six years, reaching 44 trillion gigabytes of data by 2020 [4] that can be stored and analysed to give unprecedented insights at macro and micro scales, allowing to understand and predict the global trends, the growing of data markets and individual behaviors. According to many sources, electronics will be embedded in everything and will enable the monitoring and control of every aspect of the current world, blending both the physical, e.g. Internet of Things, and digital worlds in an unimaginable way [5]. The pervasiveness of information and communication technology and ubiquity of digital infrastructures means that the digital civilization is now a fact of life. Some examples illustrate this: digital media, digital social relations, critical infrastructures, services, surveillance, industrial control, government, intelligent transport systems, and smart cities, amongst others. The introduction of ICTs into many aspects of everyday life has led to the development of the modern concept of information society. This development brings great opportunities and improvement to the daily life. However, this is accompanied by new and serious threats. Essential services such as water and electricity supply now rely on ICT [2]. Cars, traffic control, elevators, air conditioning or telephones also depend on the smooth functioning of ICT. Attacks against information infrastructure and Internet services now have the potential to harm society in new critical ways. On-line fraud and hacking attacks are just examples of computer-related crimes that are committed on a large scale every day [1]. The financial damage caused by CC is reported to be enormous and the damage per enterprise caused in USA only exceeds USD 15 million [6]. By some estimates, revenues from CC were outstripping the illegal trade in drugs for the first time in 2007 [2]. These estimates clearly demonstrate the importance of understanding CC and of developing effective prevention and protection methods and tools. As the major difference between CC and CT is in its dominating motivation [7], CC is generally committed for individual, personal reasons such as personal gain or personal revenge. CT attacks may have the same results and use the same methods, but the motivations are usually different. Such motivations may be aimed to destabilise an institution or country, or to intimidate a population into changing its government's behavior [8]. In that context, analysts and legislators are facing the problem of understanding the motivations of persons who carry out a cyberattack when trying to classify it and determine how the perpetrators should be prosecuted. These distinctions have a clear significance for justice and law enforcement, despite the use of similar techniques, methods and approaches for committing attacks. The techniques and some of the results are usually identical to certain instances of CC, as fundamentally any attack consists of individuals or groups seeking either to disrupt or take over communications and information systems or to extract information by tapping a wire. A key concept in this context is the "advanced persistent threat", frequently employed in espionage and cyberwarfare to continuously monitor and extract data from specific targets, using a set of stealthy and continuous hacking processes. Such long-lasting attacks require the capability, resources and intent and are thus commonly seen as requiring the resources and motivations of governmental agencies.

In the last decade many definitions appeared for CC, for example Hartel [9] defines CC as a behaviour in which computers or networks are a tool, a target, or a place of criminal activity [10]. This includes as a subject the information security, namely techniques to prevent or detect attacks on information assets, but the issue is much broader because it also includes such topics as the use of computers to commit "traditional" crime. For these reason the Global Cybersecurity Agenda [1] has seven main strategic goals, built on five work areas: (1) Legal measures; (2) Technical and procedural measures; (3) Organizational structures; (4) Capacity building; and (5) International cooperation.

It is possible that CC will become nothing special in the future. Something similar has happened before, with the introduction of new technology: The industrial revolution urbanised crime, which the law enforcement of the day was unable to cope with [11]. This eventually led to the introduction of the modern police force. We may expect also that the information revolution, especially if the speed of change is considered, will have a significant effect on law enforcement too in fighting against CC. However, before CC is subsumed by the definition of crime, there are some significant challenges to be met. For example, the Lockard's exchange principle [12] which is the foundation of forensics, does not seem to apply to CC scene investigations. In addition, the existing technical infrastructure of the Internet has a number of weaknesses, such as the monoculture or homogeneity of operating systems. Solutions, technical and strategic measures need to be developed to prevent attacks and develop countermeasures, including the development and promotion of technical means of protection, as well as an adequate and sufficient legislation allowing law enforcement to prevent and fight CC effectively [13].

2.2 Complexity and Interdisciplinarity

The complexity in the prevention of CC and the fight against cybercriminals is based on several dimensions originating from the characteristics of the current digital world. CC knows no borders. The crime site where the attack happens is independent of the presence and location of the attackers. There is no need for the criminals to be present at the same location as the target. As the location of the criminal is usually different from the crime site, many cyber-offences are transnational by nature. International CC offences affect more than one country, and the protocols used for data transfer on the Internet are based on optimal routing if direct links are temporarily blocked. Even when the domestic transfer processes within the source country are limited, data can leave the country, be transmitted over routers outside the territory and be redirected back into the country to its final destination. On the other hand, because no general control instruments exist, users are able to use filter circumvention technologies to send encrypted anonymous communication out of the country. As the number of people connected to Internet is growing, there is also a simultaneous increase of the number of offenders. As a consequence, any estimation of the number of offenders or people who use the Internet for illegal activities [2] is rather difficult. The increasing number of offenders causes difficulties for law enforcement agencies, as currently there is no possibility to automate the CC investigation process. Another problem is the short life of the data vital for tracing offences, especially in cases when cloud infrastructure is involved. The data are deleted after a short time. The short time available for investigation is problematic as the traditional mutual legal assistance regime often takes time to organise. Offenders may also include third countries in their attacks to make the investigation more difficult. Due to the complexity of the field, CC is by definition a multidisciplinary field as, among others, it makes use of mathematics, engineering, economics, medical science (psychology), sociology, criminology, law and public management [9].

3 Challenges and Threats

3.1 On-Line Anonymity and Data Protection

One of the major challenges of the fight against CC is the on-line anonymity as many Internet services are designed in such manner to make the identification of offenders difficult. The possibility of anonymous communication is either a by-product of a service, or is offered with the intention to avoid disadvantages for the user. Some examples are public access terminals, public wireless networks, prepaid mobile phone services that do not require registration, storage capacities for homepages, anonymous communication servers and remailers. Offenders can use several tools to hide their identities, such as fake mail addresses, or use free mail servers. In order to protect user privacy, several countries support the principle of anonymity, as is the case with the EU. The data protection applied to protect information from access by unauthorized people uses encryption technologies as a key technical solution. However, the same technology is used by offenders, making it difficult for law enforcement agencies to break the encryption and access the data. The recent case with encrypted data in an Apple iPhone and the US Agency's request to the manufacturer to reveal the encryption key which was refused is a good example. The availability of encryption technologies and their use by criminals are challenges for forensic investigators and law enforcement agencies [14].

3.2 Challenges and Technical Aspects of Data Sharing

Cyber-attacks happen in all types of organizations and individuals. They can start in many different places, including any device connected to the Internet. This is highly problematic in the modern digital society where devices such as copy machines are hooked up to the Internet in order to update themselves, report usage, install software, etc. Having all these devices connected to the Internet increases the exposure and vulnerability to CC [14]. In addition, it makes information sharing between the victims and the law enforcement bodies more difficult. Due to so many targets on the Internet sharing information, among the investigation instructions and law enforcement agencies the request by stakeholders for an effective sharing of information for fighting CC is obvious. There is an urgent need to create an orderly way of looking for threats and reporting the facts found in case of committed crime in a standard and understandable way for all involved. The implementation of countermeasures, for example the intrusion detection systems (IDS), which are part of the network hardware and software, requires maintenance and updating with recent developments [15]. Information about this should be shared as well. The systems used for monitoring and tracing should be adaptive and will need to have some level of self-awareness, self-learning and self-explanation to be able to address a moving target, such as CC criminals. Some predictability will be needed based on the shared data collected from different sources that will essentially allow the understanding of crime scenarios and learning from past wrong decisions. Information sharing should be implemented also by building new awareness and methods that enable the crime trends to be recognized in their early sprouts.

3.3 Illegal Content and Underground Market

The Internet is becoming the main instrument for the trade and exchange of material containing child pornography. The major reasons for this development are the speed and efficiency of the Internet for file transfers, its low production and distribution costs, and its perceived anonymity. Pictures placed on a webpage can be accessed and downloaded by millions of users worldwide. One of the most important reasons for the "success" of web pages offering pornography or even child pornography is the fact that Internet users are feeling less observed while sitting in their home and downloading material from the Internet. The same applies to hate speech and racism and xenophobia-motivated propaganda on the Web. The problem in that context is that not all countries criminalise hate speech [2]. An additional problem is the appearance of the Dark Web, i.e. overlay networks which use the public Internet but require specific software, configurations or authorization to access. The dark web includes marketplaces trading in mainly illicit products and services, such as drugs, software exploits (e.g. Trojan horses, botnets), network attacks offered as a service, and weapons. In addition to services such as fraud, this illegal marketplaces offer illegal and ethically disputed pornography, phishing and scam services and tumblers for Bitcoin services [16]. One of the major features of the dark web is the obscuring of the originating Internet Protocol (IP) address of its users via Tor protocol applications. The nature of activity of the Dark Web explains why little research exists related to this challenge.

3.4 Big Data, Abundance of Information and Analysis

Data sets on the Internet are growing rapidly, partly because they are increasingly gathered by cheap and numerous information-sensing mobile devices, aerial (remote sensing), software logs, cameras, microphones, radio-frequency identification (RFID) readers and wireless sensor networks. The world's technological per-capita capacity to store information has roughly doubled every 40 months since the 1980s [17]; as of 2012, every day 2.5 Exabyte (2.5×10^{18}) of data is created [18]. The abundance of data and information within the ICT systems raises several issues related to cybercrime. This includes the protection of Internet privacy, international government cooperation, passenger name record transfers, anti-terrorism developments, freedom of information, Internet censorship, e-Identity systems, corporate governance, the appointment of privacy regulators, cross-border data flows, data retention, judicial process, government consultation procedures, information security, national security and aspects of roughly a hundred technologies and technology applications ranging from video surveillance to DNA profiling. However, sophisticated solutions for their analysis and the successful removal of the potential appearance of false answers may contribute to the development of effective cyber intelligence features, by exploiting the huge potential of currently available as well as emerging information management technologies. Emerging technologies and new analytic techniques on big data are crucial for a better understanding of the criminal strategies and the anticipating trends and they will become crucial for the prevention and fight of CC.

3.5 Human Elements

In many instances the weakest point in the ICT system's defences is the human element. CC attacks are made possible by the fact that the current security technology was developed only with an aim to protect the ICT systems, and the consideration of how real users react when exposed to malicious attacks to their assets or privacy was neglected. Developing effective protection and system defences requires an understanding about how users behave and what traits of their behaviour make them and the systems vulnerable. Understanding the aspects of human psychology exploited by criminals will enable the building of robust systems able to resist most of the known CC attacks [19]. Research into victims' issues, their rights and policy recommendations will enable the voice of victims to be transferred to government and criminal justice agencies and will contribute to the changes of the legislation and policies affecting victims and witnesses.

3.6 Challenges in Anticipating a New Generation of Cybercrime

One of the appearances of crime without borders is its spreading through the Internet, causing CC cases in all manners of appearance. This is seen as an emerging spreading phenomenon that appears and will appear in the future in different shapes and scenarios. Cybercrime is a high-profit and low-risk endeavour. A successful fight against it requires a compendium of methods for preventing and combating this type of crime [20]. The expected occurrence of new CC will be caused by not yet forecasted, not yet foreseen crime related to the auxiliary structure of the free Europe, enabling the free transport of people, goods and capital. This addresses a cross cutting new challenge where several EU and member state bodies could become partners to be aligned but also confronted with the impact of travelling criminals causing high impact or high volume crime, or will be confronted with new ways of fraud and threats without any physical travelling. Crime fighting and prevention are usually implemented in traditional ways. The low flexibility of these methods is a risk that needs to be addressed. The adaptability for new solutions is low due to the hierarchical structure and fixed and insufficient budgets. The fight against crime and crime prevention will require flexible and fast measures and resources and justly discussions on competence and ethical rules. Solutions need to be developed for "real case scenarios", recognisable for policy makers, but above all for the leaders of law enforcement agencies. In that context the following is needed:

- Forecast and understanding of fast-appearing or potential new crimes,
- Technologies that can sufficiently anticipate new trends, upcoming crimes and potential threats. The challenge and objective of using new technologies for discovering what are the rapidly evolving trends, enabling the development of new mobile and flexible methods for identifying group structures and alliances, multi-crime and different crime activities. Their use should allow the understanding and detecting of the dynamics of potential threats and crimes in a sufficiently anticipatory manner in order to be able to act in time and appropriately.

4 Missing Elements and Solutions

This section presents the missing elements and solutions required to be developed to cope with the challenges of CC and CT described above. The elements and solutions have been identified on the basis of past and on-going research activities in the field and by applying the COURAGE gap analysis methodology. The sources of information included EU projects with topics addressing CC and CT and their repositories, the IEEE Explorer, SCOPUS, Google Scholar and ProQuest databases, and organizations, such as Europol, ENISA, UNICRI, OECD, and ITU. The outputs of this analysis are later combined with the results of Chap. 3 in defining the elements of the Research Roadmap presented in Chap. 16.

4.1 Fraud Prevention Techniques

Fraud is defined as an act of deceit to gain an unfair advantage. For an act to be legally considered fraud, the attacker needs to knowingly communicate false information to the victim, and the act must affect the victim in a negative way. Computer fraud refers to "*acts involving interference with or illegal accesses to a computer system or data with the intent of deceitfully or dishonestly obtaining money, other economic benefit or evading a liability, as well as to acts involving interference with a computer system or data in way that results in the creation of inauthentic computer data*" [3]. It thus uses electronic resources to present fraudulent or misrepresented information as a means of deception [21].

Methods to counter computer fraud can be divided into methods to detect computer fraud, and methods to prevent it. The former concentrate on analysing the system and user behaviour and detect fraud by searching for anomalies or certain deceitful characteristics. While these methods are already heavily used in some business areas, they are still an intensive research topic. Prevention of computer fraud concentrates on a fast response when detecting fraudulent actions to avoid (further) losses, as well as on policies, education and awareness, and technologies that prevent fraud related threats to be realized.

Many fraud scandals in recent years and statistics [22] show that the means to counter computer fraud are still lacking effectiveness, and that fraud detection and prevention methods are still an open field for research. In this subsection several challenges of the techniques that are used for fraud prevention are described together with missing related technical solutions, in particular the required solutions for effective and efficient protection against malware, data protection, authentication, and fraud prevention of digital currency.

Efficient and Effective Protection Against Malware

An important step in preventing computer related fraud is to protect against malicious software or malware, which is the top cyber threat [23]. Malware is becoming increasingly sophisticated, intelligent, versatile, available, and is affecting a broader range of targets and devices [22]. The increasing use of smart devices, e.g. smart phones, constitutes an opportunity for malware to steal information such as online banking login credentials and account information as well as other data stored on mobile devices [24]. Infected mobile devices are also targets for ransomware (e.g. Locky, TeslaCrypt, Simplocker) and have the potential to act as an infection vector for other platforms and devices [22].

Malware detection mechanisms are either signature-based, detecting patterns of known malicious behaviour, or anomaly-based, detecting anomalous activities within a system. Both mechanisms have certain issues. Signature-based mechanisms cannot detect previously unknown threats, while the anomaly-based ones often have a high degree of false positives. The efficiency and effectiveness of the mechanisms is also challenged by sophisticated evasion techniques that make malware detection and analysis harder [25]. Evasion techniques can be VM-aware, sandbox-aware or debugger-aware [26], and can complicate the detection and analysis of malware in virtual security environments (virtual machines) or prevent it from deploying or running in a sandbox environment [27]. Malware, such as the UpClicker Trojan, is able to detect the context and act accordingly, for example to remain silent in case of absent activities [28]. Advanced techniques for information hiding, e.g. malware traffic, by means of steganography or through hidden channels are also expected in the near future [29] and require proper discovery technologies.

More effective and efficient protection technologies for resource-constrained devices such as mobile phones and tablets are required [30], as well as the improved detection by correlating and analysing a broader set of features from the system and network. The solutions should also be able to perform malware analysis on-line and in a non-intrusive fashion [31].

Data Protection

Cryptographic algorithms are the basic security mechanisms for protection against illegal data modification, forgery, and disclosure. A number of symmetric and asymmetric algorithms exist [32] that provide different degrees of protection against specific types of attackers, such as individuals, organizations and intelligence agencies. The security level provided depends on the selected algorithm, key sizes, parameters, usage mode, as well as implementation details [32].

While the properly implemented and used standard cryptographic algorithms can ensure an adequate security level for most of the legacy and future systems, several issues still exist. Those issues are mostly a result of the deployment of protection measures in emerging constrained environments (e.g. Internet of Things) and the new computing possibilities in the future, especially the ones expected from quantum computing.

The NIS WG3 report [33] identifies the following three main research challenges regarding cryptographic algorithms, which are also relevant for the area of computer fraud prevention:

- ultra-light algorithms for systems and devices with constraints in, for example, computational power, memory, and energy, such as sensors, moving objects and other lower-resource devices,
- ultra-high-speed algorithms,

- public key algorithms that ensure long-term security, in particular the algorithms that cannot be broken when quantum computing reaches the level of practical usability.

Several recent and past security incidents, e.g. the Heartbleed bug in the OpenSSL library, the POODLE flaw in the TLS protocol, or the FREAK weakness in some implementations of SSL/TLS, have shown the importance of an adequate design and implementation of network security protocols. Cyber criminals can exploit any flaws of the protocols to obtain illegal access to computer systems and confidential data, such as private keys, login credentials and other private data, which can be then used to impersonate a legitimate user and commit fraud.

Existing network security protocols also face different issues in constrained environments, such as the Internet of Things, low-power wireless sensor networks or ad-hoc wireless networks of moving objects with low resource capabilities. Lightweight security protocols need to be developed for those environments at network and transport layers. Security mechanisms are also required to protect end-to-end communications, and to address cross-layer security aspects [34].

From the aspect of law enforcement, data protection mechanisms and secure protocols can be seen as a technical barrier obstructing the efficient and effective fight against computer fraud. While end users use encryption algorithms to protect their data and prevent fraud, cyber criminals can exploit these to cover the traces of criminal activities [22]. An example of the use of secure network protocols for illegal activities is the use of Tor and I2P (Invisible Internet Project) networks to provide anonymity in drug marketplaces, such as the Silk Road. New peer-to-peer networks that host the command and control infrastructure are more resilient and create additional difficulties for the disruption or taking down of botnets [22]. Law enforcement is seeking new solutions to be able to gather, access and decrypt digital evidence of CC and CT activities more easily, as well as to identify offenders using anonymization technologies.

Authentication Techniques

Strong authentication methods facilitate computer fraud prevention. Despite the research on alternative authentication mechanisms in the past years, there has been little change for users in practice [35]. People still use passwords that create too much of a burden and are plagued with security and usability problems. Users choose weak passwords that can be easily broken, even if stored in a protected form. This becomes a problem especially in the cases when attackers steal millions of them from large service providers' databases [36]. Advanced and more secure authentication mechanisms need to be used by default to prevent cyberattacks or minimize their effect, and the mechanisms should be combined (multifactor authentication) in a way that is acceptable to the end users and provides a higher level of security. Also, the number of explicit authentication events for the user has to be reduced in authentication mechanisms, and advanced technologies for implicit authentication of users developed [35].

Additional research is also required for stronger authentication mechanisms for mobile systems, constrained environments and clouds. Examples of such mechanisms are the graphical authentication for touchscreen devices, biometric authentication for mobile phones, for example the Android face unlock and iPhone fingerprint unlock [37].

Fraud Prevention and Digital Currency
Digital currency, being a sequence of bits, may be copied much easier than paper-based currency. Developing mechanisms to protect from such copies and/or fraud in general are still required for the digital currency to succeed and for confidence in digital financial systems to be developed. Methods and tools that will provide the user with strong security including some level of control over their data usage (assuring transparency on who is using what and for what purpose), while providing protection of their privacy, are needed. These tools should be able to verify who has access to the user data, and revoke this access if desired (assuming that this does not conflict with any local law) [38].

Both types of digital currency (the centralized Web and Perfect money, and the decentralized Bitcoin and Darkcoin) continue to evolve and with them the entire criminal economy [22]. The current processing power is still not sufficient for an easy decryption of the used cryptographic mechanism for digital currency creation. The development of quantum computers can contribute so this will become hypothetical. Novel cryptographic models thus need to be developed, as well as more efficient traceability tools and forensic tools for the file formats of digital currency wallets and accounts.

4.2 Operational Standards for Data Sharing

Collaboration between stakeholders such as law enforcement, public institutions and industry has been recognized as an important step in the fight against CC. However, collaborative actions in the field of CC data sharing are not trivial and easy to achieve. The heterogeneity in goals, strategies, and approaches on how stakeholders manage security issues, as well as how different sectors, for example critical infrastructure, energy, finance and banking, or public administration, manage data sharing and information exchange, must be taken into account. Companies often do not share incident related data because they are afraid their reputation would be damaged or they would lose their competitive advantage against other companies. Given the transnational nature of CC activities, different legislative frameworks in different countries make the issue even more challenging.

Several intra-sector and cross-sector initiatives have already been established to improve the sharing of cybersecurity incidents on the level of the EU and globally [39]. However, despite those initiatives, the approach for efficient knowledge sharing that would allow for a secure interoperability and collaboration between national and international bodies operating in the prevention of CC and CT is still missing. The lack of incentives from the private sector, primarily to share information on network information security issues, has been identified as an issue. As such, the scope for improving the incentivisation of cooperation and also practical mechanisms for increasing the level of information sharing between the public and private sectors remains a key area for research. Efforts are needed for the standardization of formal representations of threats, attacks and CC incidents. Some of the problems were elaborated in several initiatives (e.g. Mitre's STIX specifications [40] and approached in the ACDC project Centralized Data Clearing House data schemata [41]. Standard protocols for threat/incidents data exchange have also been proposed, e.g. Mitre's TAXII specifications [42]. However, more work is required for an efficient provision of shared knowledge between law enforcement agencies and other stakeholders. Solutions are also missing in the following areas:

- Global standard of CC information representation/exchange formats;
- Standardization of APIs for information sharing among the shareholders;
- Models for CC and CT attacks/incidents behaviour patterns.

Finally, dynamic and semantically annotated databases/repositories of known vulnerabilities for an automatic detection of vulnerabilities in source code would be helpful. Current repositories of known vulnerabilities are kept up-to-date, but in practice when reviewing the code the checking must be done manually. There are no automated methods for matching/finding patterns in the code that are already present in the repositories of vulnerabilities [43]. Also, the information in the repositories could be used for predicting, at design time, the likelihood of including a vulnerability or security flaw in the implementation code.

4.3 Solutions for Dealing with Illegal Content, Dark Web and Virtual Cybercrime

In CC, computers and computer networks can be a tool, a target or a place of criminal activities. Places vary from mobile devices, personal computers, web servers, clouds and companies' private networks to virtual worlds, social networks and parts of Internet known as Darknet or Dark Web, accessible only by the previously mentioned anonymous communication protocols such as Tor.

The biggest portion of the Darknet seems to be devoted to illegal activities, such as stolen goods, drugs, weapons and information selling, exchange of illegal content, for example content related to child pornography, child-sexual abuse, and illegal financial transactions [16]. The technologies needed to fight those activities include technologies for exploring the Darknet, detecting and monitoring criminal activities and identifying criminals in the dedicated servers of the Darknet, and seizing illegal content. The missing solutions should provide (1) monitoring of social sites to detect message exchanges containing new Darknet domains, (2) marketplace profiling for collecting information about sellers, users and the kinds of goods exchanged, (3) locating and mapping hidden services directories by deploying nodes in the distributed hash tables, and (4) monitoring hidden services of newly added sites. New investigation approaches are also needed for decentralized marketplaces such as the OpenBazaar, a BitTorrent-style peer-to-peer network [29].

The usefulness of virtual worlds and mixed reality environments in many different fields was proven by several R&D projects and other research (e.g. GALA Network of Excellence in Serious Games). Use of well-designed mixed reality makes the actors feel that they are immersed in cyberspace. Unfortunately, virtual environments are not immune to CC activities, as shown by the increase of such activities in the past years [44]. It is estimated that millions of dollars in virtual goods are stolen in virtual worlds. Virtual worlds face also other types of criminal activities and offences, such as money laundering, extortion, stalking, or hate speech. Normative frameworks to deal with virtual crime need to be developed, including (reputation-related) offences against avatars.

4.4 Information Management of Big Data

Big data is data characterized by high volume, high variety, high velocity, low veracity and high value. Here, variety refers to different formats of structured and unstructured data, velocity to the speed of data change, and veracity to the data quality. In the cybercrime domain, big data is used both by law enforcement and criminals. On one side, emerging technologies and new analytic techniques on big data are crucial for a better understanding of criminal strategies, anticipating trends and preventing and fighting cybercrime. On the other side, criminals use big data analytics to increase the value of stolen data [22].

Big Data Collection, Processing and Use for the Detection and Prevention of Cybercrime and Cyberterrorism

Big data mining and analysis represent important techniques for the identification of potential CC threats and trends, criminal and terrorist group structures and different crime activities. The technique should enable understanding and detecting the dynamics of the threats and activities in a sufficiently anticipatory manner in order to be able to act in time and appropriately. Big data analysis should thus add predictive and proactive capabilities to the fight against CC.

Solutions are also required that can quickly provide sense based on big data to an investigator and do not leave room for misinterpretation of the analysis results. Misunderstandings can be caused by an improper use of big data for predictive analytics, especially by equalling correlation with causality. Standardised procedure and best practices are therefore needed by law enforcement for properly conducting Big Data-related investigations and interpreting results [22]. A better understanding can also be facilitated by adequate visualization techniques that are scalable in visually representing massive amounts of data from heterogeneous and distributed data sources, and capable of rendering these in real time.

Privacy Protection Issues in Big Data Management

Big data can include vast amounts of personal data collected through various sensing devices to gain insights about individuals and their environment. Personal information or personal data that need to be protected are any information relating to an individual who can be identified, directly or indirectly. An important requirement of big data management is thus the protection of personal data, and finding a balance between the protection of privacy and the use of advanced data correlation and intelligence capabilities for cybercrime prevention, for example when conducting automatic mass video analysis. New solutions are also needed for a safe anonymization, aggregation, and deletion of stored data in a way that prevents de-anonymization and de-aggregation. The solutions should be sensitive to the contexts in which the data is considered private. They (in particular the ones that have some proactive properties) should be capable during their use to be aware of privacy issues and must be capable to control the information found or discovered in order the disclosure of private information to be minimal [38].

4.5 Human-Centred Solutions

Past experience has shown that technical solutions for prevention of and protection against CC are often too complex to use for non-experts, not convenient or not applicable for certain groups of users, and potentially privacy-intrusive. It is therefore of big importance that technical solutions in this field are human-centred, usable and able to protect user privacy.

Usability Issues
The literature review shows that additional study is needed to provide security and privacy services and mechanisms that are user friendly, without discarding the consideration of the security capabilities and performance of the service or the mechanism. The research gap needs to be tackled both from the technological and psychological points of view. Law enforcement officers need more usable and simpler tools for their daily work and investigations. Hibshi et al. highlight a number of usability issues that need to be taken into consideration when designing and implementing, for example, digital forensics tools [45]. The issues include the consistency, information overload and non-intuitive interfaces. Usability is especially critical here because misunderstanding that leads to false interpretations may impact real-life forensic cases [45].

Privacy Issues
Despite existing privacy protection services and privacy principles that should be followed when designing and developing services and systems that process personal data, e.g. privacy by design, different research issues still exist. Bettini and Riboni identified various technical, legal, user experience, and economical challenges related to privacy protection in pervasive systems [46]. From the technical point of view, they for example miss tools able to integrate and present the information about an individual held by adversaries, as well as more accurate models of adversary knowledge about a user. The proposed research directions for mobile participatory sensing include impact assessment of sensor reading combination and correlation on the user's privacy, as well as the provision of composable privacy solutions [47].

Privacy-friendly authentication and authorization mechanisms are also missing in wide deployment. STORK 2.0 has built an infrastructure for the use of strong authentication by means of national eID credentials in the EU for secure cross-border services, such as e-banking, e-health, e-education, and e-commerce [48]. The authentication mechanism is extended with the privacy-friendly use of business attributes for authorization purposes. Privacy protection is added to the authentication and authorization procedures by anonymous credentials and some other authentication mechanisms, such as privacy-preserving attribute-based credentials [49].

4.6 Harmonization of Terms in Cybercrime and Cyberterrorism

The current terminology used among law enforcement and other stakeholders was found to be ambiguous [7]. The definitions and the topics are overlapping. Despite the high perceived levels of awareness around the general concept of terrorism, there is little consensus towards an internationally agreed definition of CT [50]. Despite numerous attempts towards establishing a common definition for CT, none have resulted in a common, agreed international consensus on the issue [51].

The absence of an equal representation of subject areas, the definition of terms and the different taxonomy proposed in the field are identified as a problem by academia, law enforcement agencies, and by entities representing legal and ethical organizations and critical infrastructures [10]. Such an absence of harmonization can cause problems at all levels, from first response and research, right through to policy formulation and the development of legislative frameworks. Clear and logical definitions in the area of CC and CT are necessary to understand, measure, and fight CC and CT. It is necessary to have a robust framework in which different aspects of CC/CT can be classified, categorized and explained within the context and the meaning. Finally, a clear definition and an exhaustive taxonomy that may lead to metadata specification are necessary.

5 Conclusion

Identifying challenges and missing solutions is an essential step in the design of a comprehensive research agenda in any domain. This chapter focused on the field of CC and CT and the contemporary challenges and solutions, due to the very nature of CC and CT the primary focus of the chapter was on technical features. The presented results facilitate a better understanding of the challenges one faces in the prevention of and the fight against CC and CT, as well as the technical elements and solutions that one still requires to be able to cope with those challenges.

Acknowledgement. The research leading to these results has received funding from the European Union Seventh Framework Programme (FP7-SEC-2013) as the COURAGE project under grant agreement no 607949.

References

1. International Telecommunication Union (ITU): Understanding Cybercrime: Phenomena, Challenges and Legal Response (2012). http://www.itu.int/ITU-D/cyb/cybersecurity/docs/Cybercrime20legislation20EV6.pdf
2. International Telecommunication Union (ITU): Understanding Cybercrime: Guide for developing countries (2011). http://www.itu.int/ITU-D/cyb/cybersecurity/projects/crimeguide.html
3. United Nations Office on Drugs and Crime (UNODC): Comprehensive Study on Cybercrime (2013). http://www.unodc.org/documents/organized-crime/UNODC_CCPCJ_EG.4_2013/CYBERCRIME_STUDY_210213.pdf
4. Bisson, P., Martinelli, F., Granadino, R.R. (eds.): Cybersecurity Strategic Research Agenda (2015). https://resilience.enisa.europa.eu/nis-platform/shared-documents/wg3-documents/
5. Hui, S., Jiafu, W., Caifeng, Z., Jianqi, L.: Security in the internet of things: a review. In: 2012 International Conference on Computer Science and Electronics Engineering, Proceedings, pp. 648–651 (2012)
6. Anderson, R., Barton, C., Böhme, R., Clayton, R., van Eeten, M.J.G., Levi, M., Moore, T., Savage, S.: Measuring the cost of cybercrime. In: Böhme, R. (ed.) The Economics of Information Security and Privacy, Chap. 12, pp. 265–675. Springer, Heidelberg (2013)
7. Sims, D., Ghernaouti, S.: A report on taxonomy and evaluation of existing inventories. D2.1, E-CRIME deliverable (2014). http://ecrime-project.eu/
8. Koops, B.J.: The internet and its opportunities for cybercrime. In: Manual, T.C., Herzog-Evans, M. (eds.) vol. 1, pp. 735–754. WLP, Nijmegen (2010)
9. Hartel, P., Junger, M., Wieringa, R.: Cyber-crime Science = Crime Science + Information Security, University of Twente, Version 0.15 (2010)
10. Newman, G.R.: Cybercrime. In: Krohn, M.D., Lizotte, A.J., Penly Hall, G. (eds.) Handbook on Crime and Deviance, pp. 551–584. Springer, New York (2009)
11. Newman, G.R., Clarke, R.V.: Superhighway Robbery: Preventing E-Commerce Crime, pp. 8–9. Willan Publishing, Uffculme (2003)
12. Brenner, S.W., Clarke, L.L.: Distributed security: preventing cybercrime. John Marshall J. Comput. Inf. Law **XXIII**(4), 659–667 (2005)
13. Helfgott, J.B.: Criminal Behaviour Theories, Typologies and Criminal Justice, pp. 4–18. SAGE Publications, Thousand Oaks (2008)
14. Lipson, H.P.: Tracking and Tracing Cyber-Attacks: Technical Challenges and Global Policy Requirements for Next-Generation Internet (2002). http://www.sei.cmu.edu/reports/02sr009.pdf
15. Oehemen, C., Peterson, E., Dowson, S.: An organic model for detecting cyber-events. In: CSIIRW 2010 Proceedings of the Sixth Annual Workshop on Cyber-security and Information Intelligence Research, Article No. 66. ACM, New York (2010)
16. Moore, D., Rid, T.: Cryptopolitik and the Darknet. Survival **58**(1), 7–38 (2016). doi:10.1080/00396338.2016
17. Hilbert, M., López, P.: The world's technological capacity to store, communicate, and compute information. Science **332**(6025), 60–65 (2011). doi:10.1126/science.1200970
18. Boyd., D., Crawford, K.: Six Provocations for Big Data, A Decade in Internet Time: Symposium on the Dynamics of the Internet and Society (2011). http://papers.ssrn.com/sol3/papers.cfm?abstract_id=1926431

19. Victim support. https://www.victimsupport.org.uk/more-us/policy-and-research/
20. Horizon 2020, Secure Societies Advisory Group, Strategic Input for 2016-2017 Workprogram, April 2015, Private communication (2015)
21. Kunz, M., Wilson, P.: Computer Crime and Computer Fraud. University of Maryland, College Park (2004)
22. European Cybercrime Centre (EC3), Europol - The Internet Organised Crime Threat Assessment 2014 (iOCTA) (2014)
23. Marinos, L.: ENISA Threat Landscape 2014: overview of current and emerging cyber-threats. ENISA (2014)
24. Choo, K.-K.R.: The cyber threat landscape: challenges and future research directions. Comput. Secur. **30**, 719–731 (2011)
25. Marpaung, J.A.P., Sain, M., Lee, H.-J.: Survey on malware evasion techniques: state of the art and challenges. In: 14th International Conference on Advanced Communication Technology (ICACT) (2012)
26. Ortega, A.: Your malware shall not fool us with those anti analysis tricks. AlienVault Labs (2012)
27. Arntz, P.: Sandbox sensitivity. Malwarebytes unpacked (2013). https://blog.malwarebytes.org/intelligence/2013/02/sandbox-sensitivity/
28. Singh, A.: Don't Click the Left Mouse Button: Introducing Trojan UpClicker. FireEye Blog (2012)
29. European Cybercrime Centre (EC3), Europol - The Internet Organised Crime Threat Assessment 2015 (iOCTA) (2015)
30. Suarez-Tangil, G., Tapiador, E.J., Peris-Lopez, P., Ribagorda, A.: Evolution, detection and analysis of malware for smart devices. IEEE Commun. Surv. Tutorials **16**(2), 961–987 (2014)
31. Chen, P., Desmet, L., Huygens, C.: A study on advanced persistent threats. In: De Decker, B., Zúquete, A. (eds.) CMS 2014. LNCS, vol. 8735, pp. 63–72. Springer, Heidelberg (2014)
32. Agency, E.U., for Network, Information Security (ENISA): Algorithms, key size and parameters report - 2014 (2014)
33. Kert, M., Lopez, J., Markatos, E., Preneel, P.: State-of-the-art of Secure ICT Landscape (Final, Version 1), NIS Platform, Working group 3 (WG3) (2014)
34. Granjal, J., Monteiro, E., Sá Silva, J.: Security in the integration of low-power wireless sensor networks with the internet: a survey. Ad Hoc Netw. **24**, 264–287 (2015)
35. Sasse, M.A.: "Technology should be smarter than this!": A vision for overcoming the great authentication Fatigue. In: Jonker, W., Petković, M. (eds.) SDM 2013. LNCS, vol. 8425, pp. 33–36. Springer, Heidelberg (2014)
36. Mirante, D., Cappos, J.: Understanding password database compromises. Polytechnic Institute of NYU, Technical report TR-CSE-2013-02 (2013)
37. Bhagavatula, C., Ur, B., Iacovino, K., Kywey, S.M., Cranor, L.F., Savvides, M.: Biometric Authentication on iPhone and Android: Usability, Perceptions, and Influences on Adoption. USEC 2015 (2015)
38. European Union Agency for Network, Information Security (ENISA): ENISA Report on Strategic Research Agenda, draft v02.63 (2014). https://resilience.enisa.europa.eu/nis-platform/shared-documents/wg3-documents

39. European Union Agency for Network and Information Security (ENISA): ENISA cybersecurity Information Sharing: An Overview of Regulatory and Non-regulatory Approaches (2015). https://www.enisa.europa.eu/activities/cert/support/information-sharing/cybersecurity-information-sharing/at_download/fullReport
40. MITRE: Structured Threat Information eXpression (STIX) specification (2014). http://stix.mitre.org
41. Advanced Cyber Defence centre (ACDC) (2016). https://www.acdc-project.eu/
42. MITRE: Trusted Automated eXchange of Indicator Information (TAXII) specifications (2014). https://taxiiproject.github.io/
43. Torres, R., Gallego-Nicasio, B., Zanetti, R.: Initial set of research activities listed to meet gaps. CAPITAL (cybersecurity research agenda for privacy and technology challenges) D3.1 deliverable (2014)
44. Adrian, A.: Beyond griefing: virtual crime. Comput. Law Secur. Rev. **26**(6), 640–648 (2010)
45. Hibshi, H., Vidas, T., Cranor, L. Usability of forensics tools: a user study. In: Sixth International Conference on IT Security Incident Management and IT Forensics, pp. 81–91. IEEE (2011)
46. Bettini, C., Riboni, D.: Privacy protection in pervasive systems: state of the art and technical challenges. Pervasive Mob. Comput. **17**, 159–174 (2015)
47. Christin, D.: Privacy in mobile participatory sensing: current trends and future challenges. J. Syst. Softw. (2015). doi:10.1016/j.jss.2015.03.067
48. Klobučar, T., Gabrijelčič, D., Pagon, V.: Cross-border e-learning and academic services based on eIDs: case of Slovenia. In: eChallenges 2014: 29–30 October, 2014 Belfast, Ireland. Dublin: IIMC: = International Information Management Corporation, 9pp (2014)
49. Camenisch, J., Dubovitskaya, M., Enderlein, R.R., Lehmann, A., Neven, G., Paquin, C., Preiss, F.-S.: Concepts and languages for privacy-preserving attribute-based authentication. J. Inf. Sec. Appl. **19**(1), 25–44 (2014)
50. Record, J.: Bounding the Global War on Terrorism. Strategic Studies Institute (2003). http://oai.dtic.mil/oai/oai?verb=getRecord&metadataPrefix=html&identifier=ADA419754
51. Jarvis, L., Nouri, L., Whiting, A.: Understanding, locating and constructing cyberterrorism. In: Chen, T.N., Jarvis, L., Macdonald, S. (eds.) Cyberterrorism: Understanding, Assessment and Purpose, pp. 25–41 (2014) doi:10.1007/978-1-4939-0962-9

The Never-Ending Game of Cyberattack Attribution

Exploring the Threats, Defenses and Research Gaps

Piotr Kijewski[1(✉)], Przemyslaw Jaroszewski[1], Janusz A. Urbanowicz[1], and Jart Armin[2]

[1] NASK/CERT Polska, Warsaw, Poland
{piotr.kijewski,przemyslaw.jaroszewski,alex}@cert.pl
[2] Cyberdefcon, Hove, United Kingdom
jart@cyberdefcon.com

Abstract. In this article we approach the problem of attributing a cyberattack to real world actors, and the social context of the problem. The basic premise is that while it is socially acceptable to assign attribution of cybercrime after the act, society expects law enforcement to attribute the possibility of cyberterrorist acts to perpetrators in advance, and to disrupt them in the making. This blends the cyberattack attribution problem with the much wider problem of fighting terrorism and organized cybercrime, far beyond the limits of "cyber" understood as the fifth domain of warfare. The main contribution of the paper is identifying research gaps and attributing complexities derived from key problems such as offline criminal activity, as well as practical difficulties in researching cybercrime and cyberterrorism. To get to those conclusions, we analysed the attribution problem from the point of view of the perpetrator, using the SWOT methodology, which gave us insight on tactics of cyberattacks that give the most protection against attribution and prosecution, which led us to identifying current research gaps.

Keywords: Cyberattack attribution · Cybercrime · Cyberterrorism · Attack attribution · Threat intelligence · Organized crime · Espionage · Activism · Counterintelligence · Research gaps · Privacy

1 Attribution of Cyber Threats

Attribution is a core aspect of fighting cybercrime (CC) and cyberterrorism (CT) [1]. Without finding out who the actor behind an attack is, we are limited to mitigation of the technical aspects of the attack. On the other hand, if the attack can be attributed to a specific actor, individual or group, the potential to counter future cyberattacks from the same source is presented through targeted use of the information gathered by offline investigators, such as police work, political negotiations, intelligence gathering, covert operations and other available means.

B. Akhgar and B. Brewster (eds.), *Combatting Cybercrime and Cyberterrorism*,
Advanced Sciences and Technologies for Security Applications,
DOI 10.1007/978-3-319-38930-1_10

Attribution of an attack is also of paramount importance in a much broader context than through mitigation and prevention. Only if we know who the real attackers are, can we understand the real nature of the attack, take their motives into account, estimate the resources that can be deployed in an attack and thus determining its scale, and establish the political or financial motivations behind the attack. Correct attribution in cyberespionage and cyberwar cases establishes the nation state actor and allows deployment of either counterattack or effective counterintelligence techniques. It also must be noted that only high-certainty attribution of an attack can justify retaliatory actions under international law (for example, justifying invoking Article 5 of the NATO Treaty, as Internet can be considered the fifth theatre of war – "cyber") [2]. This is to firmly establish the responsibility of the offenders, before a response is deployed [3].

In CC, in police investigation and judicial proceedings, correct attribution allows a case to be built against an attacker, with the possibility of correlating it to related cases based on crimes of the same actors or their co-conspirators.

2 Methodology of the Study and Organization of the Paper

The methodology used to organize this paper is as follows. First of all, we identify the societal, economic and legal driving forces behind the need for cyberattack attribution. We then explore in general how these factors play out in today's world (current view) and look into the foreseeable future. We then attempt a deeper dive into the subject matter, by exploring its context in CC and CT. For this purpose, we perform a SWOT analysis of CC and CT attribution from an attacker's point of view. For both aspects we look into current threats and defenses, with the ultimate purpose being the identification of current gaps in tackling attribution that could serve as a basis for future research.

3 Definitions of Cybercrime and Cyberterrorism

For the purpose of this study we adopt a broad definition of the terms CC and CT, as understood by the CyberROAD project consortium [4].

CC encompasses two forms of criminal activities:

- *The use of computer systems to enable traditional forms of criminal activity (e.g., child pornography, money laundering);*
- *The use of a computer system to launch a cyberattack (an action against the integrity, confidentiality or availability of computer data, systems or network).*

Cyberterrorism, on the other hand, encompasses three forms of terrorist activities:

- ***CT attacks****, the possibility to use electronic means/information technologies to perpetrate attacks, whose dimension threatens human lives, may cause huge* *damage, challenging and jeopardizing the State security based on democracy and the rule of law. Such attacks have a political-ideological, ethnical and/or religious nature and motivation;*

- ***CT perpetrated by terrorists**, such as defacement of sites, disturbing the regular functionality of services as TV Channels and other infrastructures. These attacks may have a great impact on society holding the potential to disturb the organization of the societies;*
- ***Use of Internet by terrorists**, the use of internet/information technologies for terrorist purposes, like propaganda, financing, communication, recruitment, plotting, indoctrination, radicalization, logistics, planning, training, material dissemination, etc.*

4 Driving Forces

During our research, we identified the following classes of forces driving the needs for cyberattack attribution:

- Societal
  - Attribution is a crime deterrent.
  - Placing blame on a particular actor may be used for political advantage.
  - Attribution with certain level of confidence is required to justify actions against individuals, non-aligned political groups, and nation-states.
- Economic
  - Correct attribution is necessary for prioritisation of actions against particular actors.
- Legal and law enforcement
  - Identification of a criminal is a step towards securing punishment.
  - Attribution justifies and helps in effective use of monitoring (focused on specific groups and actors).
  - Counterintelligence purposes - the need to identify and disrupt state-level threat actors.

4.1 Current View of the Attribution Problem

When confronted with a threat, attribution is usually not on the top of the priority list. Unfortunately, unlike in a physical world, digital evidence is highly volatile, and easy to procure in a way that would make investigators look the wrong way. Attribution of cyber-related activities and attacks is therefore a complex and difficult task, requiring both skills and experience, which are not easy to gain [5]. Evidence used for attribution in cyberspace is based on several types of sources. For network-based attribution this would include server logs, netflow data, headers etc., and the key information here are network addresses (specifically, IPv4 addresses in most cases). Theoretically, tracing the malicious activity to its source should be possible, provided that activity logs are preserved at each point. However, this reasoning is oversimplified because identifying a source network address is not the same as identifying the device that was using it at a given time, let alone the individual behind it. Moreover, the assumption of availability of traces is almost never true.

Networks were not designed with attribution in mind. Data such as IP addresses, TTLs etc. are included in packets for the purpose of effective traffic delivery, not to provide accountability or forensic capabilities. Hence, they do not provide an adequate level of confidence to be solely used for attribution. Moreover, malicious actors can easily cover their tracks using network-level anonymization tools such as anonymous proxy, VPN services or the Tor network. Cyber-criminal activity is often staged, using numerous systems as stepping-stones. Those systems usually fall into different administrative domains, are located in different countries, with different jurisdictions. In practice, this makes effective gathering of evidence from the entire traffic path unfeasible. A good example of an alleged staged attack was a security breach in Lockheed Martin in 2011, where attackers supposedly leveraged vulnerabilities in RSA SecureID system, obtained in an earlier attack against this security vendor [6].

Other methods of attribution include analysis of malware and other artifacts in a compromised system. Language indicators or other characteristic strings may be often found in the code. However, such traces are inconclusive. First of all, such indicators may easily be planted by the author for distraction. More importantly, large parts of code are sold, stolen or exchanged for further reuse, so its authors may be unrelated to malicious actors who used it later.

Non-technical traces may also aid in the attribution process. Attackers' behavior and motivations can give a hint about their background and origins. Common criminals are usually not discriminative regarding their targets and are looking for fast ways to make money (such as encrypting data for ransom, stealing banking credentials or harvesting emails). On the other hand, state- or industry-sponsored actors are more likely to deploy stealth, and their attacks closely focused on particular targets. Again, this attribution strategy may yield inaccurate results, as trade secrets, design projects and other confidential data increasingly become a target for criminals.

Cybercriminal profiling can add valuable information to the task of digital forensics and help prevent unnecessary analysis of data that brings nothing to the investigation. Profiling and behavior analysis can be used to good effect to reduce response times to CC events, helping in the identification of organized crime groups (OCGs). An ongoing project into hacker profiles is provided by UNICRI (United Nations Interregional Crime and Justice Research institute) [7].

An often over-looked area, which is able to provide crucial leads on attribution of cybercriminals can be provided by analysis of the role played by the hosting provider within a specific cyberattack. This is not so easy when VPNs, TOR or similar anonymity services are used to facilitate an attack but, still, it is not impossible to successfully investigate such a scenario. Hosting providers may be unwitting victims of cybercriminals too, but in employing known best practices it is possible to reduce this risk. With the aid of the necessary legal processes, hosting providers can sometimes lead an investigation straight to the attackers or, at least, provide sufficient information from which to identify them.

It must be noted that attribution is a process highly susceptible to cognitive biases. As stated before, it is impossible – or at least very difficult – to collect all possible evidence. Hence, it is tempting to draw conclusions based only on what is collected even when the evidence is inconclusive and alternative reasoning could be proposed. It is especially easy to fall victim to the confirmation bias - looking for pieces of evidence that support early assumptions, or disregarding alternative interpretations. A good example is a vulnerability in Juniper's routers disclosed in December 2015. The vulnerability introduced backdoor functionality to cryptographic functions. In the context of increased pressure from many governments to gain side-channel access to encrypted transmissions, many researchers announced that the backdoor – due to its sophistication – also must have been planted by some government body, most likely NSA. However, there is no hard evidence to support this conclusion [8].

4.2 Future View

Cyber threats evolve rapidly and in agile ways, unmatched by the development of digital forensics and other mechanisms supporting attribution of cyberspace threats and activities. Focus is shifting towards proactive surveillance, allowing governments access to as much data as possible – with longer retention periods, relaxed procedures for requesting access to information and electronic eavesdropping, government-level decryption keys etc. Such an approach inevitably leads to conflicts with advocates of privacy, civil rights and freedom, and is generally not greeted with enthusiasm. At the same time, it is not a guarantee of success with attribution (even less with prevention) of CC.

It is essential to make efficient use of available methods, by improving legal and organizational environments to facilitate quick exchange of information between international law enforcement organisations. Operational information on malicious activities (such as IOCs – Indicators of Compromise, "digital fingerprints" of an attack or malware infection) should be routinely exchanged. This may be fostered by tightening cooperation with CSIRTs (Computer Security Incident Response Teams) and other researchers, who already engage in data exchange on daily basis. However, certain players – such as law enforcement agencies, intelligence agencies or anti-terrorist forces – have greater capabilities and should be encouraged to share their intelligence regarding cyber threats with their international counterparts. In addition to technical analysis, a path that also proves to be useful in attribution is following the money. Here, existing methods of fighting economic crimes and money laundering can be put into effective use. Some technical changes may be proposed to the design of Internet protocols and services in order to build accountability into them. For example, it was proposed to distribute IPv6 addresses in an organized way, where each country would be given its own prefix for further allocation. However, such changes have limited positive effects (e.g., they do not address staged attacks in any way) which may be outweighed by technical and organizational difficulties associated with their implementation. More sophisticated methods are needed, such as using beacons (cookies, honeytokens [9] etc.) to track back stolen data.

It is a simple axiom that all CC, cyberattacks, and Internet badness is hosted from somewhere and by someone. The introduction of best practices can be used to help improve accountability which in turn encourages efforts to eliminate, or at least reduce, the vulnerabilities that cybercriminals take advantage of. Hosts and service providers can take a proactive stance here and, in addition, increase their efforts to uncover the source of a variety of nefarious practices. Furthermore, there needs to be more research into how scientific models and algorithms can be applied to cybercriminal profiling to increase the effectiveness of an integrated holistic approach. Bringing together real-life knowledge about the conditions that create cybercriminals to a mathematical approach may lead to better methods of attribution in the future.

5 Attribution of Cybercriminal Acts

In this section we look into more detail into attribution of CC specifically.

5.1 SWOT Analysis

To start our deeper dive into attribution, we perform a SWOT (Strengths, Weaknesses, Opportunities and Threats) analysis of the attribution problem, assessed from a perpetrator's (cybercriminal's) perspective:

Strengths

- In CC it is relatively easy to anonymize oneself online.
- In CC it is relatively easy to create fake trails online.
- Inadequate international cooperation & legal framework.
- Hosting providers turning a 'blind-eye' or unable to detect nefarious practices.

Weaknesses

- Bulk traffic monitoring is carried out by governmental agencies.
- Offline means of obtaining information (HUMINT).
- Extortion attempts may be subverted by third parties (easy impersonation).
- Re-use of known malware & attack tool code, which can be detected via stylometry.

Opportunities

- It is relatively easy to frame somebody else (same concept as false flag but different language).
- Easy access to CC tools (e.g., github) makes analysis of code reuse difficult.
- Digital anonymity tools and services.
- Open access to current cyber-vulnerabilities, blacklists, what methods & tools are currently detected (in order to develop counter measures).

Threats

- Monitoring & threat analytics from law enforcement organizations and private companies ("threat intelligence").
- Progress in work on attribution of authors of code and other forensic techniques.
- Possible analysis of relationships between tools.

5.2 Current Threats

Current CC attribution-related issues are of a fundamental nature – it is practically impossible to determine the perpetrators of a professional CC coming from the analysis of attack artifacts. It is common to associate different attacks using forensic analysis [10], but for most of the high-profile cases it is impossible to determine the acting perpetrators, getting them apprehended and tried for the crime. It is usually only for the crimes perpetrated by the least technically sophisticated attackers that the perpetrators can be identified and apprehended. For the technically competent ones, it is only their mistakes in keeping operational security or their personal weaknesses as the need to brag, can lead to their identification, as the cases of Silk Road [11] and Blackhole Exploit Kit [12] authors demonstrate. Professional cybercriminals can even designate unwilling co-conspirators that can easily take the fall – for example, those who act as money mules. These are traceable and reachable by the police and take the blame for the crimes, while those who orchestrate the scheme and reap most of the gains roam free. Even researching the criminal infrastructure gives only minimal hints to the whereabouts and identities of CC operators, as the global services IT market makes it very easy to deploy the infra-structure in almost any place in the world. Unsurprisingly, this is similar to fighting against organized crime, where only the low-level "foot soldiers" are typically apprehended and tried. This is not in a small part because of an overlap in organized crime and CC, which in turns makes attribution of CC the same as identifying members and actions of organized crime rings. Thus, for attribution of organized CC, one must proceed with attribution of a whole criminal operation in general, which is equivalent with investigating the whole organization.

Another problem in the sphere of CC attribution is that apprehending the identified cybercriminals usually requires global, coordinated cooperative law enforcement results requiring lots of legal and operational resources. In a similar manner to offline crime, this requires risky investigative techniques – controlled, warranted, undercover infiltration of the organization and active exploitation of CC infrastructures to identify its owners and operators [13].

Last, but not least, attribution is crucial to establishing whether the attack is an act of CC or rather one of CT [14]. Depending on whether the attacker is a profit-driven criminal, a nation-state or a political organization, the attack can be recognized as CC, act of cyberespionage (also a crime) or war, or as an act of CT [15]. This is especially difficult, taking into account that we lack most of the background information on actors orchestrating the observed attacks, and most of the assumptions about their organizational structures and grouping them into campaigns and organizations are presumptions based on analysis of technical artifacts [16]. The artifacts need not only be obtained by analysing the victims' environments, but also by extensively analysing the cyberattack command and control infrastructure and back office.

5.3 Current Defense

Many of the current defences that are utilized to approach attribution & digital forensics, are also controversial in many EU countries. Traffic analysis and DPI (Deep Packet Inspection) cover some aspects of the CC committed and thus at the level of attribution and analysis of tools, tactics and procedures of the attacker. Databases of security events and archives of previously published internet contents and proper-ties of Internet hosts, passive DNS monitoring databases, Internet Archive and Google Cache allow looking at past clues about connected devices, published web pages and IP addresses. This is an immense help for an analyst untangling the clues of cyberattack infrastructure – the possibility to reference the site's past, even if the infrastructure was taken down by the perpetrators.

The second defence is police work, offline investigative techniques, same as applied to solving offline crimes, such as financial analyses, HUMINT methods (detainment and interrogation, undercover infiltration of criminal groups, or establishing a protected witness program deal with one of the perpetrators), and correlating physical and Internet surveillance.

Another defense is "active measures" – which includes, for example, direct action against attacker's exposed infrastructure ("hacking back"). While being the gray area of the law or outright considered illegal, this controversial method has been used by some to gain information on perpetrators and additional information on other victims [17].

5.4 Future Threats

Possibly the biggest threat against attribution of future CC is using the Internet of Things as an intermediary in perpetrating crimes. Currently the criminals need human beings in the roles of mules and witting, or unwitting, co-conspirators and accessories to a crime. Connecting everything to the Internet will reduce the need of human involvement in criminal operations exactly as it is reducing it in business and military operations. Furthermore, reducing the human involvement will also reduce the crime's forensic footprint and thus, the possibility of attribution. For example, if the potential victim of a kidnapping is traveling in a manned car, the perpetrators need a group of well-trained and armed physical operatives, a getaway car and some bait (like another car to crash into the victim's car); on the other hand, if the victim is traveling in an autonomous car, the kidnappers only need to take over the car systems and victim's mobile phone (to make him or her unable to call the police). This already can be done over the Internet and utilizing online techniques of ensuring one's anonymity, thus avoiding the possibility of easy attribution.

The second threat against attribution of CC attacks is the proliferation of anonymity protocols and tools such as Tor and I2P, progress in malware obfuscation techniques and side-channel attacks [18].

It is debatable whether encryption tools are helping cybercriminals. Law enforcement representatives repeatedly insist that mass-market cryptography tools are a hindrance in investigating and prosecuting CC and CT [7]. However, the actual level usage of encryption in actual CC is a subject for discussion and analysis. While cryptography is used as a countermeasure to forensic analysis of CC malware, it is often Tor and VPN solutions that are used to protect data in transit and hide the identity of the perpetrator. This stays in stark contrast with law enforcement position, stating that endpoint device storage encryption is the most harmful to investigations. This discrepancy is constant in the discourse concerning fighting both CC and CT [19].

Advancements in turnkey crimeware as a service infrastructure (CaaS) are a major potential threat. Current commercial malware requires a significant amount of skilled work to set-up for CC operations. As the developments of commercial malware mimic those of legitimate business IT services, we can expect some sort of turnkey cloud-based malware to appear on the criminal market, making distinguishing operators of a given campaign much harder, if not entirely impossible.

Finally, there is a danger that the chase for CC attribution will lead to mass surveillance of online users, which could be abused for other purposes [20].

5.5 Future Defense

Defenses against the attribution problem will be sought in the progress of forensic analysis and reverse engineering. New areas in those fields of knowledge will be forensic analysis of autonomous devices, anonymity protocols and tools to automatically perform the analyses in mass numbers and to automatically correlate the resulting data, as it is shown that human analysts are not up to the task, and the amount of data to be correlated and processed will only increase.

The solution to those shortcomings is using artificial intelligence and big data analytics methods to sift through the vast amount of data, to correlate them and to find CC-related items among the collected data and traffic in real time.

Progress should be made in the area of actionable intelligence sharing and international and inter-organizational information sharing. A legal framework for enabling this is required as many institutions that possess forensic data on threats are unwilling, or legally unable, to share their findings with other branches of law enforcement. This lack of coordination and information sharing should be re-mediated at both national and international level to spur information sharing among the respective stakeholders.

Another possible defense is retaliatory attack – "hacking back". There are already some proposals to make that explicitly legal, at least for selected actors [21]. As we discussed in 2.4, such active measures are controversial.

Last, but not least, the new possibilities of autonomic Internet-connected objects (Things) make it obvious that for autonomous Things that can operate in the physical space (like drones) a registration and tracking system is necessary to make sure that restrictions on their activities are enforced and attributable to a given device's operator.

6 Attribution of CT Acts

In this section we look into more detail into attribution of CT specifically.

6.1 SWOT Analysis

Strengths

- No toolset or framework to perform systematic & meaningful attribution.
- Despite data collection, agencies keep information to themselves or do not recognize their importance to act in a preventive manner (for political or formal reasons).
- Hosting providers turning a 'blind eye' to suspicious practices.

Weaknesses

- Difficulties in establishing and maintaining long-term operational security (OPSEC).
- Potentially difficult to prove authorship of certain acts (which may defeat part of the purpose of a CT act).

Opportunities

- It is easy to plant false flags.
- Easy access to CC tools (github etc.) makes analysis of code reuse difficult.
- Use of anonymity tools & services, and defeating stylometry (i.e. statistical analysis of variations in literary style between one writer or genre and another) or HUMINT (i.e. intelligence gathered by means of interpersonal contact) based investigation (e.g., Anonymouth [22]).

Threats

- Growth in private companies providing SIEM (Security information and event management) services i.e. "threat intelligence".
- Progress in work on attribution of authors of code and other cyber-forensic techniques, e.g. stylometry [23, 24].
- Possible analysis of relationships between tools.

6.2 Current Threats

Unlike CC, which is focused on making money through illegal means, CT is about sending a message, especially through mass- and social-media. The message is to further a political goal and also to incite fear of repeated attacks, to spread the message further and to recruit followers who will identify with the message and with the perpetrators, and who will become a base to expand the perpetrators' operations.

The second crucial difference between CT and CC that affects attribution is that it is socially acceptable to mitigate CC effects after an act was committed (for example, by reimbursing money to a victim of a bank fraud), while it is both politically and socially unacceptable for the government to expect the citizens to be subjected to terroristic acts with mitigation proceeding after the act. Thus attribution of a CT act should also include attributions of possible attacks and attacks as they are proceeding, extending the problem of attribution into areas of threat and anti-terrorism intelligence.

While a substantial focus has revolved around the activities of nation states or groups based in China, Russia, and Iran, recent discoveries have revealed the capabilities of Western nations. Many have argued that clandestine digital operations are a logical, even desirable part of modern statecraft. The step from digital espionage to its use in CC or CT is, however, a small one. Commercially written, offensive software from EU based companies like FinFisher and Hacking Team has been sold to repressive regimes and non-aligned groups under the guise of "govern-mental intrusion" software [25]. There have been several examples of cyberattack and digital espionage tools, that were created for this modern statecraft, being stolen, leaked, or found their way into the hands and use of CC and potentially CT [26,27].

Nation state/group hacking operations are frequently well funded, difficult to attribute, and rarely prosecuted even if substantive evidence can be discovered. While efforts have been made to counter this problem, proof is hard to find and even more difficult to interpret correctly. This creates an ideal basis for propaganda, and incorrect attribution. For some actors it is also common to utilize tools, tactics and procedures of cybercriminals to deliver the strike, as the recent BlackEnergy APT attack on Ukrainian power grid has shown [28].

Due to the nature of Internet attacks, acts of CT may include attack classes rarely or never used by cybercriminals and vice versa, thus making forensic analysis of attack artifacts from CT attacks a slightly different area of knowledge than for CC attacks. Examples are attacks against industrial control systems, not commonly carried out for direct financial gain.

Another threat that is much more likely in CT than in CC attacks is the perpetrators leaving false clues to point at another actor – including false flag operations used as a political provocation [29].

A strong political message coming from the law enforcement organizations links network encryption with terrorism and CT [7]. This connection is debatable, as in no major terrorist incident were their communications observed to be encrypted.

6.3 Current Defense

CT attribution uses methods similar to investigation of CC augmented with the possibilities given by a state apparatus aimed at fighting terrorism in general. Its prerogatives are much wider in both operational and legal aspects, thus giving bigger leverage against the potential attackers. Recent improvements, with the introduction of newly adapted techniques within cyber-forensics investigations have shown promise in improved attribution. For example:

- Within a recent exercise it was possible to identify up to 80 % of of users of one major 'Anonymous' forum through the use of various methods including stylometric analysis [30], Latent Dirichlet allocation [31] (a technique to explain data similarity) and the authorship attribution framework Jstylo [32]. Stylometry uses linguistic information found in a document to perform authorship recognition [33].
- Although the Sony hack made major headlines and even President Obama directly highlighted attribution to North Korea, later a stylometric analyst was able to establish over 20 references that the authors of the hacking tools used Russian language, and that it was nearly identical to the early hack of Aramco in Saudi Arabia [34,35].

Overall current events show that operational and technical means of detecting terrorist and CT threats are insufficient in the areas of recognizing and correlating actionable intelligence. However, within the technological sphere there has been steady progress from the identification and classification of CC and cyberattack tools.

6.4 Future Threats

As in the case of CC, the biggest future threat against attribution of future CT is using the Internet of Things as intermediary in perpetrating the crimes. Currently the terrorists need human beings in the roles of operatives and witting or unwitting co-conspirators and accessories to a terror act. Connecting every device in the world to the Internet will reduce the need of human involvement in terrorist operations exactly as it is reducing it in business and military operations. Furthermore, reducing the human involvement will also reduce the act's forensic footprint and thus, the possibility of attribution.

The second future threat to CT attribution is proliferation of anonymity tools and protocols, and proliferation of poorly secured internet-connected devices that can be used in a terrorist act while not retaining any access information to be used for later forensic purposes. Note here the contrasting viewpoint, that the chase for attribution may lead to mass surveillance, which could be abused for other purposes [20].

As active measures may gain popularity and legal standing, it will be possible to trigger a retaliatory attack against a false attacker by hiding the original attacker's identity and leaving clues leading to an intermediate victim that will be attacked in retaliation. A primitive version of this technique, called "joe jobbing" was used in first spamming attacks against e-mail and Usenet users [36].

Table 1. Attribution research gaps identified by the study

Cyber-terrorism/ cyber-crime?	Threat (future)	Defence (current)	Defence (future)	Research gap
Both	Mass surveillance considered danger to modern society	Privacy-enabled internet services, privacy controls in internet services	Privacy mechanism for using internet and sharing information that do not compromise the possibility of criminal act attribution	Research on privacy and anonymity technologies that do not limit possibilities of attribution
Both	Inadequate information sharing mechanisms between parties. With a lack of clarity, differences, and misunderstandings between EU countries relating to privacy, traffic monitoring, data storage & analysis	Ad hoc, provisional, information sharing platforms, often informal	(International) legal frameworks & formalized sharing platforms	Analysis of international, and inter-EU country data protection - how & what can we legally share this data, for cyber forensics
Both	Poor level or inconsistent attribution with attack tool analysis, e.g. malware reverse engineering & methodologies such as stylometry. With linguistic obscuration	Early stage attack tool analysis, e.g. malware reverse engineering & methodologies such as stylometry, yara etc.	Defeating attacker obscuration and advanced digital forensics tools e.g. stylometry	Cyber intelligence gathering of malware & attack tool behaviour, signatures, and AI based linguistic analysis
Cybercrime	Framing others	Police work following the money or motives	Evolution of methods of fighting organized cybercrime	Research on tools, tactics, procedures of organized crime

continued

Table 1. *continued*

Cyber-terrorism/ cyber-crime?	Threat (future)	Defence (current)	Defence (future)	Research gap
Both	Wide proliferation of easy to use cyber-crime/offensive kits	Active countermeasures disseminating back-doored/subverted cybercrime tools, active infiltration of tool development markets	Refined fingerprinting of tools aided by contextual attack information, infiltration of crimeware development markets	Next generation of analysis, fingerprinting tools with context
Cybercrime	Lack of enforcement of internet-wide policing standards	Voluntary implementa-tion/enforcement of actions against cybercriminals	Robust legal frameworks for ensuring coordinated actions against cybercriminals	Creation of world-wide policing standards
Cyber-terrorism	False flag operations	Intelligence: who benefits politically from such an action, who has the capabilities	Evolution of "threat intelligence" (incl. mapping of potential attackers ahead of attack)	Research on tools, tactics, procedures of cyberterrorism
Cyber-terrorism	Difficulty in identifying meaningful information on upcoming threat (information overflow: needle in haystack problem of finding what is important)	Ad-hoc analysis tools for intelligence analysis, tip-offs, HUMINT	Artificial intelligence, machine learning, big data applied to threat intelligence	Intelligence has access to great volume of information but lacks of tools to identify the most meaningful
Cyber-terrorism	Lack of knowledge where cyberterrorism comes from (root cause)	Current cyber aspect of war on terrorism - investigation of known terrorism suspects	Understanding motivation of cyberterrorists, enabling profiling for early identification of radicalization	Research on cyberterrorism motiva-tions/root cause

6.5 Future Defense

Similarly to CC, future defence in the attribution problem will focus on progress of cyber-forensic analysis and reverse engineering, linguistic analysis (stylometry), especially in the new areas of autonomous devices, and anonymity protocols. Further research and development should be done to create effective tools to perform the analyses in large numbers and to automatically correlate the resulting data, using artificial intelligence and big data methods, to achieve the real-time analysis and correlation capability.

The same capability should be employed to intelligence and surveillance data gathered while expanding the influence of anti-terrorism legal frameworks into the cyber domain, as it is shown that human analysts are not up to the task, and the amount of data to be correlated and processed will only increase.

Much progress is needed in the area of actionable intelligence sharing and international and inter-organizational information sharing. A legal framework for enabling this is required as many institutions that possess the threats forensic data are unwilling or legally unable to share their findings with other branches of law enforcement. This lack of coordination and information-sharing should be re-mediated at both national and international level to spur information sharing among the respective stakeholders.

Last but not least, the new possibilities of autonomic Internet of Things (IoT) make it obvious that for autonomous devices that can operate in the physical space (for example drones) a registration and tracking system is necessary to make sure that restrictions on their activities are enforced and attributable to a given device's operator.

7 Identified Research Gaps

Based on the previous discussion and analysis, we can now move on to identifying research gaps for future work in the area of cyberattack attribution. This is carried out by comparing the potential future threats to attribution with current and future defences. Overall, we found perhaps not surprisingly, significant overlap between attribution challenges in CC and CT. Of the 10 different research gaps we identified, 5 were characteristic of both CC and CT. Our results are summarized in (Table 1).

8 Conclusions

The game of cyberattack attribution is complex with a multitude of constantly evolving, often-conflicting elements and interests at play. As of now, there is no single method or sure methodology for its establishment. It is quite probable that this is an intrinsic property of the problem and there always will be some degree of uncertainty involved.

In the study, we have identified the following key points concerning the cyberattack attribution problem and research gaps in the field:

- The problem of attribution is crucial in combatting cyberattacks, yet the available research on the problem mostly focuses on narrow technical aspects of forensic analysis.
- In some aspects, the problem is equivalent to investigating entire CC and CT networks.
- Unlike offline crime or terrorist activity, Internet perpetrators have much more opportunities and means for misguiding investigations. Such actions are much more common than in offline crime, as they are a viable deterrent of investigation and prosecution.
- The attribution of CT acts are expected to be performed before the act rather than after, to mitigate it. This is a significant difference from CC, where after-the-fact mitigation is more socially acceptable, probably due to CC being a crime against property.
- We found the subject area to have multiple research gaps that should be pursued to improve the certainty of attribution. This includes a mix of political, social, legal, organizational and technical issues that need to be explored in depth.
- Some aspects of the attribution problem are heavily politically loaded as they may be interpreted as a call for mass surveillance of Internet users.
- More research is needed on ensuring users privacy online without compromising possibility of attribution of criminal acts.

References

1. Schneier B.: Attack attribution and cyber conflict. https://www.schneier.com/blog/archives/2015/03/attack_attribut_1.html. Accessed 11 Jan 2016
2. Tsagourias, N.: Cyber-attacks, self-defence and the problem of attribution. J. Confl. Secur. Law **17**(2), 229–244 (2012). http://papers.ssrn.com/sol3/papers.cfm?abstract_id=2538271. Accessed 03 Feb 2016
3. Healey, J.: Beyond Attribution: Seeking National Responsibility inthe Cyber Attacks. Atlantic Council Issue Brief. http://www.atlanticcouncil.org/images/files/publication_pdfs/403/022212_ACUS_NatlResponsibilityCyber.PDF. Accessed 19 Jan 2016
4. CyberROAD Consortium: https://www.cyberroad-project.eu. Accessed 11 April 2016
5. Buchanan, B., Rid, T.: Attributing cyber attacks. J. Strategic Stud. **38**(1-2), 4–37, doi:10.1080/01402390.2014.977382. http://www.tandfonline.com/doi/10.1080/01402390.2014.977382. Accessed 03 Feb 2016
6. Moscaritolo, A.: RSA confirms Lockheed hack linked to SecurIDbreach. http://www.scmagazine.com/rsa-confirms-lockheed-hack-linked-to-securid-breach/article/204744/. Accessed 11 April 2016
7. From Encryption to Failure of Traditional InvestigationInstruments, Freedom From Fear Magazine, UNICRI.it. http://f3magazine.unicri.it/?p=343. Accessed 25 Jan 2016
8. Constantin, L.: Juniper's VPN backdoor: buggy code with a dose ofshady NSA crypto. PC World. http://www.pcworld.com/article/3017803/security/the-juniper-vpn-backdoor-buggy-code-with-a-dose-of-shady-nsa-crypto.html. Accessed 12 April 2016

9. Pouget, F., Dacier, M., Debar, H.: Honeypot, Honeynet, Honeytoken: Terminological issues. Research Report RR-03-081, InstitutEurecom. http://www.eurecom.fr/en/publication/1275/download/ce-pougfa-030914b.pdf. Accessed 12 April 2016
10. Linfeng, Z.: Effective techniques for detecting and attributingcyber criminals, Iowa State University. http://lib.dr.iastate.edu/cgi/viewcontent.cgi?article=2935&context=etd. Accessed 18 Feb 2016
11. Mullin, J.: Sunk: How Ross Ulbricht ended up in prison for life. Ars Technica. http://arstechnica.com/tech-policy/2015/05/sunk-how-ross-ulbricht-ended-up-in-prison-for-life/. Accessed 12 April 2016
12. Krebs, B.: Who is Paunch. Krebs on Security. http://krebsonsecurity.com/2013/12/who-is-paunch/. Accessed 12 April 2016
13. Brown, C.S.D.: Investigating and prosecuting cyber crime: forensic dependencies and barriers to justice. Int. J. Cyber Criminol. **9**(1) (2015). http://www.cybercrimejournal.com/Brown2015vol9issue1.pdf. Accessed 18 Feb 2016
14. Carr, J.: A critical review of tom rid and ben buchanan's attributing cyber attacks. Digital Dao. http://jeffreycarr.blogspot.com/2015/01/a-critical-review-of-tom-rid-and-ben.html. Accessed 18 Feb 2016
15. Mejia, E.F.: Act and Actor Attribution in Cyberspace. http://www.au.af.mil/au/ssq/digital/pdf/spring_2014/Mejia.pdf. Accessed 18 Feb 2016
16. Carr, J.: Responsible attribution: a prerequisite for accountability. The Tallinn Papers, CCDCOE. https://ccdcoe.org/multimedia/responsible-attribution-prerequisite-accountability.html. Accessed 18 Feb 2016
17. Kovacs, E.: Researchers Hack Infrastructure of Iran-Linked CyberSpies http://www.securityweek.com/researchers-hack-iran-linked-spy-groups-infrastructure. Accessed 16 April 2016
18. Armstrong, H.L., Forde, P.D.: Internet anonymity practices incomputer crime. Inf. Manage. Comput. Secur. **11**(5), 209–215 (2003)
19. Schneier, B.: FBI and Apple's encryption. Schneier on Security. https://www.schneier.com/blog/archives/2015/09/fbi_and_apples_.html. Accessed 18 Feb 2016
20. EFF: Mass Surveillance Technologies. https://www.eff.org/issues/mass-surveillance-technologies. Accessed 18 April 2016
21. Anthony, S.: UK government quietly rewrites hacking laws to give GCHQ immunity. Ars Technica. http://arstechnica.com/tech-policy/2015/05/uk-government-quietly-rewrites-hacking-laws-to-grant-gchq-immunity/. Accessed 12 April 2016
22. Bennett, L.: This Computer Program Turns Famous Writers Into Anonymous Hacks. https://newrepublic.com/article/114112/anonymouth-linguistic-tool-might-have-helped-jk-rowling. Accessed 08 April 2016
23. Brocardo, M.L., Traore, I. et al.: Authorship verification forshort messages using stylometry. Dept. of Electr. & Comput. Eng., Univ. of Victoria - UVIC, Victoria. http://ieeexplore.ieee.org/xpl/articleDetails.jsp?reload=true&arnumber=6705711. Accessed 15 April 2016
24. Caliskan-Islam, A., Yamaguchi, F., Dauber, E. et al.: When Coding StyleSurvives Compilation: De-anonymizing Programmers from Executable Binaries. http://www.princeton.edu/~aylinc/papers/caliskan-islam_when.pdf. Accessed 12 April 2016
25. Marquis-Boire, M., Marschalek, M., Guarnieri, C.: Big Game Hunting: The Peculiarities in Nation State Malware Research. https://www.blackhat.com/docs/us-15/materials/us-15-MarquisBoire-Big-Game-Hunting-The-Peculiarities-Of-Nation-State-Malware-Research.pdf. Accessed 08 April 2016

26. Pi, P.: Unpatched Flash Player Flaw, More POCs Found in Hacking Team Leak. http://blog.trendmicro.com/trendlabs-security-intelligence/unpatched-flash-player-flaws-more-pocs-found-in-hacking-team-leak/. Accessed 08 April 2016
27. Kafeine: CVE-2015-5119 (HackingTeam 0d - Flash up to 18.0.0.194) and Exploit Kits. http://malware.dontneedcoffee.com/2015/07/hackingteam-flash-0d-cve-2015-xxxx-and.html. Accessed 08 April 2016
28. Lipovsky, R., Cherepanov, A.: BlackEnergy trojan strikes again: Attacks Ukrainian electric power industry. We Live Security. http://www.welivesecurity.com/2016/01/04/blackenergy-trojan-strikes-again-attacks-ukrainian-electric-power-industry/. Accessed 12 April 2016
29. Schneier, B.: FBI and Apple's encryption, Schneier on Security. https://www.schneier.com/blog/archives/2015/09/fbi_and_apples_.html. Accessed 25 Jan 2016
30. Goswami, S., Sudeshna, S., Mayur, R.: Stylometricanalysis of bloggers' age and gender. In: Third International AAAI Conference on Weblogs and Social Media (2009)
31. Blei, D.M., Ng, A.Y., Jordan, M.I.: Latent dirichlet allocation. J. Mach. Learn. Res. **3**, 993–1022 (2003)
32. JStylo-Anonymouth software. https://psal.cs.drexel.edu/index.php/JStylo-Anonymouth. Accessed 08 April 2016
33. Thegift83: Up to 80 % of Anonymous Users Can Be Identified By Using Linguistic Software. http://www.techfleece.com/2013/01/09/up-to-80-of-anonymous-users-can-be-identified-by-using-linguistic-software/. Accessed 12 April 2016
34. Perlroth, N.: New Study Adds to Scepticism Among Security Experts That North Korea Was Behind Sony Hack. http://www.nytimes.com/2014/12/20/world/fbi-accuses-north-korean-government-in-cyberattack-on-sony-pictures.html. Accessed 12 April 2016
35. Novetta Threat Research Group: Operation Blockbuster, Unraveling the Long Thread of the Sony Attack. https://www.operationblockbuster.com/wp-content/uploads/2016/02/Operation-Blockbuster-Report.pdf. Accessed 12 April 2016
36. Joe job: Wikipedia, The Free Encyclopedia. https://en.wikipedia.org/w/index.php?title=Joe_job&oldid=686605265. Accessed 12 April 2016

Emerging Cyber Security: Bio-inspired Techniques and MITM Detection in IoT

Michał Choraś[1,2](✉), Rafał Kozik[1,2], and Iwona Maciejewska[3]

[1] ITTI Sp. z o.o., Poznań, Poland
{michal.choras,rafal.kozik}@itti.com.pl
[2] University of Science and Technology, Bydgoszcz, Poland
{chorasm,rkozik}@utp.edu.pl
[3] DFRC AG, Bern, Switzerland
iwona@dfrc.ch

Abstract. The major goal of this chapter is to overview and present selected emerging technologies for cybersecurity. In the first part we show the practical realisations of the bio-inspired concepts for cybersecurity. We do not focus on discussing the bio-inspired techniques on a high and abstract level, but we focus on our own practical developments. We want to present concrete solutions with the magazine-like language understandable to all readers. Our goal is to prove that the bio-inspired techniques can be really implemented to protect networks and that the readiness level of such technology is constantly increasing. In this chapter, we present and focus on our own results and give references to our past and on-going cyber security projects where we successfully implemented different nature-inspired solutions.

Keywords: Cybersecurity · Anomaly detection · MITM detection · Bio-inspired techniques · Genetic algorithm · Internet of Things (IoT)

1 Introduction

The motivation of our work and results, presented in this chapter, come from the current needs to protect computer networks from cyber attacks, cybercrime (CC) and cyberterrorism (CT). As always in the history of the world, when a technology is created and evolves, it can be used for good and criminal purposes. The same happens with the quick evolution of the communication networks and software applications which are now often the target of so called CC and cyberattacks.

What is more, currently it is very difficult, even for large and wealthy organisations (such as big industrial companies, banks, public administration), to counter and eliminate cyber attacks. Of course, the same challenge (in even greater extent) applies to smaller organisations (e.g. SMEs) or citizens.

The important question in our civilisation has always been: what can be done to ensure effective security? The general (independent on application or domain)

B. Akhgar and B. Brewster (eds.), *Combatting Cybercrime and Cyberterrorism*,
Advanced Sciences and Technologies for Security Applications,
DOI 10.1007/978-3-319-38930-1_11

view on the security chain is as follows: analyse and understand the context and situation; look for vulnerabilities; analyse threats and manage risks; observe the situation using sensors and monitoring capabilities (also humans); collect data; analyse data and, on the basis of the data processing and analysis; either detect danger and attacks, or decide that there are no attacks happening at a given moment. The last stage is the reaction and remediation, and of course, there are different reactions available depending on situation, capabilities and legal aspects. This chain of action is illustrated in Fig. 1.

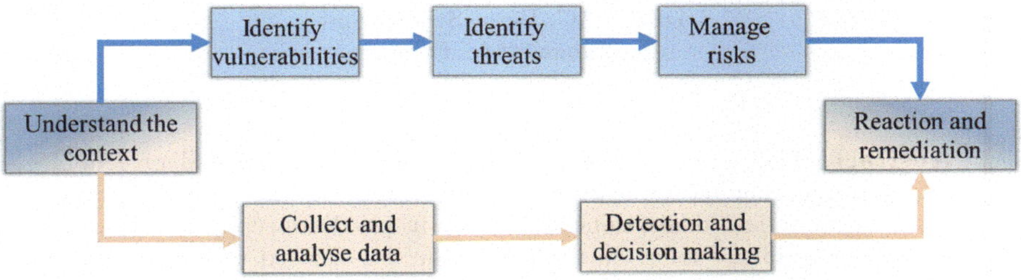

Fig. 1. Cyber security chain of actions

2 Bio-inspired Techniques for Cybersecurity

There are plenty of methods to analyse the collected data. Hereby, we will focus on the bio-inspired techniques for network protection. As for the decision making, there are two possible approaches based on detecting certain patterns in the observed data. Let us assume that all the needed and desired sensors, probes, monitoring devices are installed and operational – then what can be done with the collected data?

The first approach is to learn patterns of 'evil' (e.g. cyberattacks or terrorist attacks) and detect those patterns. The second approach is to learn the pattern of normal and safe state, and detect abnormalities (also called anomalies or outliers) that do not fit to the normal and typical patterns. Those two approaches apply not only to computer networks but also to security in general, e.g. in domains such as counter-terrorism, analysis of bank transactions, urban safety, etc.

If we knew the modus operandi and patterns of the terrorists and criminals, then they would be easier stopped once effective monitoring is applied. However, the terrorists never promised to use the same ways and modus operandi, and they always want to 'outsmart' those responsible for security and safety. Law enforcement agencies train officers not to look for the biased patterns (e.g. white vans) but to analyse context, look for anomalies and think out-of-the-box.

As for the networked systems, current cybersecurity solutions can be classified as signature-based and anomaly-based. Typically, signature-based solutions are installed and widely used on personal computers, intrusion detection systems, etc. For deterministic attacks it is fairly easy to develop patterns that will clearly identify particular attack. The drawback of signature-based solutions is that, since there are no signatures (patterns) of the future attacks, new cyberattacks and so called zero-day exploits cannot be detected and mitigated.

The reality is that hackers and cyber terrorists/criminals never promised to use the same attacks, tools, means and worms. For instance, top-ranked application layer attacks such as SQL Injection attacks or XSS (Cross Site Scripting) are often used due to their diversity, complexity, and availability of obfuscation techniques. Therefore, signature-based approaches are not efficient and anomaly-based solutions are needed. Of course, the typical drawback of anomaly-based approach is that such solutions produce significant number of false alarms. In other words, not all of the detected anomalies are signs of terrorist or cyber-attacks, and the context needs to be understood while making decisions (e.g. rapid growth on network traffic to certain service might not be the sign of Distributed Denial of Service (DDoS) attack, it can be the start of selling tickets to important sports events or concerts).

The evolution of species is based on the ongoing battle between the predator and the victim – in such a battle the victim learns to avoid and protect themselves from predator (either biologically or in behaviour), while the predator has to improve (skills, behaviour or biological characteristics) to catch the victim.

In the rest of the chapter, we will present how the bio-inspired optimisation as well as mimicking the behaviour of the living organisms can be practically realised to enhance cybersecurity of computer networks and systems.

It is worth to notice, that other researchers work is only shortly mentioned, while we focus on our own solutions that have been implemented in practice in the prototype systems and projects for computer network protection.

2.1 Bio-inspired Methods for Cybersecurity – Practical Examples and Implementations

We live in a world of information that is ruled by information theory; on the other hand we also live in a natural world bounded by laws of physics. These two worlds present common analogies and similarities visible at macro and micro levels. For instance particles and data have similar statistical properties (uncertainty, entropy, etc.) that can be measured using common tools. This fact is heavily exploited by variety of the optimisation techniques like simulated annealing, stochastic climbing, particle filtering, etc. Also the macro scale of our physical world (interaction between organisms, complex mammals' brains capable of multimodal perception or evolution of species) inspires variety of large scale genetic and evolutionary-based optimisation or swarm/ant colony optimisation techniques.

We can also observe many similarities between computer networks and biological organisms, especially when it comes to communication and the security of telecommunication systems. Even the term "viruses" has been 'borrowed' from life sciences to highlight the behaviour resemblance [1]. As for the cyber defence and protection, there are also examples of solutions that are inspired by biology. Some of the methods include artificial neural networks, swarm optimisation methods, ant colonies, collective intelligence, artificial immune systems, and genetic algorithms.

In this chapter we analyse and discuss different bio-inspired techniques applied for cybersecurity domain. We focus on our own practical implementation and during the analysis we make references to results of our past and running projects related to cyber security.

The discussed cybersecurity implementations (in the following subsections) use the bio-inspired algorithms for different purposes such as:

- to optimise some cost functions (e.g. to find IDS rules),
- to leverage collective intelligence and distributed properties (cooperative behaviour of social insects),
- to mimic the behaviour of living organisms (e.g. defence mechanisms).

2.2 Practical Implementations of the Bio-inspired Optimisation Techniques Applied to Cyber Security

Is this subsection we will present two practical implementations of the bio-inspired techniques:

- genetic algorithm for SQL injection attacks detection,
- genetic algorithm for detection of anomalies in HTTP requests.

A variety of the Evolutionary Algorithms (EAs) that mimic the biological evolution or social behaviour of the living organisms have been successfully used over the last decades to find near-optimum solutions to large-scale optimisation problems. Most commonly used ones are: genetic algorithms, particle swarm, ant-colony, firefly algorithm, or shuffled frog leaping.

A practical implementation is proposed in [2], where authors proposed a Genetic Algorithm (GA) based technique to learn IF-THEN rules of the firewall from the historical data. The authors first extracted the relevant features describing TCP/IP connections using the principal component analysis, and then they encoded the rules as chromosomes within the typical GA framework. In [3] authors used the genetic algorithm to enhance the effectiveness of the fuzzy-classifier for detecting the insider threats.

To give the example from our own research and implementations: in one of our recently finalised research project (called SECOR [4]) devoted to investigating innovative anomaly detection methods, we proposed a novel method for SQL Injection Attack detection based on the genetic algorithm (GA) for determining anomalous queries.

SQL Injection Attacks are relatively easy to perform and hard to detect or prevent. In order to perform injection attack, an attacker sends text, which exploits the syntax of the targeted interpreter, therefore almost any source of data can be an injection attack vector. In result, injection can cause serious consequences including data loss, corruption, and lack of accountability or denial of access. These factors cause the growing popularity of such form of cyber attacks.

Our proposed solution exploited genetic algorithm implementing a variant of social behaviour of species, where the individuals in the population explored the lines in the log-files that were generated by the SQL database. In our model, each individual delivers a generic rule (which was a regular expression) that describes the visited log line.

The proposed algorithm is divided into the following steps:

- *Initialisation:* The line from the log file is assigned to each individual. Each newly selected individual is compared to the previously selected in order to avoid duplicates.
- *Adaptation phase:* Each individual explores the fixed number of lines in the log file (the number is predefined and adjusted to obtain reasonable processing time of this phase).
- *Fitness evaluation:* The fitness of each individual is evaluated. The global population fitness as well as rule level of specificity are taken into consideration, because we want to obtain the set of rules that describe the lines in the log file.
- *Cross over:* Randomly selected two individuals are crossed over using algorithm for string alignment. If the newly created rule is too specific or too general, it is dropped in order to keep low false positives and false negatives.

In our work, we used the modified version of the Neddleman-Wunsch algorithm [5], originally invented to find the best match between DNA sequences. In order to find correspondence between those two sequences, but also for any text strings such as the logs analysed here, it is allowed to modify the sequences by inserting the gaps. For each gap (and for mismatch) there is a penalty while the award is given for genuine matches. The fitness function, that is used to evaluate each individual, takes into account the effectiveness of the particular regular expression (number of times it fires), the level of specificity of such a rule and the overall effectiveness of the whole population. The level of specificity indicates the balance between number of matches and number of gaps. This parameter enables the algorithm to penalise these individuals that try to find general rule for significantly different queries like SELECT and INSERT.

The SQL detection results for our method were better when compared to those obtained with standard signature-based solutions like SNORT or Apache SCALP. SNORT is a widely deployed IDS system that uses set of rules that are used for detecting web application attacks (signature based approach). Apache SCALP is an analyser of Apache server access log file. It is able to detect several types of attacks targeted at web application. The signatures have a form of regular expressions that are borrowed from the PHP-IDS project.

Our results are significantly better especially in terms of the detection rate and still comparable when it comes to false positives.

The second practical implementation from our different project is the method to use genetic algorithm to find an alignment of common segments in the consecutive request/packets, in order to develop evolutionary-based anomaly detection method for web layer attacks detection [8]. In this work, we focused on detecting cyber attacks and anomalies in HTTP protocol. Our method works as an additional cyber security measure protecting the WWW server against cyber attacks. The current implementation works as a passive analyser that analyses the HTTP streams. Therefore, the proposed algorithm operates on a server side where the web application is deployed. It intercepts the HTTP(S) traffic generated by client web browser. Through the proxy server it is possible to split the HTTP streams (in order to process them simultaneously) without affecting the quality of the web service.

In our work we analyse and classify the content of HTTP requests. We represent the structure of the payload by means of tokens. The token of HTTP request is defined as the sequence of bytes that are common for all the requests sent to the same resource. There could be several tokens identified for one request. Tokens are used to identify delimiters of those regions of the requests sequences that are likely to be related to the data provided by the client sending that request. Hence, this allows us to identify possible points where malicious code can be injected. It is out of scope of this chapter to show how tokens are generated. However, once we have tokens, we practically apply the genetic algorithm to align them for further processing and decision. In order to build HTTP request model, we need to identify the right subset of tokens and their order.

In order to address tokens alignment problem, we may formulate it as discrete knapsack problem: "given a set of items, each with a mass and a value, determine which item to include in a collection so that the total weight is less than or equal to a given limit and the total value is as large as possible". In our case single token represents item. We assign the value to each token (in current implementation we favour longer tokens over shorter) and mass which represents the position of token in a sequence. The limit, in our case, is determined by analysed sequences.

To solve this optimisation problem, we proposed to use genetic algorithm with classical binary chromosome encoding schema and one point crossover. The chromosome in our algorithm represents candidate solution and it is a string of bits (1 indicates that given token is taken to build the structure of request, while 0 is used to reject given token). The genetic algorithm is used in the following manner:

- The population is initialised randomly. The chromosome length is determined by the number of tokens identified during the extraction procedure.
- The fitness of each chromosome is measured. Individuals are ordered by fitness values.
- Two chromosomes are selected randomly from the population.
- Selected chromosomes are subjected to crossover procedure.
- The procedure is terminated (and individual with the best fitness is selected) if maximal number of iterations is exceeded, otherwise it goes back to step 2.

Further in our system, once the tokens are identified, we describe the sequences between tokens using their statistical properties and apply machine-learning algorithms to decide if the requests represented by tokens are anomalous or not.

The proposed method containing practical realisation of genetic algorithm achieves satisfactory results, better than state of the art methods, on a benchmark CSIC'10 database [6].

2.3 Practical Realisations of Techniques Mimicking the Behaviour of Living Organisms

The second group of the bio-inspired methods include mechanisms that mimic the defence techniques adapted by living organisms. One of these techniques is called the Moving Target (MT) strategy and aims at providing security through the system diversity [7,8]. It is achieved by changing various system properties (system configuration). For instance, in [9] authors used genetic algorithm to address the problem of uniform and deterministic configuration (e.g. of computing clusters, databases farms, etc.). In the proposed approach authors modelled the configuration of single computer as a chromosome and used the evolutionary approach to identify new possible configurations. Other MT strategies may include dynamic IP addresses translation or techniques to fool the network scanners [8,10].

Another recent strategy inspired by the nature, is to use heterogeneous multimodal sources of information and to correlate them for improved decisions. The multimodal perception of physical world that is exhibited by mammals' brains is also used as a guidance when prototyping machine learning algorithms. For instance, living organisms use different heterogeneous sources of information (touch, smell, etc.), in order to reduce the uncertainty of single source and to better identify objects, threats or to estimate more accurately the position with respect to the environment. The same phenomenon also applies when it comes to pattern matching, objects detection or identification, data mining and machine learning. As it is explained in [11], there is no single pattern recognition algorithm that is suitable for all the problems. In fact, each classifier has its own domain of competence. The reason why the researchers are focusing on an ensemble of classifiers is the fact that combined classifiers: (i) can improve the overall effectiveness of recognition, (ii) can be easier deployed in distributed systems, (iii) allow overcoming the initialisation problem of many machine-learning methods (e.g. k-means, tree learner, GMM).

We followed such bio-inspired approach in practice in one of our previous works [6], where we used several techniques adapting the idea of ensemble learning. One of the challenges of producing the ensemble of classifiers is the diversity problem. Although, the formal definition does not exist, it can be intuitively perceived as correlation and similarity of classifiers results. For instance, if the outputs produced by the pool of classifiers are similar, those will have poor diversity, thus we may not expect performance improvement.

According to [12], there are the following methods to improve diversity, namely:

- to use different partition of the data to train the classifiers,
- to exploit local specialisation of given classifiers,
- to use different sub-set of feature.

In order to address the first and the second approach, we applied boosting and bagging techniques.

For the last one, we applied random selection of features subspace. In our approach we selected two types of classifiers that build the ensemble, namely:

- *Decision Stump (DS):* machine learning model that is a decision tree with the single level. For example, if subsequent features are considered in one-class classification problem, this machine learning technique will produce a threshold.
- *Reduces Error Pruning Tree (REPTree):* machine learning technique that uses pruned decision tree. REP Tree algorithm generates multiple regression trees in each iteration. Afterwards, it chooses the best one. It uses regression tree adapting variance and information gain (by measuring the entropy). The algorithm prunes the tree using back fitting method.

Our experiments conducted on publicly available benchmark database show that ensembles of weak classifiers can achieve better results than classical approach using single classifier [6].

Moreover, we have explored the advantages of data heterogeneity and multimodality in order to detect cyber attacks conducted in the application layer. The same way as living organisms use different senses to identify and avoid threats, we may use different sensors to detect wide variety attack targeting web applications. For instance, we may deploy the sensors and firewalls in different layers of TCP/IP protocol stack, but we may also deploy different detection techniques at the same layer (e.g. we can combine anomaly-based attacks detection with signature-based detection). Our experiments showed that this technique (for instance using simple weighting between sensors) can lead to significant improvement of the detection effectiveness.

2.4 Practical Realisation of the Collective Intelligence and Distributed Properties

The third group of the bio-inspired methods include techniques that mimic the collaborative strategies of social insects such as bees, ants, fireflies, etc. For instance, in [13] authors proposed a system adapting ant colony to identify a potential cybersecurity attack against smart meter deployments. In [14] authors combined ant colony optimisation with cybersecurity scanners to identify vulnerabilities in the networks in more effective way.

We have applied such bio-inspired approach to design and develop the Federated Networks Protection System [15].

Our motivation was that the successful cyber attacks are considered as a threat for military networks and public administration computer systems. Therefore, the goal of the Federated Networks Protection System, developed in the SOPAS project, is to protect public administration and military networks which are often connected into Federations of Systems. While adopting the concept of federation of networks and collective intelligence, the synergy effect for security can be achieved.

In our approach, we use the capability of the federated networks and systems to share and exchange information about events in the network, detected attacks and proposed countermeasures. Also in our case, the collective intelligence concept refers to a set of different independent systems, which are not centrally managed, but cooperate in order to share knowledge and increase their security.

Of course, as in nature, the important factor for implementation of such approach is trust. Trust of the networked systems has to be managed by administrators and decision-makers following certain procedures.

The general architecture of the Federated Networks Protection System is presented in Fig. 2. It consists of several interconnected domains, which exchange information in order to increase their security level and the security of the whole federation. Different subnet works are arranged in domains, according to the purpose they serve (e.g. WWW, FTP or SQL servers) or according to their logical proximity (two networks closely cooperating with each other). In each of the domains, a Decision Module (marked as MD in Fig. 2) is deployed. Each MD is responsible for acquiring and processing network events coming from sensors distributed over the domain.

If the attack or its symptoms are detected in one domain, the relevant information is disseminated to other cooperating domains so that the appropriate countermeasures can be applied.

All Decision Modules within the federation can also interact with each other and exchange security information. The information about network incidents, such as attacks in one domain, may be sent to different Decisions Modules in order to block the attacker before the attack takes place in another domain. Communication between domains and Decision Modules is based on P2P (Peer-to-Peer) protocol in order to increase communication resiliency and enable data replication. Moreover, for decision-making, we proposed a semantic approach to network event correlation for large-scale federated intrusion detection system.

In our experiments, we showed that the proposed system can for example correlate various network events from different layers and domains (traffic observations and application logs analysis) in order to detect Injection attacks (e.g. SQL Injection Attacks) on the public administration web services. As a result of attack detection, the Decision Module creates reaction rules and sends it to MD in another domain. Therefore, the same injection attack targeted at the other network can be prevented [15].

It is relevant to note that Decision Modules are central units in their own domains, but are treated as advanced cooperating sensors by other domains. Of

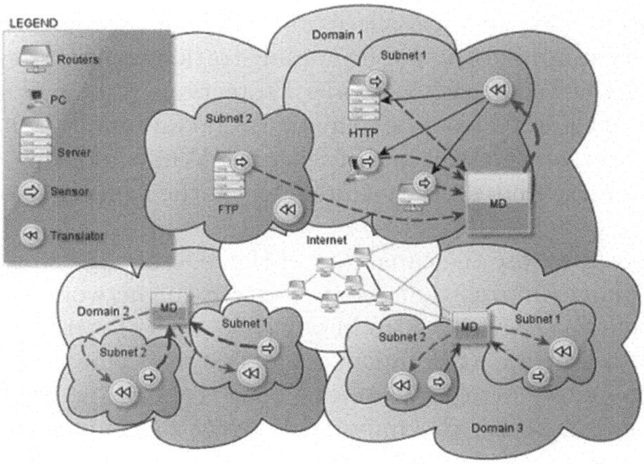

Fig. 2. The concept of multi-domain collective intelligence collaboration for cyber security and networks protection

course, the decision and reaction can be different in each domain (even for the same attack or event) depending on internal policies and legal requirements.

3 Man the Middle Detection in IoT

3.1 IoT Ecosystems

The most recent report of European Union's Agency for Network and Information Security: ENISA Threat Landscape 2015 (ETL) states clearly that threat agents have increased the sophistication of their attacks and their tools. On the other hand, and unsurprisingly, ETL overview of current and future cyber-threats is a mirror image of the evolving digital technology environment.

The most notable fact is cyber-threats: attack patterns and tools developed in the past, which were targeting PCs, have now migrated to mobile ecosystems.

Accordingly to the aforementioned ENISA report the drivers behind this trend are:

- Proliferation of Internet of Things (IoT) devices in home environments, as well as an increased role of wearables in the area of health.
- The increased use of wireless connectivity among devices is of all kind and in various sectors.
- Interaction of all components with mobile and cloud platforms is a key of their architecture design.

IoT can be also seen as a special case of Cyber Physical Systems (CPS), a connecting point between the cyber and physical worlds, where interconnected devices deal with some physical events.

The traffic on mobile data networks had increased spectacularly over the last years and it is foreseen wireless data traffic will continue to grow more than 60 % a year for the next years, meaning by 2017, monthly mobile data traffic will reach 11.2 Exabytes per month! Moreover an enormous growth of Wi-Fi Internet access in both public and private spaces, logically leads to the expectation of ubiquitous connectivity. And WiFi 2.0, the next step, will allow mobile devices to automatically join a Wi-Fi subscriber service whenever the user enters a hotspot area.

5-G network technologies, currently under development, will drive the next network revolution leading to "ambient internet" – Internet access present everywhere and essential for everyone and everything, people and objects.

From the point of view of IoT ecosystems cyber security the weakest part are the communications, making IoT especially vulnerable for Man In The Middle (MITM) attacks.

3.2 IoT and Communications Security

Man in the Middle (MITM) devices were originally developed to steal IMSI (International Mobile Subscriber Identity), therefore sometime they are called IMSI Catchers. The operational mode of MITM is based on the premise the mobile devices do prefer the strongest communication cell signal in vicinity to maximize its own power consumption. Today MITM, besides its classical applications such as mentioned before IMSI stealing, tracking mobile devices, deliver geo-target spam, etc., can be also used to perform more sophisticated cyber attacks, such as interception (getting data), DOS (locking data) or deception (false data).

Taking into account IoT ecosystems deal with certain complexity due the convergence of various and heterogeneous platforms and applications, storage and management systems, resulting system-of-systems, any cyber attack of the weakest component would easily exploit in amplification effect along the entire chain of these interrelated components.

MITM typically introduces irregularities in the network layer that give hints for an educated observer. These irregularities could be any of following [16]:

- usage of off-band frequency,
- cell IDs are very static,
- changes in base station capabilities,
- incomplete network parameters,
- RF jamming,
- sudden absence of encryption,
- cells that suddenly appear (with very good signal quality) for a short period of time and disappear afterwards.

3.3 Methods for MITM Detection

Nowadays there are two groups of methods to detect MITM, called IMSI Catcher (ICC): either needs to run on mobile/IoT device (mICC) or a dedicated stationary device (sICC). It is important to state, in both methods, ICC needs to be able to generate and maintain its own database, but the mobile application cannot assume the online access is guaranteed during the possible attack.

Strong encryption of IoT devices could be also useful against interception and deception but IoT devices do not have the processing and storage capacity that is required for such an encryption. The result is that IoT device is unable to detect a man in the middle attack in general (unless the attack is very simple) and a single device can do nothing against the DoS.

The best method to detect and to overcome this type of attacks is by monitoring the frequency band, locating and identifying the legitimate transmissions and checking any unknown, so suspicious one. Basically detection of abnormalities (as those listed in the previous subsection) in the spectrum can indicate the existence of a nearby MITM.

3.4 MITM Detection– Practical Examples and Implementations

One of the proposed solutions is DFRC[1] MITM Detector by based on the technology of spectrum monitoring.

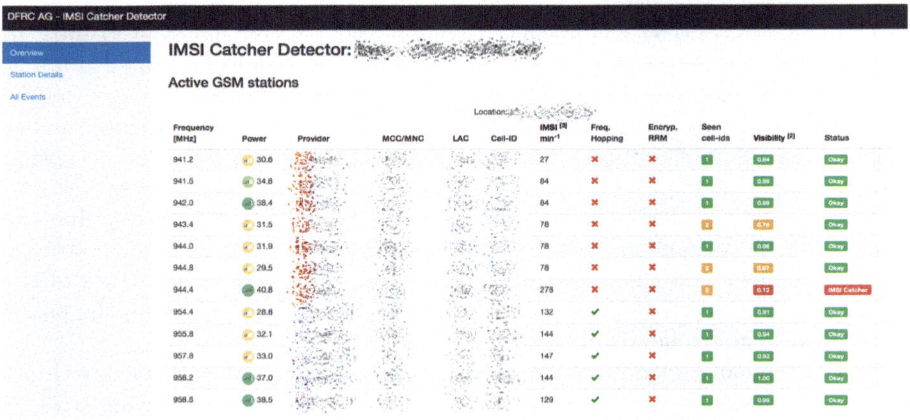

Fig. 3. Overview control panel

MITM Detector is fully autonomous, low cost, simple to install and capable to highlight in near real-time the possible existence of IMSI catchers in the nearby areas: through its online password-protected platform, it delivers to the end user, over a secure connection, the actual situation of the active stations around the target location; it marks the suspicious base stations and reports the ones that prove to be IMSI Catchers.

---

[1] DFRC AG is a high-tech Swiss SME, with its core business focused on analysing, understanding and finding new information in location database.

Main identifiers for each base station (frequency, power, network carrier, cell ID, etc.) are described in an overview control panel, as depicted in Fig. 3, where all details are continuously reported.

Moreover, the user can go to any of the scanned base stations and check in real time all the relevant parameters for this particular base station (Fig. 4)

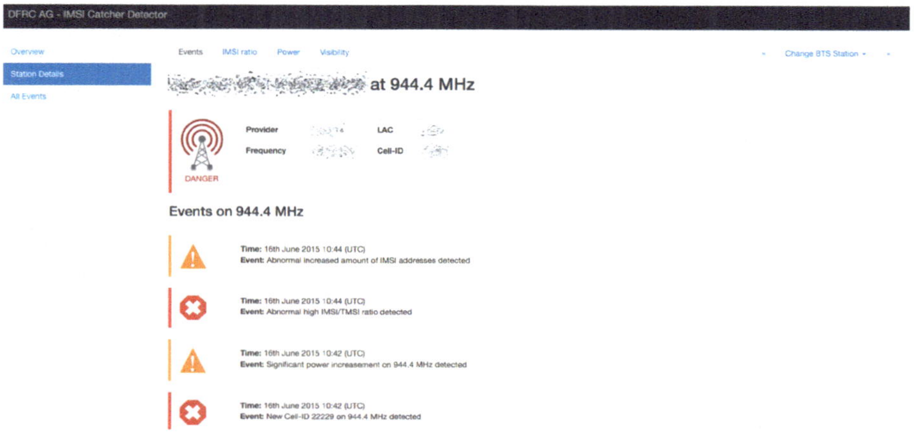

Fig. 4. Detailed BS information

MITM Detector provides so the clear indication of MITM presence, defining also its approximate position. This information is forwarded to the competent authority, which in order to efficiently counter the attack, would need to check physically the suspicious location and find the attacking device.

4 Conclusions

In this chapter, we summarised our own results and implementations related to cybersecurity solutions that exploit techniques inspired by the nature. We showed how those techniques can be practically implemented for cyberattack detection, anomaly detection and protection of computer networks.

We have investigated and presented the practical solutions for the evolutionary-based optimisation techniques and the techniques that mimic social behaviour of species. The proposed genetic algorithms improve detection of SQL injection attacks and anomalies within HTTP requests. Similarly, the proposed ensemble of classifiers and correlation techniques allow for the improved networks protection.

We believe that the bio-inspired techniques will further find many applications in cybersecurity domain since, as proven, the readiness of such technology has increased and practical implementations are possible.

On the other hand, our results and implementations related to cybersecurity solutions for IoT cybersecurity are presented from the perspective of secure communications. We showed how those techniques could be practically implemented for cyber attacks detection, abnormalities and protection of IoT ecosystems.

In particular, we have developed and presented the stationary MITM detector, based on spectrum monitoring techniques, which aim to improve significantly the IoT systems protection.

In spite of the reduced scope and impact of attacks on IoT performed so far, it is just a matter of time to witness attacks affecting more than one component (cascade effect) of IoT systems, by just single points of failure and the weakest link exploitation in IoT ecosystems.

Finally it is important to highlight, that besides our internal research and development, there are many others developments taking care of IoT security, providing insights over IoT vulnerabilities and attack methods.

Creating awareness among users, who in case of IoT are, and will be, mostly ordinary citizens, with a very general cybersecurity culture, remains the first requirement towards achieving IoT effective protection.

Acknowledgement. The research leading to these results has received funding from the European Union Seventh Framework Programme (FP7-SEC-2013) as the CAMINO project under grant agreement no 607406.

References

1. Mazurczyk, W., Rzeszutko, E.: Security - a perpetual war: lessons from nature. IEEE IT Prof. **17**(1), 16–22 (2015)
2. Bankovic, Z., et al.: Improving network security using genetic algorithm approach. Comput. Electr. Eng. **33**(5–6), 438–451 (2007)
3. Bin Ahmad, M., et al.: Using genetic algorithm to minimize false alarms in insider threats detection of information misuse in windows environment. Math. Prob. Eng. **2014**, 12 (2014). Article ID 179109
4. Choraś, M., et al.: Correlation approach for SQL injection attacks detection. In: Herrero, Á., et al. (eds.) International Joint Conference CISIS'12-ICEUTE'12-SOCO'12. AISC, vol. 189, pp. 177–185. Springer, Heidelberg (2013)
5. Needleman, S.B., Wunsch, C.D.: A general method applicable to the search for similarities in the amino acid se1quence of two proteins. J. Mol. Biol. **48**, 443–453 (1970)
6. Kozik, R., Choraś, M.: Adapting an ensemble of one-class classifiers for web-layer anonaly detection systems. In: Proceedings of 3GPCIC, Cracow, IEEE Press, pp. 724–729 (2015)
7. Fink, G.A., Haack, J.N., McKinnon, D., Fulp, E.W.: Defense on the move: ant-based cyber defense. IEEE Secur. Priv. **12**(2), 36–43 (2014)
8. Okhravi, H., Hobson, T., Bigelow, D., Streilein, W.: Finding focus in the blur of moving-target techniques. IEEE Secur. Priv. **12**(2), 16–26 (2014)
9. Lucas, B., et al.: An initial framework for evolving computer configurations as a moving target defense. In: Proceedings of the 9th Annual Cyber and Information Security Research Conference (CISRC) (2014)

10. Kewley, D., et al.: Dynamic approaches to thwart adversary intelligence gathering. In: Proceedings of the DARPA Information Survivability Conference & Exposition II (DISCEX 2001), vol. 1, pp. 176–185 (2001)
11. Nguyen, H.T., Torrano-Gimenez, C., Alvarez, G., Petrović, S., Franke, K.: Application of the generic feature selection measure in detection of web attacks. In: Herrero, Á., Corchado, E. (eds.) CISIS 2011. LNCS, vol. 6694, pp. 25–32. Springer, Heidelberg (2011)
12. Wozniak, M. (ed.): Hybrid Classifier. SCI, vol. 519. Springer, Heidelberg (2014)
13. McKinnon, A.D., et al.: Bio-inspired cyber security for smart grid deployments. In: 2013 IEEE PES Innovative Smart Grid Technologies (ISGT), pp. 1–6, 24–27 February 2013
14. Chhikara, P., Patel, A.K.: Enhancing network security using ant colony optimization. Global J. Comput. Sci. Technol. Netw. Web Secur. **13**(4), 19–22 (2013)
15. Choraś, M., et al.: Information exchange mechanism between federated domains: P2P approach. In: Herrero, Á., et al. (eds.) Int. Joint Conf. CISIS'12-ICEUTE'12-SOCO'12. AISC, vol. 189, pp. 187–196. Springer, Heidelberg (2013)
16. Dabrowski, A., et al.: IMSI-Catch Me If You Can: IMSI-Catcher-Catchers. In: Annual Computer Security Applications Conference (ACSAC) (2014)

Cyber Situational Awareness Testing

Joel Brynielsson[1,2](✉), Ulrik Franke[3], and Stefan Varga[2,4]

[1] FOI Swedish Defence Research Agency, 164 90 Stockholm, Sweden
joel.brynielsson@foi.se
[2] KTH Royal Institute of Technology, 100 44 Stockholm, Sweden
[3] SICS Swedish Institute of Computer Science, Box 1263, 164 29 Kista, Sweden
ulrik.franke@sics.se
[4] Swedish Armed Forces Headquarters, 107 85 Stockholm, Sweden
stefan.varga@mil.se

Abstract. In the cyber security landscape, the human ability to comprehend and adapt to existing and emerging threats is crucial. Not only technical solutions, but also the operator's ability to grasp the complexities of the threats affect the level of success or failure that is achieved in cyber defence. In this paper we discuss the general concept of situation awareness and associated measurement techniques. Further, we describe the cyber domain and how it differs from other domains, and show how predictive knowledge can help improve cyber defence. We discuss how selected existing models and measurement techniques for situation awareness can be adapted and applied in the cyber domain to measure actual levels of cyber situation awareness. We identify generic relevant criteria and other factors to consider, and propose a methodology to set up cyber situation awareness measurement experiments within the context of simulated cyber defence exercises. Such experiments can be used to test the viability of different cyber solutions. A number of concrete possible experiments are also suggested.

Keywords: Situational awareness · Measurement technique · Experimental design · Cyber defence exercise

1 Introduction

In cyber security it is seldom straightforward to get a sense of the threat landscape as a whole in order to really know "what is going on"[1]. Still, to understand an immediate threat or a detected attack not only in itself but also in terms of the surrounding threats and its strategic implications will most likely be the key to effectively be able to deal with more elaborate forms of cyber threats. To understand the roots and causes underlying a threat and to be able to put this

[1] To know "what is going on" is a phrase used by Endsley [12] in order to provide an informal and intuitive definition of the situational awareness concept.

B. Akhgar and B. Brewster (eds.), *Combatting Cybercrime and Cyberterrorism*,
Advanced Sciences and Technologies for Security Applications,
DOI 10.1007/978-3-319-38930-1_12

information in an overall cyber arena context, is what cyber situational awareness[2] (CSA) is about. Such CSA will help the decision-maker/analyst to better understand the organisational implications, and how to assess and act given that a threat or an attack has been detected. As identified in previous work [20], CSA is considered to be the part of situational awareness which concerns the "cyber" environment, whilst at the same time acknowledging that acquiring and upholding CSA requires that external factors concerning, e.g., the physical environment, the political dimension, etc., need to be taken into account.

The cyber threat is omnipresent in today's connected world, and the necessity to uphold a high level of CSA naturally follows in many operational applications. Examples include the importance for IT departments to be able to distinguish between "background noise," e.g., attack attempts with slim chances of success, and more advanced attempts with potentially severe effects, and for intelligence personnel to understanding a cyber attack strategically in terms of its political implications. Related to the sought for operational CSA capacity, it follows that the ability to acquire and maintain a high level of CSA is also something that ought to govern educational endeavours. Moreover, the usefulness of solutions for tackling the cyber threat—be it technology, processes, or policies—is also closely related to CSA since the level of CSA that a solution provides, is a measure of its usefulness. As a consequence, it is important to develop reliable and valid measures of, and ways to measure, CSA so that, e.g., relevant training goals can be stated and cyber solutions can be evaluated.

The present paper presents an overview of existing situation awareness measurement techniques, and exemplifies how these techniques can be used for CSA measurement. The paper is structured as follows. Section 2 introduces the reader to the area of CSA and provides the necessary background regarding situational awareness. Then, Sect. 3 reviews the area of situational awareness measurement, and discusses measurement design from a cyber perspective. Next, Sect. 4 discusses experiment design considerations in general and how to perform measurement through using cyber defence exercises (CDXs) in particular, which is followed by a practical example of how to setting up a CDX for being able to train for a diversion attack. Finally, Sect. 5 concludes the paper.

2 Background

The purpose of this section is to frame the concept of situation awareness and its development. Situation awareness existed before [8] the publication of Mica R. Endsley's seminal article entitled "Toward a Theory of Situation Awareness in Dynamic Systems" [12], but a wider acceptance of the theories undoubtedly seem to have gained traction in the academic community thereafter as manifested by increasing numbers of research papers on the subject [40]. The reason for studying situation awareness, SA, in the first place is the assumption that good SA contributes to better system design, which in turn ultimately leads to

[2] In this paper we use the terms "situation awareness" and "situational awareness" interchangeably.

better decisions, actions and more successful mission outcomes. There are several proposed models for SA, but many of those appear to view the SA construct differently, and most models focus on the process of acquiring SA from the view of an individual operator as opposed to the multiple individual perspective where acquiring of shared or team SA is emphasised [45]. There are, however, theories that specifically aim to describe and measure phenomena such as team awareness, shared situation awareness and distributed shared awareness, DSA, and the like [1,44]. According to Artman [1], team members in a studied military command and control setting created SA at least by their interactions with the environment through active monitoring, negotiation with other team members, and by use of artefacts. Thus, when situation awareness theories involve groups or teams, a social dimension is also added.

According to Stanton et al. [52], three models and their associated theoretical perspectives dominate. Besides Endsley's three-level model, here: Endsley's model, there is the perceptual cycle model [50] and the activity theory model of Bedny and Meister [2]. In short, the perceptual cycle model emphasises that situation awareness is dependent on the task environment and that situation awareness is externally-directed, that goals and criteria for performance must be explicit in the environment and that the cyclic nature, as suggested by the name of the model, is due to the assumption that knowledge influences behaviour, which in turn sometimes affects and modifies the environment [50]. The activity model, which is a significantly larger construct than Endsley's model, gives that situation awareness can not be viewed in isolation, and that other behavioural concepts tied to human activity have to be understood as well [2]. To summarise, all three models of situation awareness build upon the assumption that the operator has to have a cyclic iterative interaction with the environment, but the perceptual cycle model emphasises the need for interaction with regard to perception, and the activity theory model emphasises the interplay via performed actions. We will not elaborate further on the perceptual cycle model or the activity theory model in this paper.

Endsley's model of situation awareness has found its use and gained widespread acceptance during the years as reflected in the contemporary literature, even if the scientific rigour of some of its theoretical underpinnings or different definition issues are questioned by some [4,5,19,48]. The formal definition of SA, due to Endsley [8], is that it denotes a person's "perception of the elements in the environment within a volume of time and space, the comprehension of their meaning and the projection of their status in the near future." In addition, the person, or operator, also has to have an understanding of the relevant parameters of the system itself [11].

Endsley's model emerged from the aviation domain. She submits that the above mentioned definition merely specifies the scope of the situation awareness construct, and that the elements for different aircrafts or, indeed, systems, have to be determined [10] for each domain. She also proposed a methodology, situation awareness requirements analysis, for the task of determining those elements for the air-to-air combat fighters domain [10]. Other areas for which relevant elements have been identified include, for example, en route air traffic control [15] and command of infantry platoons [35]. The proposed methodology includes the consecutive steps of conducting unstructured interviews with subject matter experts, SMEs, followed by a goal-directed task analysis in which goals, sub-goals and SA requirements to meet those goals are determined. In the next phase a structured questionnaire is submitted to another group of SMEs in order to add an objective assessment to the goals identified in previous phases. Each item is then rated depending on its criticality to reach the sub-goals. The resulting battery of questions about the identified parameters, is intended for the measurement of all three levels of situation awareness. To have a set of questions that reflects the relevant aspects of situation awareness is a critical prerequisite needed to perform further measurements of an operator's, or a team of operators', SA.

2.1 Evaluation of Cyber Threat Insight

As indicated above, situational awareness is often defined following Endsley [8] as "the perception of the elements in the environment within a volume of time and space, the comprehension of their meaning and the projection of their status in the near future." As suggested by Endsley in later work [12], this definition can be seen as delineating ascending levels of awareness ranging from (1) mere basic perception of important data, over (2) interpretation and combination of data into knowledge, to (3) the ability to predict future events and their implications.

In this paper we define cyber situational awareness to be the part of situational awareness which concerns the "cyber" environment. In other words, CSA is what enables system administrators and incident managers to swiftly and appropriately respond to cyber attacks and other incidents pertaining to their operations. However, to acquire and uphold appropriate CSA requires a full understanding of the threat in order to be able to plan strategically for appropriate actions concerning, e.g., training undertakings, possible insider threats, etc. Hence, CSA needs to be understood not only in itself but also with respect to external factors concerning, e.g., the physical environment, the political dimension, etc.

It is easy to see that lack of appropriate CSA makes victims more vulnerable to cybercrime (CC). This is all the more true today, when many crimes also have an IT aspect in them. For example, in June 2011, enterprise networks in the port of Antwerp, Belgium, were hacked by drug traffickers, so as to facilitate their smuggling operations alongside legitimate goods delivered in containers. By manipulating the dispatching of containers upon arrival, the smugglers were able to retrieve the containers holding drugs before the legitimate container owners did. The operation was exposed only when port workers started to notice containers disappearing for no apparent reason. Once the criminal operation was exposed, the police seized over two tons of cocaine and heroin, and more than a million euros [17].

Another example which is interesting to reflect upon from the perspective of CSA is the digital bank attack tactics exposed by Symantec in 2012: distributed denial of service (DDoS) attacks are no longer just a blunt tool that causes a lot of annoyance, but less harm. Rather, attackers have started to use DDoS attacks as diversions, in order to draw the attention of system administrators away from a more sophisticated attack[3]. This kind of tactic really emphasises the need not only to perceive lots of data (e.g., by means of intrusion detection systems, etc.) but also to correctly interpret it in order to predict what the adversary will do next. In other words, countering these new and sophisticated attacks hinges on proper CSA.

3 Measurement of Awareness Level

The formal definition of situational awareness according to Sect. 2.1 has gained acceptance during the years and is widely used throughout the contemporary literature. *Testing* of situational awareness, however, has not matured into an equally well-defined tool set. Endsley's definition suggests that situation awareness can be reached in a gradual manner where the understanding on higher levels to some extent depends on the awareness on lower levels, but not in a linear way [14]. To test to what extent there is an understanding of the situation in terms of these levels typically requires that specific measurement solutions are developed in order to account for the specific domain. It follows that the validity of situational awareness measurement, and of CSA measurement as a means to evaluate cyber solutions, is closely related to (1) the measurement design, taken together with (2) the application of interest.

Concerning measurement design, many more or less elaborate and valid methods to measure SA exist. Hence, to determine whether it is possible to evaluate/test a cyber solution in terms of achieved CSA then amounts to identifying whether the cyber solution, in itself or a part of it, lends itself to CSA measurement, and, if so, to identifying a suitable activity where CSA can be measured using existing SA measurement techniques. Depending on the need, this activity can, e.g., be a small-scale exercise or a full-scale CDX using an exercise design where it is possible to perform relevant training whilst at the same time evaluating to what extent the cyber solution has resulted in individual understanding of the overall cyber situation. To measure the obtained CSA the exercise is typically frozen at randomly selected times and subjects are queried as to their perception of the situation at the time (queries on specific data or data criteria). The reasoning behind the randomly selected times of breaks is that it will not be possible for the subject to mentally prepare for the queries. Hence, it needs to be stressed that SA (and thereby CSA) is a distinct and unique phenomenon which applies to individuals' mental models in a universal sense. It refers to the availability of a comprehensive and coherent situation representation of what is currently known, and which is continuously being updated based on the individual's recurring assessment of the situation.

3 http://www.zdnet.com/article/symantec-data-stealing-hackers-use-ddos-to-distract-from-attacks/.

As indicated, the three levels to be measured and distinguished between during CSA measurement consist of perception, comprehension, and projection. From a cyber security perspective, the perception level thus concentrates on the perception of cyber environment changes including, e.g., noticing an intrusion detection system alarm, whilst the comprehension level focuses on the understanding of what this actually means in terms of, e.g., a website defacement attack, a new kind of friendly user behaviour, etc. Finally, the projection level signifies a more in-depth understanding of the situation in that one is also able to make predictions concerning the forthcoming development of the situation to make informed decisions regarding how to act in order to manage the situation. For the purpose of constituting a means for assessment of cyber solutions, it is necessary that the cyber solution—be it a technical tool, a methodology, or something else—lends itself to testing with regard to understanding of some aspect of the cyber environment along the lines of perception, comprehension, and projection.

The objective for all kinds of measurement is to be able to compare an object or event with another. Stanley Smith Stevens, who made contributions to the field of measurement theory, states that it for measurement is essential that "numbers are assigned to aspects of objects or events according to one or another rule or convention" [53]. Accordingly it follows, when we have those numbers, that they have to be compared to something. For SA, the operator's SA has to be compared to, ideally, an objective truth in order to be able to rate the level of SA. Parasuraman et al. [39] claim, without further comment, that there is such a "ground truth" against which the SA can be compared, while Dekker et al. [5] vehemently argue against the feasibility of acquiring such a "ground truth" as unattainable since it requires an aperspectival, e.g., extracorporeal, objectivity. As we have established that the forms of situation awareness are highly context dependent, the question of *what* constitutes the situation, and what the relevant aspects are, therefore arises.

To address that problem, however, there are a number of techniques that are developed with specific SA target domains in mind. The techniques are asserted to inherently provide a sufficiently good "ground truth" and they also to some extent prescribe how and what to measure. Further, Salmon et al. [47] make the point that most measurement techniques are, consequently, developed in line with corresponding specific models.

According to an excellent inventory of situation awareness measurement methodologies for C^4I (command, control, communications, computers and intelligence) environments, made by Salmon et al. [46], such domains include military, aviation, air traffic control, nuclear power plants, and also a few techniques intended for generic use. Their inventory contains an analysis of 17 different measurement techniques suitable for measurement of military C^4I. One of the proposed techniques is the situation awareness requirements analysis [10], an integral part of SAGAT [9] which we will dwell further into below. Following the Salmon et al. categorisation [46], the remaining 16 techniques can be grouped into self-rating techniques, probe techniques, observer rating techniques, performance measures, process indices, and combinations thereof:

Self-rating techniques: CARS [37], MARS [34], SARS [58], SART [54], SA-SWORD [57].

Probe techniques: Sacri (freezing on-line probe) [25], SAGAT (freezing on-line probe) [9,11], SALSA (freezing on-line probe) [23], SPAM (real-time probe) [7].

Observer rating techniques: SABARS [34].

Performance measures: performance measures can be collected both by measuring explicit and implicit performance.

Process indices: eye tracker, verbal protocol analysis.

Combinations: QUASA [36], C-SAS [6], SASHA [29].

In addition, we also have CAST [22], which is designed to measure team SA. CAST can arguably be classified as a combined observer rating and performance measuring technique.

Endsley's definition suggests that ascending levels of perception, comprehension, and projection, also called level 1, 2, and 3 respectively, as derived from her definition, can be reached [14], but, as we have seen, to test to what extent those levels have been achieved often requires that specific measurement solutions are developed [47].

Endsley asserts that (good) SA can be seen as a factor that increases the probability for good performance, but does not guarantee it [11]. By measuring situation awareness, good design choices for systems can be made, which in turn ultimately increases the probability for the operator to make good decisions and avoid bad ones [13]. In order to develop useful measurement techniques she sought to ensure the validity and reliability of a technique by (1) establishing metrics that solely measure the construct that the technique claims to measure, (2) providing the required insight using sensitivity and diagnosticity measures, (3) utilising a well-balanced probing method in relation to its purpose, and (4) not substantially altering the construct during the process.

In her quest, Endsley reviewed and analysed several existing techniques. She concluded that physiological techniques such as electroencephalographic measurements as well as eye tracking are inadequate to measure situation awareness by themselves. With regards to performance measures she submits that a global performance measure may be useful for obtaining a "bottom line measure," but that performance measures otherwise are hard to conclusively tie to situation awareness as performance may be affected by many other factors than that of situation awareness [11]. Another technique, external task measures, which involves artificially changing or removing pieces of information as proposed by Sarter and Woods [48] was also deemed inadequate. She regards embedded task measurement, i.e., the measurement of specific subtasks, as a possible way to gain information that can be used to infer conclusions about overall situation assessment. An identified potential problem, though, is that the achieved SA for the measured subtask may not correspond to the level of overall SA. The observer rating technique was also discarded as being insufficient in itself to measure situation awareness because it, according to Endsley, probably does not provide an unbiased assessment of the operator's situation awareness. Further techniques were also reviewed by Endsley who eventually arrived at the conclusion that a probe technique best met her requirements, according to above, for a measurement technique. In the following we elaborate further on three selected techniques, namely SAGAT, SART, and QUASA, due to their popularity and proven validity.

A standard technique suggested by Endsley [9], is the situation awareness global assessment technique (SAGAT). As depicted above, SAGAT may be classified as a probe technique, or more specifically as a freezing on-line probe technique. SAGAT includes queries about all situation awareness requirements as discussed above, including level 1, 2, and 3 components, system functioning and status, as well as relevant features of the external environment [11]. SAGAT suggests that operators are intermittently queried concerning carefully chosen state parameters at random points of time during a dynamic situation. The SAGAT protocol prescribes that a number of questions are asked for each of the three situational awareness levels in order to determine to which degree the subject is currently aware of the situation for each level. A commonly occurring setting in which SAGAT is typically used is in a simulator, such as a flight simulator, that simulates real-life situations. For querying the subject, the simulation is typically frozen so that the SAGAT questions can be asked whilst the simulation is at rest. The underlying idea is to remove all relevant information from the operator (e.g., the operator's displays) before the questions are asked. The answers are then compared to the states of the selected variables in the simulation, and the more accurate the answer, the better. Examples of states of variables that are asked for in the context of aviation [10] include own heading, own location, aircraft heading, G level, fuel level, weapon quantity, etc. Although SAGAT is intrusive, Endsley reports that the performance during the continuation of the simulation is not affected if the probing questions are answered within, at the most, five to six minutes [11].

Another wide-spread, versatile and easy to use measurement technique for SA is Taylor's [54] situation awareness rating technique, SART. SART uses self-rating. The protocol requires the subject to rate to what degree he or she perceives (1) a demand on operators resources, (2) supply on operator resources, and (3) understanding of the situation, on a set of bipolar Likert scales. The ratings are then combined in order to provide an overall SA measurement score [16].

The quantitative analysis of situational awareness technique (QUASA) [36] is a combined self-rating and probe technique. QUASA is performed via probe statements that state a proposition as of the current state of parameters in, e.g., a simulation to which the subjects have to agree or disagree, e.g., "true or false?," thus the probe. Then, the subject has to rate to what degree of confidence the prior assessment was made using a scale with five degrees, hence the self-rating part of the technique. As a third question, the subject is then asked "Which teams will mostly answer this probe correctly?" The idea behind QUASA is to take advantage of concepts from signal detection theory, i.e., the analogue of the detection and the consecutive step of determination of the quality (of the signal). Further, QUASA aims to measure the "actual situation awareness" as acquired via cognition, and "perceived situation awareness" as sensed by metacognition. In experiments made within a military context (operational net assessments), it was shown that the technique provided insights into individual's situation awareness, but also regarding levels of sensitivity and biases in groups which may be useful information as well [36].

In a comparative study of the three situational awareness measurement techniques SAGAT, SART and CDM (Critical Decision Method, which is not further mentioned in this paper) within the context of a military planning task, it was shown that SAGAT level 2 (comprehension) showed a significant correlation relative to task performance as opposed to any other of the analysed techniques [47]. Another interesting conclusion was that no significant correlations between SAGAT and SART were found, indicating that the techniques may have measured different variables, as opposed to the stated intent not to do so, which is also the same conclusion that Endsley et al. made in a comparative analysis in 1998 [16]. Furthermore, Salmon et al. [47] make the important remark that success of SAGAT as a measurement technique is dependent on the ability to find relevant elements of situation awareness a priori, which is why they see SAGAT primarily as useful for measuring situation awareness in linear and deterministic settings.

3.1 The Cyber Domain

The U.S. Army Field Manual 3–38 entitled "Cyber Electromagnetic Activities" [56] defines cyberspace in terms of a man-made construct of systems of systems in that many small and diverse systems comprise the structure as a whole. These systems exist in the physical world. Cyberspace, which continually evolves, facilitates the use and exploitation of information, human interaction, and intercommunication through computers and telecommunication systems. Cyberspace and the electromagnetic spectrum, EMS, have converged into a global interdependent network, emphasising that the environment is not confined to a specific physical place. In order to successfully tackle cyber issues it is therefore asserted that a holistic approach involving physical infrastructure, data networks, and the EMS is suitable.

It seems, as given by the discussion hitherto, that there currently is no situation awareness measurement technique that is suitable for all domains. Although it remains to be thoroughly analysed to what extent the listed measurement techniques according to Salmon et al. [46] can be used for measuring situation awareness in the cyber domain, it is our belief that it may be fruitful to assemble components from several of the existing techniques in order to create a feasible measurement solution for the cyber domain.

Endsley's proposed situation awareness definition, i.e., a person's "perception of the elements in the environment within a volume of time and space, the comprehension of their meaning and the projection of their status in the near future" [8] may have to be carefully reconsidered because both "time" and "space" can be viewed differently in the cyber domain than in other domains, and both of these aspects are judged to be of importance in the situation awareness model construct. Temporal aspects of situation awareness are mentioned [12] and further elaborated on [13] by Endsley, where she notes that (1) the perception of time, (2) the temporal dynamics associated with events, and (3) the dynamic aspect of real-world situations, are aspects that may be considered. Spatial aspects of SA are also mentioned by Endsley [12] who points out that, in order to gain situation awareness, an operator needs to take the subsets of the environment that are relevant to tasks and goals into account.

As derived from the U.S. Army Field Manual mentioned above [56], the spatial properties of cyberspace is plainly that cyberspace is global, which makes the task of determining the outer geographical boundaries of a situation according to the situation awareness model problematic if not "everything everywhere" should be included. As the other delimiting boundary, the location of one's own system or network along with its externally facing connection point/points may be suitable.

Regarding the relevant temporal aspects to be considered in the cyber domain, we feel that it is of essence to keep several parallel time scales in mind, namely those that may be labelled near real-time, mid-term, and long-term. The near real-time perspective pertains to the time for signals to traverse through various communication systems to and from one's own system or network, and the processing time of those signals in electronic circuitry, which typically takes place during fractions of a second. The mid-term perspective may constitute the interval between minutes, e.g., updating software or applying a patch, and months, e.g., the increased user security awareness with regard to social engineering attacks. This is the timeframe in which different additional effects, other than the near instantaneous, of (cyber) actions will surface and be understood. The long-term perspective may stretch from months to years, and involves relevant aspects of the evolution of the domain itself, e.g., introduction of new (technical) protocols or changes in the governance of the internet.

In Table 1 the discussed cyber domain characteristics with regard to time and space are contrasted relative to other domains that are commonly discussed within the SA literature.

Table 1. Domain comparison with regard to geographical and temporal boundaries for situational awareness.

Domain/context	Geographical boundaries	Temporal boundaries
Tactical flight operations	The aircraft vs. the immediate vicinity of the aircraft	Start of flight mission vs. end of flight mission
Nuclear power plant process control	The power plant	Arbitrary starting point vs. continuous/infinite time
Military command and control	Own position vs. area of operations	Arbitrary starting point vs. mission/campaign time
Cyber defence	Own network vs. globally interconnected computers	Near real-time vs. continuous/infinite time

Endsley originally asserted [12, p. 50] that information reaches the operator from two sources, the real world and through an interface of a system, but later refined her assertion to include a third source [13, p. 7], the communication with team members and others, without, as far as we know, revising her SA model. Consequently, how information reaches the operator is another factor that may differentiate the cyber domain from other domains. In the cyber domain no direct observations, e.g., looking out the window, of the external physical world are feasible. All information about the state of the external environment comes mediated to the operator through artefacts or direct interpersonal communication, e.g., the status of a remote industrial control system is conveyed via sensors, a telecommunication system and displays. Details about a cyber threat may be learned through a conversation.

Drawing from another U.S. military publication, the Joint Publication 3–12 entitled "Cyberspace Operations" [55], we obtain another, functional, view of cyberspace, in terms of three layers:

1. the **physical** network layer in which the physical network components reside in the geography,
2. the **logical** network layer where nodes are interconnected, sometimes without a straightforward mapping to the other network layers, and
3. the **cyber-persona** layer, which takes advantage of the rules that apply in the logical network layer to "develop a digital representation" of an individual or entity identity in cyberspace.

We submit that all three layers have to be treated in a holistic way, but that the logical network layer is the layer that distinguishes the cyber domain from other domains the most. According to the mentioned Joint Publication 3–12, the "logical network layer consists of those elements of the network that are related to one another in a way that is abstracted from the physical network, i.e., the form or relationships are not tied to an individual, specific path, or node" [55]. Hence, the logical layer is an intangible abstraction that exists in computer memory only, that provides the cohesion between the physical hardware and the humans in the other two tangible layers.

We argue that, if made known to operators, a well-performed situation awareness requirements analysis, perhaps using the above mentioned viewpoints as a basis, may be viewed as an educational effort. The resulting hierarchical goal structure, subsequently, can also be used to inform the operator and drive her or his data collection. However, it has recently been shown that availability of an increased volume of additional task-relevant information does not result in significant effects with regard to mission performance [33]. Therefore it is rather the right information than the amount of information that counts. Endsley [13] concur based on her statement that more data does not equal more (relevant) information. Besides being a component in SA measurement, the information can also provide input to system designers who can design better systems, thus contributing indirectly to greater situation awareness and better mission outcomes.

To further expand the scope, yet other dimensions that constitute the cyber domain, besides a cyber security perspective, may be added. Recently Rid and Buchanan proposed a framework, named the Q Model, that covers multiple dimensions and levels of presumably attainable knowledge in an article that discussed cyber attack attribution [42]. We assert that the proposed framework is also useful for identifying required elements for CSA. The quite extensive and coherent model contains three functional, not necessarily hierarchical[4], levels: the tactical/technical level that mainly deals with the questions "What?" and "How?," the operational level that mainly deals with the question "Who?," and the strategic level dealing with the "Why?" The model proposes several specific questions for each level concerning, e.g., technical modus operandi, attacker characteristics, and involved organisations, but also some questions related to predictive knowledge concerning, e.g., attacker intent, second-order effects, and so forth. In short, we notice that the Q Model levels have a striking resemblance to Endsley's three-level model as discussed throughout this paper, in that they also show an ascending complexity using three levels. We feel that the proposed Q Model bears promise to be used to further the understanding of the cyber domain, and specifically contribute to the development of CSA and its associated measurement techniques.

Concerning measurement design for cyber, however, it should for the purpose of this paper suffice to mention that many more or less elaborate and valid methods to measure CSA can be developed using, e.g., SAGAT that can be adapted to suit different domains, along with other awareness measurement techniques such as QUASA, as a basis. In general, we propose the development and use of an SA measurement technique that is constructed specifically for the cyber domain, taking into account relevant elements, as mentioned above, combined with the measurement of bottom-line mission performance. We are well aware of that it is questionable if performance measures, mainly external measures, contribute to the measurement of situation awareness per se, but assert that it is indeed useful to measure performance as related to the mission goals, which is the ultimate rationale for having (good) situation awareness in the first place. The correlation between the level of CSA and the overall mission performance can also be used to gain second order insights.

4 Experiment Design

From a cyber threat perspective it is not easy to "know your enemy." Attackers typically possess a number of varying skills, have complex motives, might be organised in teams, etc. Moreover, the defending organisation's computer infrastructure is often complex and distributed, which makes knowing one's own environment a nontrivial task. It is in this context a cyber threat management solution needs to be evaluated, and this assessment needs to take the actual

[4] In military theory, the hierarchical war levels consist of the (lowest) tactical, operational, strategic, and political (highest) levels.

understanding of the cyber threat into consideration rather than solely evaluating the extent of being able to successfully make use of physical protective measures.

As a basis for measurement, the previously mentioned awareness levels proposed by Endsley serve as a baseline. That is, for any cyber management solution there is an underlying bigger picture that can be more or less understood, and for tackling, e.g., CC and/or cyberterrorism (CT) strategically it will be beneficial to have an understanding that to the greatest extent possible makes it feasible to understand the cyber threat not only in terms of mere perception of attacks but also in terms of working knowledge regarding the ulterior motives of the attack, additional attacker profiles, how to predict future attacks, how to devise new forms of training, etc.

Depending on the nature of the cyber threat of interest and the chosen measurement scheme according to Sect. 3, questionnaires or simpler simulations might suffice for situational awareness measurement in some situations whilst in other cases a more elaborate solution that can account for a higher degree of realism is required. In the following we elaborate on and suggest the use of CDXs that are adapted to accommodate possibilities for performing measurement, thereby testing the level of developed CSA.

4.1 Cyber Defence Exercises

CDXs are today being undertaken at regular intervals with relevant personnel participating in an environment that provides for a good level of realism. As an example, during the "cyber defence exercise" in the U.S., the participating schools are tasked to design and implement a computer environment providing a number of services which the participants are later supposed to defend from cyber attacks that are initiated by the "red force" of hackers which are in reality provided by the NSA [38]. The "Baltic Cyber Shield" exercise provides a similar example where six teams from across northern Europe were tasked to defend critical infrastructure networks from a group of professional penetration testers [26].

As indicated, a CDX provides an environment which can be tailored to resemble a relevant cyber threat arena, which can be further used for obtaining additional insight regarding true hacker motives. For a CDX to provide relevant higher-level data concerning a cyber threat, the CDX needs to be designed in a way that puts the cyber threat in focus and lends itself to observing the relevant aspects. The remainder of Sect. 4 discusses possible CDX setups, and the way to gain CSA insight through using both qualitative and quantitative observations. The main idea is to carefully insert suitable activities within the CDX in order to bring about a behaviour that can be observed and that makes the CDX participants engage in the cyber activity that the cyber threat management solution focuses on. As an example, setting up a honeypot of a suitable kind might attract certain types of attackers. The attacker behaviour can then be observed and used for determining the user's characteristics. In the long run, a number of such observations can turn, e.g., a stereotypical "script kiddie profile" into a more well-informed understanding of the attacker that can later play an integral role for analysing the overall organisational threat and the strategic measures that ought to be undertaken according to a higher CSA level.

4.2 Games

It is known that forensic psychology can be of great assistance to CC investigation [30], which assumes realistic hacker profiles and personality characteristics to be an important means for cyber defence and, hence, for informing CSA. Whilst many theories regarding hacker motives indeed abound, these are seldom based on actual empirical data and it is unclear whether the current knowledge is at all representative. Notable exceptions exist, though, with the "honeynet project"[5] being an interesting initiative where honeypots are placed on the internet to allure hackers in order to learn about their methods. The knowledge gained from the honeypots is used for raising awareness through issuing "know your enemy papers" where people can gain insight regarding the development of cyber threats and the measures that ought to be undertaken. From a pedagogical viewpoint, some insight regarding hacker behaviour has been gained through hands-on training within specifically designed isolated computer labs which the students are able to use as a playground for trying out various security related tools in a secure fashion. Although a number of successful initiatives have been reported on [3,24,28,43], these still remain fairly small-scale and are typically dependent on specific individuals. Full-scale exercises in terms of CDXs provide for more realism, and better chances of gaining insight that can be considered to be more relevant from a CSA perspective.

It is important, however, to consider both the limitations and the strengths of this claim. Following Raser [41], we distinguish between four criteria for the validity of gaming as a research tool: psychological reality, structural validity, process validity, and predictive validity.

For some cyber threats, these criteria are relatively easy to meet. If the objective is to find the success rate of remote code execution attacks as described by Holm et al. [27], then the exercise environment can be set up accordingly, and whenever a remote code execution attack is performed by the red team, the simulation environment ensures structural validity (operating systems, communication protocols, etc., all work just like in reality), process validity (finding vulnerabilities, using exploits, obtaining privilege escalation, etc., all work just like in reality), and predictive validity (what works in the simulated environment works in reality—if the real systems are configured just like the simulated ones). As for psychological reality, this cyber threat requires only that participants, once in a while, actually attempt to perform a remote code execution attack.

For other kinds of threats, however, the criteria are much more demanding. As noted by Sommestad and Hallberg [51], "the incentives that real attackers or defenders act upon" appear difficult to assess in exercises or competitions. The requirement for psychological reality now becomes prohibitive, as it more or less

[5] https://www.honeynet.org/.

requires the participants to actually *be*, say, ideologically or financially motivated. Indeed, not even economic incentives for the participants are certain to make them financially motivated since they "may make competitive choices not because they want to maximise their point totals, but because they want to beat the other person" [49]. There is, however, a middle ground. Even if questions regarding the psychology of attackers are beyond our reach, questions about their actions *given their incentives* are not. And the incentives of the game can be set to reflect motivation structures found in the IT security literature, gained from questionnaires, inspired by expert assessments, etc.

In the following, we consider a few examples of possible game setups, constructed to measure various aspects of CSA. Each game assumes an ongoing CDX with at least two opposing teams:

Benefits from eavesdropping. The team under attack (blue team) is given access to the communication channel(s), e.g., IRC, of the attacking team (red team). In the basic setup, the blue team has to manually read all the information in person, and take appropriate defensive measures. In more advanced setups, traffic is either pre-processed to highlight terms of interest or fused with other information sources. All of these setups can either be real-time, or lagged by a number of minutes. These setups can be compared to a baseline of no IRC access. In this way, the relative benefits of eavesdropping on the opponent can be measured. If enough trials are conducted, quantitative measures such as time-to-compromise or probability-of-compromise can be elicited. *This scenario measures the value of CSA for defence.*

Targeting with social network analysis. One team is given the ability to partially disrupt the IRC communications of the opponent. In one setup, the team can inhibit the IRC communications of a random member of the opposing team. In a more advanced game, the team has a software tool that displays the social network of the opposing team along with the centrality of each member. The team can then make a more informed decision regarding which IRC communications to disrupt. *This scenario measures the value of CSA for attack.*

Information overload. In this game, the blue team is attacked and is fed with accurate information about this attack, but is also simultaneously fed with a significant amount of irrelevant information. Variants include overloads aimed at single decision-makers, or overloads crafted to make several people in the team all slow down at a time. Quantitative measurements from this scenario include delays in decision-making, delegation of decisions and shutting down certain inputs (measures taken from Libicki [31]). *This scenario measures the extent to which competent information management and fusion tools offer remedies to information overload.*

Insider threat. In this game, the team is subject to an attack from one of their own. The individual is covertly given this task as part of the exercise setup. As noted by many authors, the insider cyber attack is a significant threat. In one setup, there is no system dedicated to detecting insiders. In another setup, an insider detector such as ELICIT [32] is employed. Additional setups would fuse ELICIT with information from other sensors. *This scenario measures the value of CSA for insider detection.*

Value of honeypots. The team under attack is allowed to configure a honeypot within their network, in order to learn from red team attacks on it. In one setup, the honeypot is monitored in real-time. In another setup, historical data from previous exercises is used instead. *This scenario measures the relative value of historical attack data vs. honeypot data for CSA.*

Automatic hypothesis monitoring. Computer network defence is not only about real-time operational measures, but also about risk analysis and planning beforehand. In this game, the team is allowed to identify high-level attack plans against their own systems before the exercise starts. They also build a threat assessment model with indicators (detectable with sensors at their disposal) allowing the model to provide a continuous threat assessment throughout the exercise. *This scenario measures the value of model-based threat reasoning for CSA.*

Service level agreements. Situational awareness is important not only during IT service operation, but also in the procurement and planning phases. In this game, the team does not fully control all of its IT infrastructure. Rather, some services are "bought" from a service provider, and the team must procure service level agreements regarding guaranteed restore times (e.g., service X is always restored within five minutes for \$1,000 or within one hour for \$100) before the actual exercise starts. With a limited budget, they must prioritise—some services must be deemed more important than others. In the baseline setup, no historical information is available. In subsequent setups, historical data from previous exercises is made available to the team, allowing more informed decisions. With the advent of cloud services and the notion of SOA, such decision scenarios are rapidly becoming increasingly relevant, but recent research suggests that decision-makers do not always make rational choices in SLA decision-making [21]. *This scenario measures the value of CSA regarding the past when making management decisions for the future.*

Aggressor identification. Four different teams at different locations participate in the exercise. One of the teams is secretively selected to be the aggressor and will during the exercise attack a randomly selected team, possibly hijacking resources from the other teams for the purpose. The task of the attacked team is to identify the aggressor using cyber information fusion techniques, optionally including help from the other teams. *This scenario measures the value of CSA for attribution.*

These examples have shed light on the interplay between specific cyber threat management scenarios and CDXs. The cyber threat specifics is required for proper incentive structures in exercises to be set up. The exercises can then serve to evaluate the level of CSA with respect to a specific cyber threat solution through conducting exercises where relevant and realistic courses of action for different attacker types are operationalised through using appropriate exercise incentives. Such behavioural information can be both qualitative, e.g., common modi operandi for espionage, and quantitative, e.g., the relative detection rates of ideological attackers compared to insiders.

4.3 Principles for Cyber Situational Awareness Measurement

In this section we discuss the differences between SA measurement experiments for the cyber domain and other domains, and highlight some important aspects to take into account for measurement of CSA. As an experiment platform, the cyber range not only enables the simulation—its computers and networks, real or simulated, are also an integral part of the system that includes the *subject* for training, experiment or measure in the cyber domain. For other domains the computers and networks are used as instruments of the simulation, but for cyber purposes the computers and the networks are at the same time the tools that are used by the operators.

It must be remembered that SA is measured on the operator, even if complex CDXs are used as a backdrop. The operator, or operators, work in an environment with all available means that we have at our disposal to execute the (cyber) mission, e.g., specific arrangements of hardware and software (a technical setup). The operators perform work in work processes. They may also have different degrees of organisation. We call this socio-technical system the *cyber solution*. By measuring the SA of the operators we ultimately aim at improving the cyber solution, be it with new and faster computers, novel pieces of software, new configurations of the software, improved visualisation techniques, or better work processes. Depending on the need, the measurement experiment can be conducted through, e.g., small-scale exercises or full-scale CDXs using exercise designs where it is possible to perform relevant training whilst at the same time evaluating to what extent the cyber solution has resulted in individual understanding of the overall cyber situation.

As discussed, information reaches the cyber operator in two ways, through artefacts via telecommunication systems, and via direct communication. Therefore, the cyber solution is of utmost importance. The cyber solution determines to what degree the operator *can* perceive, and consequently comprehend and predict future events.

Given the above we assume that the performance of the cyber solution is dependent on, and will vary with, at least three different factors: (1) how information is presented to the operators, e.g., how the technological portions of the cyber solution is configured which in turn will affect the operator's CSA, (2) the work processes, and (3) the properties of the operators themselves (including knowledge, experience, cognitive abilities etc.) We assume that, in all cases and experiments, these are the factors that affect the CSA of the operators and the levels of performance relative to the mission. We therefore assume that if we change one or more of the factors, the technical setup, e.g., the configurations of firewalls and intrusion prevention systems, etc., or the work processes, e.g., the order of which tasks are carried out, or the operators, e.g., novice or expert operators, the CSA and the performance will vary. (Alas, as noted in Sect. 3, the relation between these factors and the resulting CSA is not perfect, but subject to both random and systematic errors, making measurement more challenging.)

Noticing that SA measurement is highly context dependent according to the previous discussion, we emphasise the distinct properties of the cyber domain with regards to the missions and the cyber solutions as well as the importance of testing relevant measures in a carefully crafted game (e.g., a cyber range simulation) to be integral parts of the experiment design. Accordingly, we propose the following elements and associated criteria to be used for guiding CSA experiment design:

Mission. Existence of a clearly defined cyber mission that is realistic and attainable. Its expected outcome has to be measurable. If applicable, spatial and temporal boundaries are to be specified.

Cyber solution. Arrangements of hardware and software (a technical setup), the operators, and their associated work processes.

Metrics. Relevant metrics for (1) SA (given by an SA requirements analysis), and (2) performance (implicit and explicit "bottom-line").

Game. Simulation with a realistic scenario, planned sequence of events, and injects that provide a controlled environment.

In addition we propose using several suitable measurement techniques that are adapted to the cyber domain, e.g., domain-specific SAGAT and QUASA techniques, and both explicit and implicit measures of performance.

To make this more concrete, consider the following example from the banking domain. Nowadays most banks offer online services, e.g., internet access to their product portfolio of financial services, to customers. According to press reports the HSBC bank was struck by a distributed denial of service, DDoS, attack against their web services in January 2016[6]. These kinds of attacks, which are often carried out with the aim to intimidate or damage the reputation of its target organisations, may cause disruptions to online services for legitimate users. In other words they affect the availability of information. According to the same source, HSBC has been hit several times in the past as well, including the end of 2012. Now, expanding the view of this incident, we may add that during the approximate same time period, in the winter of 2012 and spring of 2013, other web sites belonging to other large financial institutions were also attacked by DDoS attacks, including Bank of America, Chase, Citigroup, JP Morgan, Wells Fargo, and others[7]. Furthermore, other types of malicious activity were also detected in conjunction with some of the DDoS attacks. More precisely, attempts to gain unauthorised access and carry out unauthorised transactions that are likely precursors and indicators of fraudulent wire transfers were detected. Data breach and information manipulation of this kind is an attack on the confidentiality and integrity of information.

[6] http://www.telegraph.co.uk/finance/personalfinance/bank-accounts/12129786/HSBC-online-banking-fails-again-after-succumbing-to-cyber-attack.html.

[7] http://www.cnet.com/news/cybercrooks-use-ddos-attacks-to-mask-theft-of-banks-millions/.

Some time before the incidents mentioned above, in September 2012, the U.S. Federal Bureau of Investigation, FBI, issued a warning of a new *modus operandi* for cyber criminals: that "DDoS attacks were likely used as a distraction for bank personnel to prevent them from immediately identifying a fraudulent transaction, which in most cases is necessary to stop the wire transfer" [18]. In other words, the FBI warned that they had observed DDoS attacks being used as diversion manoeuvres by criminals to cloak other more severe types of CC.

As an interpretation of these events in terms of the concept discussed within this paper, we assert that the level 1 understanding (perception) of the situation, according to Endsley, is about detecting the existence of malicious activity in the network. Level 2 understanding (comprehension) is about drawing conclusions about the types of attacks (DDoS and unauthorised access) and their immediate implications. Level 3, i.e., the higher-order understanding (projection), would be to draw conclusions about the specific *modus operandi*, i.e., the use of DDoS as a diversion manoeuvre for the purpose of hiding other attacks. In concrete terms, such insight can contribute to the prioritisation of the work of the IT (security) department to primarily focus on preventing unauthorised access attempts (even if drowned in a simultaneous DDoS attack), and not divert critical manpower to mitigate the effects of the DDoS (a less critical mission goal). For a bank we assume that the protection of the confidentiality and integrity of customer data takes precedence over the goal of protecting the availability of services (though both are important).

Our hypothesis is that it is indeed possible to defend the network (cyber mission) with only a first or second level appreciation of the cyber situation, but that it is possible to do it *even better* with additional third level insights.

4.4 Sample Cyber Situational Awareness Experiment Setup

As elaborated on throughout this section, a good way to perform measurements of CSA is within the context of CDXs. By convention the active (trained) participants of a CDX are named the blue team and the red team. The blue team, normally the primary training audience, is assigned for defensive tasks, while the red team is assigned to be the offensive attacking team. The best way to perform CSA measurement of a blue team, is by controlling the activities of the red team to the fullest extent possible in order to provide uniform conditions in several consecutive experiments, i.e., to rigidly script the attacks with regards to sequencing and timing. In this way it is possible to isolate the measured variable reasonably well. In such a case, however, the training effects for the red team are close to non-existent. Furthermore, if the activities of the attackers are fully scripted there is a risk that the blue team questions the psychological reality of the simulation [41] and that the exercise becomes static and deterministic (see Sect. 3), and is experienced as artificial.

Instead we suggest performing CSA measurements during the regular execution of CDXs (e.g., for training purposes). By giving red teams a certain degree of autonomy, a more dynamic interplay with the blue team(s) can be achieved. Through managing the red teams using a combination of loosely formulated tasks and an incentive structure (as mentioned in Sect. 4.2), possibly combined with direct instructions, both training effects and good conditions for measurements can be achieved for both blue and red teams. Cyber ranges generally have excellent data collection capabilities that enable extensive post-action analysis.

Using the banking CC case mentioned in Sect. 4.3 as a an example, we propose and discuss a possible CSA measurement experiment setup according to the principles in Sect. 4.3 as follows:

Mission. We would have one red team, and four blue teams. The cyber mission is to detect and prevent CC by protecting the information assets of the bank with regards to confidentiality, integrity, and availability. Sub-goals and subtasks include, e.g., continuous monitoring of network perimeter, matching of known malware parameters with incoming traffic, detecting suspicious network activities, logging and analysing activities on the internal network, stopping ongoing access attempts, etc.

The **cyber solution** is the computers and networks, hardware and software, that the bank has globally. The cyber solution includes the IT departments with their IT security functions and, specifically, the organisation, the personnel and the associated work processes that govern these functions. The mission has to be carried out continuously.

Metrics for availability is uptime/downtime of services. Other metrics, for confidentiality and integrity, are hard to define and measure directly. Implicit metrics can include, e.g., number of detected scans, number of refused connections, as well as quantifications of other kinds of attempts.

Game. As part of the game the red team would be given an incentive structure that awards high scores for fraudulent wire transfers. The red team would also be directly instructed to perform a DDoS attack as a diversion prior to a subsequent attempt to gain authorised access for the purpose of doing the wire transfer.

In this case it would be interesting to investigate, e.g., what changes in the cyber solution that would be required to enable the blue teams to focus on detecting and ultimately deflecting the attempts to gain unauthorised access, whilst under a distracting DDoS attack.

To gain a baseline we would instruct the red team to carry out the DDoS and the illicit transfer attacks as described. We would stop the simulation intermittently and ask the blue teams' questions according to the SAGAT and QUASA protocols. Level 1 questions would include, e.g., "What activity did you observe in the network?" Level 2 questions would include: "Which activities are hostile?," "What are the characteristics of those hostile activities?," and "How are the attacks carried out?" Level 3 questions would include, e.g., "Why are we attacked?," and "What will happen next?," for all four blue teams. At the same time we would record up-time of services (explicit performance) as well as successes or failures of the illicit transfers from customer accounts.

Next, we would test *changes* in the cyber solution to determine what might, and what might not, affect CSA and performance. A plethora of possible experiments can then be undertaken to test any number of ideas, such as, e.g., changes in firewall rules, changes in intrusion prevention system (IPS) calibration, changes in hardware, changes in software configuration, changes in information presentation, giving additional information to operators (e.g., FBI warnings, introducing bi-hourly briefings for operators for the purpose of information sharing, etc.) The changes would then be introduced to two of the teams and the simulation resumed. In further measurements the differences in CSA and performance, if any, between the teams can be used to draw conclusions about the effects of the implemented changes.

5 Conclusions

Based on the notion of situational awareness and its use for determining the level of cyber insight in terms of so-called cyber situational awareness (CSA), this paper has served to provide the foundation for developing suitable measurement techniques to be used for testing to what extent a person or a team has been able to acquire and/or maintain CSA. Being able to perform such measurement is critical for making it possible to test, e.g., to what extent training goals have been met, if a technical solution provides the sought for insight, whether a security process is capable of providing strategic insight, etc.

Although the notion of situational awareness and its role as a unique phenomenon has gained acceptance during the years, the way to measure situational awareness has been widely debated and many views exist. Also, measurement is naturally dependent on the domain, which by necessity requires that tailor-made protocols are being developed for the respective applications of interest. Hence, the development of the principles for CSA measurement that have been presented and exemplified in this paper have been based on (1) an overview of a few current situational awareness measurement techniques, in relation to (2) an analysis of the cyber domain and its similarities and differences in contrast to other domains.

It is vital to take the experiment design into account at an early stage in order for CSA testing to provide results that are relevant and applicable to the cyber aspect of interest. Albeit simpler methods requiring less resources, such as questionnaires, could sometimes be used, more elaborate simulations will most often be required for being able to providing sufficient realism and the associated measurement validity. As a result, the basis for constructing more elaborate testing mechanisms utilising cyber defence exercises (CDXs) has been provided in the article. The obvious next step and plan for future work is to develop these principles further and to validate them during the execution of a relevant CDX.

References

1. Artman, H.: Team situation assessment and information distribution. Ergonomics **43**(8), 1111–1128 (2000)
2. Bedny, G., Meister, D.: Theory of activity and situation awareness. Int. J. Cogn. Ergon. **3**(1), 63–72 (1999)
3. Brynielsson, J.: An information assurance curriculum for commanding officers using hands-on experiments. ACM SIGCSE Bull. **41**(1), 236–240 (2009)
4. Carroll, L.A.: Desperately seeking SA. TAC Attack **32**(3), 5–6 (1992)
5. Dekker, S.W.A., Hummerdal, D.H., Smith, K.: Situation awareness: some remaining questions. Theor. Issues Ergon. Sci. **11**(1–2), 131–135 (2010)
6. Dennehy, K.: Cranfield situation awareness scale: users manual. Technical report 9702, Applied Psychology Unit, College of Aeronautics, Cranfield University, Bedford, United Kingdom, January 1997
7. Durso, F.T., Hackworth, C.A., Truitt, T.R., Crutchfield, J., Nikolic, D., Manning, C.A.: Situation awareness as a predictor of performance in en route air traffic controllers. Technical report DOT/FAA/AM-99/3, Office of Aviation Medicine, Federal Aviation Administration, U.S. Department of Transportation, Washington, District of Columbia, January 1999
8. Endsley, M.R.: Design and evaluation for situation awareness enhancement. In: Proceedings of the Human Factors Society 32nd Annual Meeting, Anaheim, California, pp. 97–101, October 1988
9. Endsley, M.R.: Situation awareness global assessment technique (SAGAT). In: Proceedings of the IEEE 1988 National Aerospace and Electronics Conference (NAECON 1988), Dayton, Ohio, pp. 789–795, May 1988
10. Endsley, M.R.: A survey of situation awareness requirements in air-to-air combat fighters. Int. J. Aviat. Psychol. **3**(2), 157–168 (1993)
11. Endsley, M.R.: Measurement of situation awareness in dynamic systems. Hum. Factors **37**(1), 65–84 (1995)
12. Endsley, M.R.: Toward a theory of situation awareness in dynamic systems. Hum. Factors **37**(1), 32–64 (1995)
13. Endsley, M.R.: Theoretical underpinnings of situation awareness: a critical review. In: Endsley, M.R., Garland, D.J. (eds.) Situation Awareness Analysis and Measurement, pp. 3–32. Lawrence Erlbaum Associates Inc., Mahwah (2000)
14. Endsley, M.R.: Situation awareness misconceptions and misunderstandings. J. Cogn. Eng. Decis. Making **9**(1), 4–32 (2015)
15. Endsley, M.R., Rodgers, M.D.: Situation awareness information requirements for en route air traffic control. Technical report DOT/FAA/AM-94/27, Office of Aviation Medicine, Federal Aviation Administration, U.S. Department of Transportation, Washington, District of Columbia, December 1994
16. Endsley, M.R., Selcon, S.J., Hardiman, T.D., Croft, D.G.: A comparative analysis of SAGAT and SART for evaluations of situation awareness. In: Proceedings of the Human Factors and Ergonomics Society 42nd Annual Meeting, Chicago, Illinois, pp. 82–86, October 1998
17. Europol: Hackers deployed to facilitate drugs smuggling. Intelligence Notification 004-2013, European Cybercrime Centre (EC3), Hague, Netherlands, June 2013. https://www.europol.europa.eu/sites/default/files/publications/cyberbits_04_ocean13.pdf

18. Federal Bureau of Investigation: Fraud alert - cyber criminals targeting financial institution employee credentials to conduct wire transfer fraud. Press release, Financial Services Information Sharing and Analysis Center (FS-ISAC) and the Internet Crime Complaint Center (IC3), September 2012. http://www.ic3.gov/media/2012/fraudalertfinancialinstitutionemployeecredentialstargeted.pdf
19. Flach, J.M.: Situation awareness: proceed with caution. Hum. Factors **37**(1), 149–157 (1995)
20. Franke, U., Brynielsson, J.: Cyber situational awareness - a systematic review of the literature. Comput. Secur. **46**, 18–31 (2014)
21. Franke, U., Buschle, M.: Experimental evidence on decision-making in availability service level agreements. IEEE Trans. Netw. Serv. Manage. **13**(1), 58–70 (2016)
22. Gorman, J.C., Cooke, N.J., Winner, J.L.: Measuring team situation awareness in decentralized command and control environments. Ergonomics **49**(12–13), 1312–1325 (2006)
23. Hauss, Y., Eyferth, K.: Securing future ATM-concepts' safety by measuring situation awareness in ATC. Aerosp. Sci. Technol. **7**(6), 417–427 (2003)
24. Hill, J., Carver, C., Humphries, J., Pooch, U.: Using an isolated network laboratory to teach advanced networks and security. In: Proceedings of the 32nd ACM SIGCSE Technical Symposium on Computer Science Education, Charlotte, North Carolina, pp. 36–40, February 2001
25. Hogg, D.N., Follesø, K., Strand-Volden, F., Torralba, B.: Development of a situation awareness measure to evaluate advanced alarm systems in nuclear power plant control rooms. Ergonomics **38**(11), 2394–2413 (1995)
26. Holm, H.: Baltic cyber shield: research from a red team versus blue team exercise. PenTest magazine **2**(5), 80–86 (2012)
27. Holm, H., Sommestad, T., Franke, U., Ekstedt, M.: Success rate of remote code execution attacks: expert assessments and observations. J. Univ. Comput. Sci. **18**(6), 732–749 (2012)
28. Jacobson, D.: Teaching information warfare with lab experiments via the internet. In: Proceedings of the 34th ASEE/IEEE Frontiers in Education Conference, Savannah, Georgia, pp. T3C/7–12, October 2004
29. Jeannot, E., Kelly, C., Thompson, D.: The development of situation awareness measures in ATM systems. Technical report HRS/HSP-005-REP-01, European Organisation for the Safety of Air Navigation (EUROCONTROL), Brussels, Belgium, June 2003
30. Kirwan, G., Power, A.: Cybercrime: The Psychology of Online Offenders. Cambridge University Press, Cambridge (2013)
31. Libicki, M.C.: Conquest in Cyberspace: National Security and Information Warfare. Cambridge University Press, Cambridge (2007)
32. Maloof, M.A., Stephens, G.D.: ELICIT: a system for detecting insiders who violate need-to-know. In: Kruegel, C., Lippmann, R., Clark, A. (eds.) RAID 2007. LNCS, vol. 4637, pp. 146–166. Springer, Heidelberg (2007)
33. Marusich, L.R., Bakdash, J.Z., Onal, E., Yu, M.S., Schaffer, J., O'Donovan, J., Höllerer, T., Buchler, N., Gonzalez, C.: Effects of information availability on command-and-control decision making: performance, trust, and situation awareness. Hum. Factors **58**(2), 301–321 (2016)
34. Matthews, M.D., Beal, S.A.: Assessing situation awareness in field training exercises. Research report 1795, U.S. Army Research Institute for the Behavioral and Social Sciences, Alexandria, Virginia, September 2002
35. Matthews, M.D., Strater, L.D., Endsley, M.R.: Situation awareness requirements for infantry platoon leaders. Mil. Psychol. **16**(3), 149–161 (2004)

36. McGuinness, B.: Quantitative analysis of situational awareness (QUASA): applying signal detection theory to true/false probes and self-ratings. In: Proceedings of the 2004 Command and Control Research and Technology Symposium (CCRTS), San Diego, California, June 2004
37. McGuinness, B., Foy, L.: A subjective measure of SA: the crew awareness rating scale (CARS). In: Proceedings of the First Human Performance. Situation Awareness and Automation Conference, Savannah, Georgia, pp. 286–291, October 2000
38. Mullins, B.E., Lacey, T.H., Mills, R.F., Trechter, J.M., Bass, S.D.: How the cyber defense exercise shaped an information-assurance curriculum. IEEE Secur. Priv. **5**(5), 40–49 (2007)
39. Parasuraman, R., Sheridan, T.B., Wickens, C.D.: Situation awareness, mental workload, and trust in automation: viable, empirically supported cognitive engineering constructs. J. Cogn. Eng. Decis. Making **2**(2), 140–160 (2008)
40. Patrick, J., Morgan, P.L.: Approaches to understanding, analysing and developing situation awareness. Theor. Issues Ergon. Sci. **11**(1–2), 41–57 (2010)
41. Raser, J.R.: Simulation and Society: An Exploration of Scientific Gaming. Allyn and Bacon Inc., Boston (1969)
42. Rid, T., Buchanan, B.: Attributing cyber attacks. J. Strateg. Stud. **38**(1–2), 4–37 (2015)
43. Romney, G.W., Higby, C., Stevenson, B.R., Blackham, N.: A teaching prototype for educating IT security engineers in emerging environments. In: Proceedings of the Fifth IEEE International Conference on Information Technology Based Higher Education and Training, Istanbul, Turkey, pp. 662–667, May-Jun 2004
44. Salas, E., Prince, C., Baker, D.P., Shrestha, L.: Situation awareness in team performace: implications for measurement and training. Hum. Factors **37**(1), 123–136 (1995)
45. Salmon, P.M., Stanton, N.A., Walker, G.H., Baber, C., Jenkins, D.P., McMaster, R., Young, M.S.: What really is going on? Review of situation awareness models for individuals and teams. Theor. Issues Ergon. Sci. **9**(4), 297–323 (2008)
46. Salmon, P.M., Stanton, N.A., Walker, G.H., Green, D.: Situation awareness measurement: a review of applicability for C4i environments. Appl. Ergon. **37**(2), 225–238 (2006)
47. Salmon, P.M., Stanton, N.A., Walker, G.H., Jenkins, D., Ladva, D., Rafferty, L., Young, M.: Measuring situation awareness in complex systems: comparison of measures study. Int. J. Ind. Ergon. **39**(3), 490–500 (2009)
48. Sarter, N.B., Woods, D.D.: Situation awareness: a critical but ill-defined phenomenon. Int. J. Aviat. Psychol. **1**(1), 45–57 (1991)
49. Schlenker, B.R., Bonoma, T.V.: Fun and games: the validity of games for the study of conflict. J. Conflict Resolut. **22**(1), 7–38 (1978)
50. Smith, K., Hancock, P.A.: Situation awareness is adaptive, externally-directed consciousness. In: Gilson, R.D., Garland, D.J., Koonce, J.M. (eds.) Situational Awareness in Complex Systems. Aviation Human Factors Series, pp. 59–68. Embry-Riddle Aeronautical University Press, Daytona Beach, Florida (1994)
51. Sommestad, T., Hallberg, J.: Cyber security exercises and competitions as a platform for cyber security experiments. In: Jøsang, A., Carlsson, B. (eds.) NordSec 2012. LNCS, vol. 7617, pp. 47–60. Springer, Heidelberg (2012)
52. Stanton, N.A., Chambers, P.R.G., Piggott, J.: Situational awareness and safety. Saf. Sci. **39**(3), 189–204 (2001)
53. Stevens, S.S.: Measurement, statistics, and the schemapiric view. Science **161**(3844), 849–856 (1968)

54. Taylor, R.M.: Situational awareness rating technique (SART): the development of a tool for aircrew systems design. In: AGARD Conference Proceedings No. 178: Situational Awareness in Aerospace Operations, pp. 3/1–17, April 1990
55. U.S. Department of Defense: Cyberspace operations. Joint Publication 3–12(R), Joint Chiefs of Staff, Washington, District of Columbia, February 2013
56. U.S. Department of Defense: Cyber electromagnetic activities. Field Manual 3–38, Headquarters, Department of the Army, Washington, District of Columbia, February 2014
57. Vidulich, M.A., Hughes, E.R.: Testing a subjective metric of situation awareness. In: Proceedings of the Human Factors Society 35th Annual Meeting, San Francisco, California, pp. 1307–1311, September 1991
58. Waag, W.L., Houck, M.R.: Tools for assessing situational awareness in an operational fighter environment. Aviat. Space Environ. Med. **65**(5), A13–A19 (1994)

Part IV: Policy Development and Roadmaps for CC/CT Research

How the Evolution of Workforces Influences Cybercrime Strategies: The Example of Healthcare

Enrico Frumento(✉) and Federica Freschi

CEFRIEL, Politecnico di Milano, Milan, Italy
{enrico.frumento,federica.freschi}@cefriel.com

Abstract. Healthcare was an early adopter of ICT with the goal of improving physicians' work. The digital revolution of healthcare started several years ago with the introduction of informatics into hospitals. Today healthcare is again at the forefront: as one of the most attacked and promising areas of exploitation for cybercriminals and cyberterrorists due to the abundance of valuable information and for its role in critical infrastructure. Patients' world also changed radically and went through an ICT revolution; nowadays healthcare operators and patients' worlds are highly digitalized, modifying how healthcare operators and patients offer and use services. This chapter, starting from an introduction to the new paradigms of the modern workforces, will introduce the concepts of Hospital 2.0, the patient ecosystem and will explore specific cybercrime and cyberterrorism threats.

Keywords: Cybercrime · Healthcare · Strategy · Information security · Cyberterrorism

1 Introduction

Today, there is a blending between private and professional lives due to the flexibility of being able to work at any time and from different locations. As a consequence, physical and virtual encounters seamlessly merge. The recent global recession directly influenced labour markets adding new paradigms, more flexibility and more mobility. Thanks to mobile and ubiquitous devices, a user can complete a task anywhere, at home, in public spaces or in traditional company offices. From a technological point of view, we are faced to the presence of a digital ecosystem: a community of people who interact, exchange information, combine, evolve in terms of knowledge, skills and contacts, in order to improve their lives and meet their own needs. Among the aspects arising from the wide adoption of mobile technologies there is the evolution of workforces, i.e., the changes in how people are accustomed to carrying out their work. Digital devices have strongly shaped the way people are working and collaborating. Figure 1 demonstrates a simplified user-centric model of the modern way of working.

B. Akhgar and B. Brewster (eds.), *Combatting Cybercrime and Cyberterrorism*,
Advanced Sciences and Technologies for Security Applications,
DOI 10.1007/978-3-319-38930-1_13

The schema has four directions, from the point of view of a single worker, that impacts their working habits: Dataspace, Enabling Technologies, Use Cases and Context. Figure 1 is a conceptual representation of the most important trends in modern workforces and it is important to explain it fully.

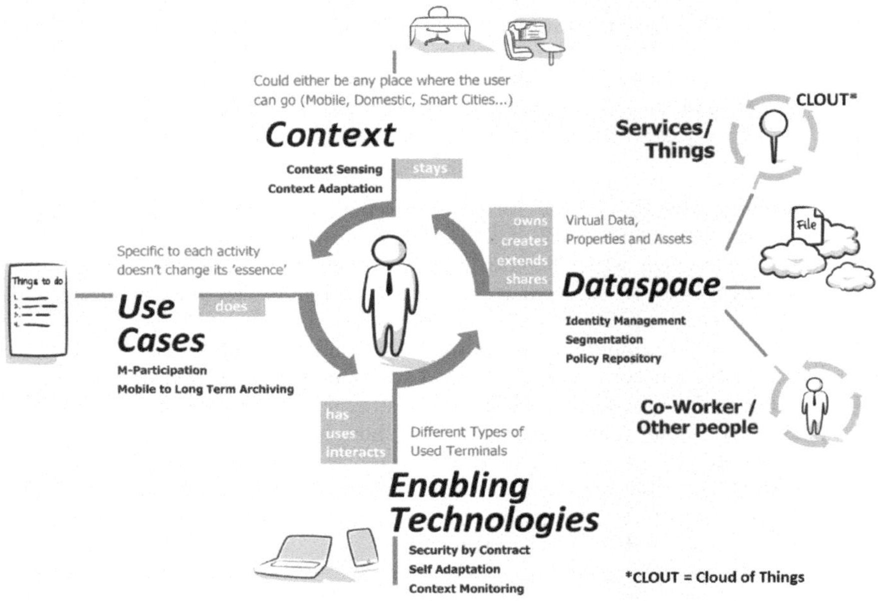

Fig. 1. Schematization of modern mobile work forces. (source: CEFRIEL)

In general, we can define a worker as a person that owns (i.e., has legal rights to access, edit and modify) a *dataspace* (also called a personal information space [1]) where all of their data are stored. He can then extend, elaborate and create new elements in this dataspace, even with the collaboration of other workers (shared dataspaces) or objects (internet of things). Thus we can define everyday working activity as a continuous process of updates to the personal dataspace. A simple definition of a working dataspace is a virtual place where data can be stored and accessed; this data can be either be strictly personal, shared or both. Nowadays trends are moving towards a complete dematerialisation of personal dataspace on centralised cloud services (no more disjoint data islands) and towards an intersection of the personal and working dataspaces [2].

To access the dataspace a worker can use several *enabling technologies* with different usability characteristics. Choosing any of these technologies is a matter of usability and ease of use for the worker. Ease of use defines how straightforward it is to perform a task, or a use case, in a specific place (context) with an enabling technology. Nowadays, the market is constantly offering new "methods" to access a user's own dataspace: smart watches are one of the most recent, but others are just around the corner, for example, the expected revolution of wearable electronics [3–5].

With reference to Fig. 1, a Use Case is the "invariant" portion of this scenario in that it is not affected by technologies and social trends. For example, a user could have written a commercial letter using different methods depending on the time period, i.e., using a typewriter machine, a video terminal with a word processor, more recently a tablet or smartphone device, or in the future wearable smart glasses that understand speech or thinking [6]. What remains the same is the requirement of writing a commercial letter.

Thanks to mobile and ubiquitous devices, a user can complete a task in anywhere: home, a public space, or in the office. It does not matter where he performs the work: only ergonomics matters (for example, carrying out a task with a laptop does not have the same ergonomics when it is performed on public transportation such as a train or compared to carrying out the same task at a desk). Therefore, sensing the *context* of a user is of enormous importance in order to adapt the enabling technologies' usability [7]. However, the context is also important for modern security solutions because it may assist, for example, in defining which data in the personal dataspace a user can access in a specific place without security problems (for example to protect his identity, privacy or with respect to company security policies). For example, a user wants to access a secured document, from a crowded place, over a data network; the system might prevent the access since from a crowded place where someone else might spy over their shoulder while they type the access password.

In this kind of environment, the essence of cybercrime (CC) is to abuse so called trust chains to steal assets. A trust chain is a trust relationship existing between two or more peers (either ICT devices or humans) that exchange assets, trusting that they will be handled correctly and that nothing intercepts them. Hence, changes in trust models and importance of assets implies changes in CC [8,9]. The future is characterised by the radicalisation of "blending life" and "immersed human" concepts.

1.1 The Healthcare Scenario

"Population ageing is a triumph of humanity but also a challenge to society" [10]. More and more people are enjoying life into their 80s and 90s.

The ageing of the population living in industrialised countries is an issue that influences the economy and the management of public and private finances allocated to health and social benefits for elderly people. Governments must be prepared to cope with modified needs and health/social expenditure. As people age, they depend more heavily upon outside support for health assessment and medical care. The current healthcare infrastructure is widely considered inadequate if it is to meet the needs of an increasingly older population. One solution is to enable ageing in place, where elderly people live independently and safely in their own homes for as long as possible, i.e., avoiding the transition to a care facility. This approach helps keeping the elderly population happy and socially connected, while reducing the strain on healthcare infrastructure. For this reason, recent studies generate substantial interest in telecare and home monitoring devices to address the health needs of senior populations, especially in rural and frontier communities. Recently, this trend led to more flexible and mobile solutions thanks to the evolution of mobile technologies and wearable medical devices.

A long-term radical change of perspective has happened in the health services in the last few years, it goes under the name of **"Patient Ecosystem"**. It consists in the evolution of the hospital as a place of care to a network of services for patients, provided in home environments, smart cities etc., through different channels and technologies.

The development of Assisted Living systems is one of the evolutionary aspects that healthcare is facing to support the creation of such an ecosystem. **"Moving to the Humans is the new wave"**, referring both to the many technological developments that have as a common characteristic to "centralise" the user (wearable systems, natural interfaces, and emotional design for user-centred innovation, etc.), and, above all, the way in which the access to services is provided.

Figure 2 reports the structure of a modern patient ecosystem (source: PRECIOUS project[1]) with four separate phases:

1. "Risk factor" data collection through a heterogeneous sampling of biometric data, performed in the houses of the patients, mobility via wearable things (mobile-health services), or in hospitals.
2. Data processing of collected information.
3. Analysis and modelling of data for the evaluation of lifestyle trends and the risk analysis
4. Feedback and responses aiming at the injection of behavioural changes in the patients either for wellness or for healthcare

Hospital 2.0, the Patient Ecosystem. Healthcare is migrating to an *Ecosystem* logic thanks to the evolution of some key technologies, such as the Body Sensor Networks, offering *integrated services*[2] (see Fig. 3).

Until a few years ago, healthcare ecosystems were perceived as limited within the hospital walls. The expected evolution relies, instead, on knocking down the localization attribute, in favour of a fully outsourced network of services. The hospital will ideally keep its traditional role for healthcare services that cannot be relocated and, will also keep being the institution where clinical competence is maintained and medical required professionalism can remotely operate.

[1] http://www.thepreciousproject.eu/.

[2] It is important to distinguish between the Services and the Ecosystems. "Ecosystem" means a network of integrated services that can interact with each other to offer the user a unique and seamless vision. Centering the vision of health services around the patient naturally leads to seamless servicing (the data are elaborated and accessed through different channels–e.g. mobile–without disruption or differences) and to a stronger control of personal data (which may be accessed through a unified ID).

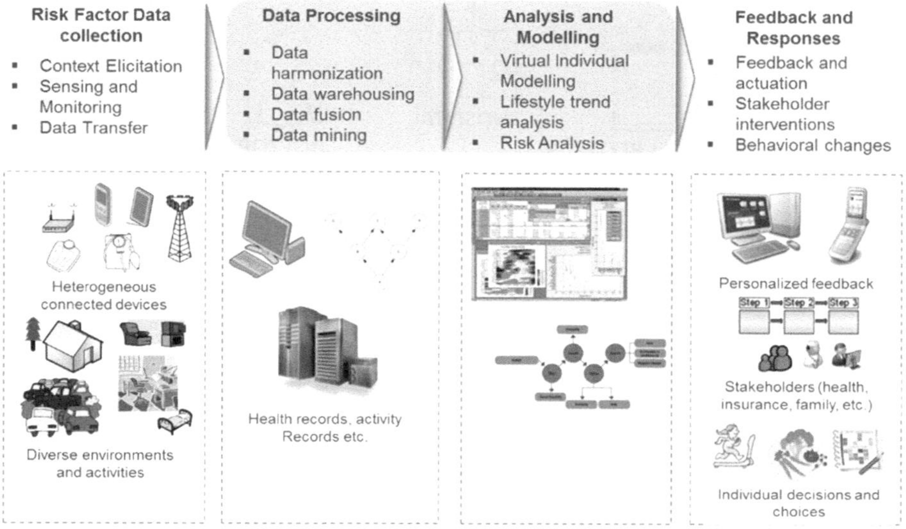

Fig. 2. Reference system for Lifestyle Management and Diseases Prevention. (source: PRECIOUS project).

Therefore, hospitals evolved from a place of care to a delocalised network of care services. The development of Assisted Living systems is only one of the evolutionary aspects of the healthcare system. The long-term radical change of perspective goes under the name of "Patient Ecosystem".

This evolution started few years ago, but it is exponentially accelerating thanks to all the following factors: the recent evolutions of mobile services, the better penetration of information technology to the patients and the increased impact of mobile wellness solutions.

Personal Information Space. The data usually collected to predict the risk of a clinical event are heterogeneous, including medium-term information (patient clinical history, exposure to environmental risk factors, and biological, therapeutic, environmental or occupational exposure) and short-term information (behavioural, biomedical signals, physical training and performance, lifestyle and diet, environmental data, social data etc.).

As a result the amount of information used to feed the data processing algorithms is huge. Extending the view above the healthcare sector it is useful to introduce the concept of personal information space or personal big-data spaces, which is the sum of our personal data, generated by the different applications or different areas of interest (see Fig. 4), often overlapping. The regulation of this dataspace (which data is allowed, who can access it, how it must be protected, when data must be deleted etc.) is one of the most problematic areas for the information security, not only in the healthcare sector.

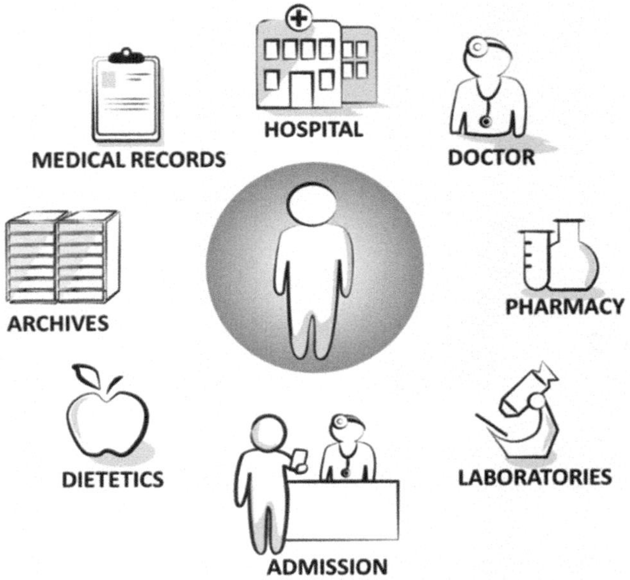

Fig. 3. Patient-centred healthcare is nowadays a service-based ecosystem.

This lack of regulation is more visible in the area of wearable devices which is one of the most critical growing sectors of the personal information space.

As a matter of fact, estimations evaluate the total wearable computing market to reach up to $34.61 Billion by 2020, growing at a compound annual growth rate (CAGR) of 20.7 % between 2015 and 2020 [11]: *The wearable computing market in this report is segmented into fitness and wellness, medical and healthcare, enterprise and industrial, infotainment, and other applications. The infotainment application is expected to grow at the highest value of $16.7 Billion by 2020. The increasing adoption of wearable devices such as smartwatches and augmented devices in the consumer market is contributing to the growth of the wearable market.*

Following this trend, healthcare operators in the US market are already experimenting the two leading ecosystems (namely Apple Health Kit and Google Fit)[3], in Europe instead this type of services is still lagging behind due to data export legislation inconsistencies.[4] The consumer wishes are already well defined,

[3] See for example: http://www.theverge.com/2015/2/5/7983707/apple-healthkit-hospitals-google-samsung.

[4] Commercial activities such as selling an end-users health information collected through the HealthKit API to advertising platforms, data brokers or information resellers is still debated, for example see these two links of 2014 http://www.forbes.com/sites/emmawoollacott/2014/08/29/will-apple-satisfy-regulators-over-healthkit-data-privacy/ and also the following post of 2015 http://www.giovannimaglio.it/articoli/lo-sviluppo-di-e-health-app reporting that privacy by design and by default criteria are not enough to cope the extreme importance of health data.

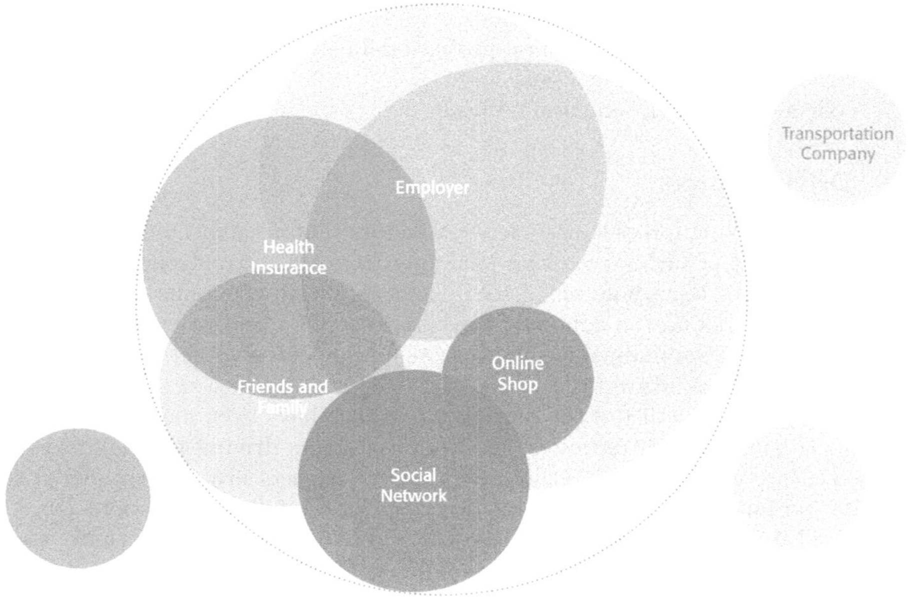

Fig. 4. An example of a personal information space or personal big-data space (Source: Talk-in-the-tower Taskforce #1 on the role of machines in our changing concepts of identity.)

a fact that operators cannot ignore anymore: Google and Apple host more than 100 thousand applications for fitness/health, downloaded by 500 million people in 2015.[5]

Mobile Health and connected devices and sensors potentially bring a large amount of data from different sources and for different purposes.

The first discriminant is to classify the data among categories based on:

1. Purpose of use:
   - health care,
   - wellness,
   - citizen profiling,
   - secondary use (e.g. clinical/non-clinical research),
   - info.
2. Level of privacy/security:
   - sensitive data,
   - non sensitive data/
3. Method of collecting of data:
   - provided by Health Professional or Health Care Service providers/Institutions,

[5] See a report from Research2Guidance: http://research2guidance.com/.

- provided by wellness professionals,
- provided by the citizen or non-professional caregivers,
- collected by Medical Devices,
- collected by "environmental" sensors.

1.2 Driving Forces

We define as driving forces the key leading factors that are expected to influence the development of future scenarios emerging from the current ones. The term scenario indicates the whole set of technological, social, economic and political conditions that define the context of cybercrime (CC) and of cyber-terrorism (CT), and the corresponding specific threats and defences, either in the present or in a hypothetical future time.

Table 1 reports the list of the driving forces that have been mentioned in the previous sections, which represent the leading aspects driving the evolution of the healthcare sector in the future years. These aspects are also subjected to exploitation for CC/CT intentions as described in the following sections.

From Table 1 it is quite evident that some of these forces are actually connected to each other, Fig. 5 reports a possible correlation of these concepts in a cause-effect diagram.

1.3 Cybercrime and Cyberterrorism Scenario in Healthcare

The main motivation for cybercriminal activities in healthcare is the financial profit made from stolen data and ransom demands [12]. Protected health information (PHI) has incredible value on the black market. A recent Ponemon Institute report on the cost of breaches found the average cost per lost or stolen record to be $154. That number skyrockets to $363 on average for healthcare organizations [13].

The modern healthcare ecosystems can be abused in different ways. In fact, hospitals have become incrementally digitalised often with complex and still largely unsolved security problems tied to the standards used, the lack of harmonization of services and problems with both roles in the hospitals and harmonizing laws among different countries (especially in Europe).

On the other side, the advanced attack techniques are becoming liquid and extremely flexible, ready to catch all the possible paths of income. For several years *the advanced attack campaigns are multi-vector, prolonged and adaptive to the defences they meet - unlike the defending side, which is inherently more rigid and structured around products and security solution silos. This siloed security approach presents an opportunity for advanced attack campaigns. While SOC (Security Operation Center) teams are occupied sifting through endless alerts and logs, with no real-time visibility and understanding of the "big-picture", attackers can exploit dead spots and misconfigurations to sneak between security policies* [14].

Lack of executive support, improper implementations of technology, outdated understanding of adversaries, lack of leadership, and a misguided reliance

Table 1. The sum of the forces driving the evolution of healthcare.

Driving force	Details
Society is getting older	Several highly technological countries are getting older and this trend is more evident in Europe. The increasing number of older people force the healthcare services to adapt both their services and care tracks
Health-care system congestion	The congestion of the healthcare infrastructures is a consequence of the increasing number of people that need to be served, in parallel some technologies such as wearable and home-automation are foreseen to mitigate the issue
Moving to the humans	Moving to the humans is the new wave, a citation that represents the new trend in healthcare, of moving data and not people
Early demission from hospitals	The foreseen increasing number of people using healthcare services pushes hospitals to increase the turnover of patients favouring the access to remote healthcare services
Assisted living system	The increasing adoption of remote assisted living systems is a consequence of the evolution of Hospitals from a place of care to a network of delocalised services
Home-care houses	The growth of the home-automation market also leads to a corresponding increase in the number of "hospitalized" houses
Pervasive health-care solutions	Pervasiveness of healthcare solutions is also a consequence of the ultra-mobile habits of people, who move more frequently, and the increasing wish of patients to continue their lives as much as possible when cured
Patient ecosystems	The interconnection of health and assistance services as well as wellness is one of the aspects of modern healthcare services, inherited from other application areas. The healthcare infrastructures are rapidly becoming a network of services
Personal big-data space	The increasing growth of our personal big-data spaces is one of the leading trends in several sectors, healthcare is foreseen to contribute to this scenario with big amounts of sensible data
Big-data analysis	Advanced data analysis is often one of the key elements that differentiate one health service from another and healthcare is just starting to mine value from the amount of accumulated personal data

upon compliance are some of the factors that result in making healthcare a very vulnerable sector to cyber-attacks [15]. In 2015, one in three Americans were victims of healthcare data breaches, attributed to a series of large-scale attacks that affected more than 10 million individuals [16]. Data summarised in Fig. 6 better clarifies the dimension of this phenomenon. Just in 2015 the number of

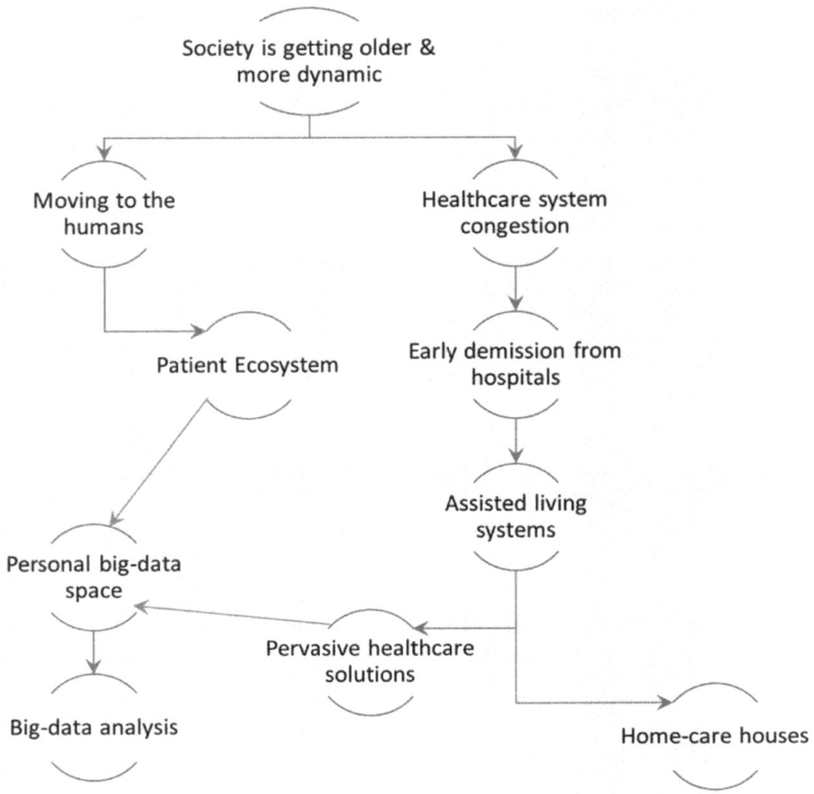

Fig. 5. A possible correlation of the driving forces in healthcare as a cause-effect tree.

breaches in healthcare was significant: what happened with the Carefirst [17] and Anthem blue cross services [18] and more recently UCLA health system [19] could happen with high probability to most European systems.

Furthermore, another key factor that according to the literature [20] that makes healthcare data breaches so prevalent is the lack of a proactive, comprehensive security systems dedicated to monitoring system irregularities. In other words, the lack of Intrusion Detection System (IDS) specialized for healthcare. Using IDS can help to identify a suspected attack and help locate security holes that gave the criminals access to your network in the first place. Without the knowledge derived from IDS logs, it can be very difficult to find system vulnerabilities, or determine if patients health data was accessed.

Healthcare is increasingly becoming a service-oriented ecosystem. This trend relies on solid market trends in the wearable industry, social needs of an ageing society and economic sustainability considerations. The number of remotely operated health services will increase in the coming years also thanks to the wider adoption of data processing algorithms based on the availability of large

Type of Breach	Individuals Affected 2014	Individuals Affected 2015
Hacking or IT Incident	1,786,630	111,803,342
Loss or Theft	7,273,157	750,802
Other	3,504,350	646,243
Total Individuals Affected	**12,564,137**	**113,200,387**

Fig. 6. In 2015, one in three Americans were victims of healthcare data breaches, attributed to a series of large-scale attacks. (source: https://www.helpnetsecurity.com/2016/01/28/why-cybercriminals-target-healthcare-data/)

personal information spaces. Nonetheless, the European regulations, granting privacy in all these aspects, are not yet adopted at state level. The discussion on privacy of the personal information space involves all these aspects:

1. applicable country law
2. relevant applied laws/directives/regulations
3. role of the third party service providers (data controller, data processor, data keeper) which often operate outside the country where the patients (the ultimate data owner) live
4. data protection/security
5. data conservation duration
6. Secondary use of data (e.g. market analysis, research, etc.)

The recently approved (April 2016) European General Data Protection Regulation (GDPR) introduces a more precise and informed control of the personal data by citizens [21].

1.4 Cybercrime

Current Threats. Security in the health sector suffers from a wider trend: increasing number of attacks to secondary markets, not primarily targeted by cybercrime until now. Health is gaining a lot of attention because it is a simpler target than financial institutions. Hospitals' security landscape is jeopardized and their employees are not as well trained [22]. This problem is getting even harder with the rise in popularity of mobile health.

Over 90 % of healthcare organisations faced a data breach in 2014 and 40 % had over five incidents in the last two years [23]. The trend up to 2014 is confirmed by a 2015 report that estimates that there were 340 % more security incidents in healthcare than the average industry [24]: "The rapid digitization of the healthcare industry, when combined with the value of the data at hand, has led to a massive increase in the number of targeted attacks against the sector".

The solutions for these trends are complex because the problem does not only involve the owner of the data (the user) and the official handler (the health services), but also external actors (e.g., the insurance sector), in some cases they are also based in foreign countries (e.g., companies selling health monitoring services through wearable bracelets, which are hosting data abroad).

Summing up the most common threats within the healthcare industry are the following:

- Physical theft/damage/loss is maybe one of the most typical cases in areas where there is the presence of very sensitive data, such as health and government. In particular, in the area of healthcare, physical theft ranks top among the breach methods [25].
- Information theft is another important element of incidents in the medical/healthcare industry. Identity theft in this sector has received particular attention by attackers [26,27]. The increase of data breaches if seen in combination with developments internet of things/wearables, makes it obvious that there is a lot of potential misuse in the area of healthcare [28].
- Targeted Attacks are among those that are exploiting more efficient social engineering techniques to facilitate data breaches. Despite not being a common attack, in hospitals the likelihood of an attack of this type is very high due to the structural and security problems of several patient ecosystems [29]. A mitigation for such attacks could be to identify the critical roles in the organisation and the estimation of their exposure to espionage risks based on their internal role and their digital footprint and shadow. Social engineering is a problem in healthcare because it is hard to identify, especially in organisations where workforce members do not always know each other. This happens despite the existence of security policies (e.g., HIPAA in the US or HITEC Act which enforces the encryption of healthcare data) and employee training programs. As recently reported by Cook [30]: *Social engineering attacks of any kind tend to be highly successful, but against an organisation with uneducated and untrained employees, these attacks are lethal, an example are the multi-faceted social engineering attacks* which combine phishing and vishing attacks and works well in healthcare.
- Threatening of the hospital users and infiltration through the external nodes. The problem of a distributed informative system like Hospitals 2.0 is that the security of the overall ecosystem is equal to the security of the weakest node. In a distributed system, like that of Fig. 3, the weak nodes are several: patients, wearable things, peripheral ambulatories, insufficient security knowledge of physicians and nurses, etc. An interesting menace comes from the abuse of patients' dataspace and medical information, for example through specialised

ransomware [31], which uses social engineering techniques against weak targets (elderly, patients etc.) [32]. Ransoms are actually a good sample of how quickly interest in CC for hospitals is growing, as recently reported by CEFRIEL [33]: "*ransomware is not actually the problem, but rather a consequence. The real problem is something happened before. The training of health operators was far from being effective and employees were not taught to correctly recognise the threat*".

Beside these problems, the modern hospitals still suffer from another class of issues addressed for decades: the existing security standards in the eHealth world, lack of on-field testing against complex real world attacks.

- Non-coherent (sometimes conflicting) standards specified by SDOs (Standard Developing Organizations) are in use. However, actions to come to a convergence among ISO/CEN, WHO, HL7, IHE and others have been undertaken by European Community and standard bodies for few years, but the work is still not complete and in large part its robustness has still to be proven against real attacks.
- Interoperability standards to allow semantically correct interoperability among Institutional Electronic Health Record (EHR), user centred Personal Health Record (PHR), mobile health/wellness applications and (medical/non-medical) devices should be defined and adopted, to allow a proper interchange of data, avoiding risks of "misunderstanding/mixing up" of concepts and data.
- Interoperability should address transactions and messages structures, document structures, adopted terminologies and code systems through the communication chain from devices to applications to EHR/PHR. The Multi Stakeholder Platform/European Interoperability Framework should consider these aspects, pushing stakeholders (SDOs, providers, etc.) to adopt compatible standards.

Current Defences. Politico [34] reports "After spending billions of dollars migrating to electronic health records, the health care industry is now looking to beef up its spending on data security". According to Politico's estimations health care organisations should spend at least 10 % of their IT budget to reach a decent cybersecurity level. Yet, the industry average is just 3 %.

The elements needed to handle the threat are the following:

- Innovative user awareness programs: the real essence of the current threats is the direct involvement of the victims into the attack's tactics (social engineering) and thus the users become an active part of the defence systems, which must be "hardened";
- Innovative mobile terminal management systems which mixes perimetral defence with pervasive awareness;[6]

[6] For example see MUSES 7th FWP EU Project (Multiplatform Usable Endpoint Security), www.muses-project.de.

- Try to mitigate the problem of security in hospitals not only promoting the adoption of best practices, which often have been developed in other application areas, like banks, but studying specific defence strategies, and also trying to foster a common culture of security
- Improving the existing solutions through the application of the known best-practices of other areas (e.g., banking)

Future Threats. The trends of CC in healthcare are summarized in the paragraph "Cybercrime", but the personal information space is getting larger and complex. This happens also thanks to the diffusion of healthcare personalized services offered though ecosystems. Moreover, the increasing adoption of mobile-health solutions (e.g., personal wearable and mobile terminals) which generates significant amounts of data is an important aspect. The personal information space is hence getting larger, weaker and its operators (e.g., both patients and physicians) do not fully understand its implications [35]. On the one hand, the general awareness of what the data sharing implies is not increasing. On the other hand, the overall general weakness of the health personal information space opens an increasing number of patients and operators to exploits [36]. The problems identified will continue to affect this sector for several years

Moreover, as with the telemedicine services, one of the most urgent driving forces is the so called **"immersion effect"**[7]: the ability for both physicians and patients, to forget the medium used to supply or receive a service. The physician must be able to concentrate on the clinical problem without worrying of any security issue, which could distract him; the patient at the same time must be confident that their data is not "abused" in any way. Currently, the way to obtain this immersion effect is to hide deeply the security issues into products, but often without real security. This is the most relevant trust chain of the health care sector but, unfortunately, it is often not present in the market solutions [37,38].

Future Defences. The evolution of defences could be described looking at the deployment EU research priorities in healthcare. The evolution of the healthcare security passes through a better harmonisation of ICT systems, hospitals services, protocols and laws.

Five areas of concern establish the priorities in the research and innovation on mobile Health:

1. Legitimisation of mobile health solutions and data. This area is necessary to increase the diffusion of mobile health solutions/applications in healthcare.

[7] Immersion effect: a generic telemedicine application should create the users immersion effect that means the physician should only think of his diagnosis without worrying about particular informatics operations that could divert his attention. Source: Committee on Evaluating Clinical Applications in Medicine. Telemedicine: A guide to assessing Telecommunications in Health Care. Marilyn J Field Editor, Division of Health Care Services.

A lot of data is collected through mobile health solutions; however, healthcare providers still do not take into account these data as a valuable source of information to be integrated with the traditional streams of healthcare data. An increased trust in mobile health solutions and on data that they are able to acquire/provide is the necessary condition for their institutionalisation in healthcare delivery services. Because of this, research should start focusing on the assessment of mobile health solutions/approaches in terms of safety, privacy, reliability and usefulness.
2. Mobile health for supporting healthcare delivery and connecting healthcare professionals. Mobile health can support and improve the processes and the services through which healthcare professionals establish and nurture their support networks. In fact, mobile health can improve not only the processes through which healthcare services are delivered within healthcare organisations, but also in the ones allowing a more robust and fruitful integration among the different providers of healthcare services within a healthcare ecosystem. The connection of different professionals – from primary and secondary care (GPs, specialists, nurses, etc.), to rehabilitative services – can enable real improvements in the delivery processes increasing their efficiency and effectiveness. Because of this, research should focus on the regulation of the information exchanged through unsecure terminals for example. The mobile terminals are exploited in different ways and methods for their secure management are still under research (e.g. MUSES EU project).
3. Mobile health for patient engagement and empowerment. Mobile health is one of the main levers to increase the level of engagement and empowerment of patients. From this viewpoint, it is necessary to understand how mobile health applications (together with their business models) have to be designed to support the delivery of patient-centred and sustainable healthcare services– allowing patients to actually securely contribute to their personal healthcare records with data collected through their (often unsecured) mobile devices.
4. Mobile health for well-being and prevention. Mobile health can widen the scope of national healthcare systems. Legitimised mobile health solutions adopted by engaged citizens can improve population well-being and strengthen disease prevention. The acquisition of data must result from safe and reliable mobile terminals and also health applications.
5. Mobile health widens the concept of "Country Specific" and "Cross Border" health/wellness services, to broader "Border Free" scenarios. A citizen may download a new mobile application while abroad, or connect to a local mobile service provider, providing data and getting data transferred in the Personal Health Record/Electronic Health Record when he is back home. These new "Border Free" scenarios should be carefully studied for legal, clinical and data interoperability implications, in a Pan-European, or even more global landscape.

The above trends influence the overall robustness of the patient ecosystem in terms of useful, secure and stable services. Concerning security, it is also of paramount importance for the European Community to foster the adoption of specific security certifications in the healthcare sector.

Most of the problem in healthcare sector arises from a lack of widely adopted secure standards and policies, which are instead a best practice in other strategic fields. The adoption of proper controls to protect the privacy and security of sensitive patient health information as well as their commitment to the healthcare privacy profession should not be a process left to the good will of single suppliers/hospitals [39]. Some certifications specific for the healthcare information security and privacy practitioners have been released[8], but the aim is to have Europe-wide accepted foundational standards to assess both information security and privacy expertise within the healthcare industry.

In general, there are some interesting security trends happening at a global level, which can have a positive effect also in healthcare:

1. *Artificial Intelligence in Antivirus systems.* Machine learning has the potential to be used to predict crimes before they happen. It is based on algorithms that, fed by many variables, can spot patterns otherwise oblivious to humans [40]. Researchers have already made use of machine learning to solve challenges in medicine, cosmology and, most recently, crime. In the cybersecurity field, some artificial intelligence techniques including heuristic technique, data mining, agent technique, artificial immune, and artificial neural network are applied in antivirus detection. It believes that it will improve the performance of antivirus detection systems, and promote the production of new artificial intelligence algorithm and the application in antivirus detection to integrate antivirus detection with artificial intelligence [41].
2. *Threat intelligence.* According to Gartner definition it is "Evidence-based knowledge, including context, mechanisms, indicators, implications and actionable advice about an existing or emerging menace or hazard to assets that can be used to inform decisions regarding the subject's response to that menace or hazard." Cyber threat intelligence has become one of the main topics in the industry in recent years [42] (look at companies such as Knowbe4, Phishme, Alienvault).
3. *New trends to increase Awareness.* According to recent researches the most interesting recipes of new awareness strategies in security are all involving the following three elements:
   - Fun (Gamification): ICTSec is boring, especially for non ICT experts. The point is how to make it fun or at least how to add on the awareness experience some pleasure mechanism;
   - Incidental learning: little step-by-step learning and knowledge improvement, for example through mini-games during the day, trying to avoid monolithic tracks;

[8] See for example the HCISPP (HealthCare Information Security and Privacy Practitioner), available at https://www.isc2.org/hcispp/default.aspx.

   - Personalization: adapt the learning experience to the stereotyped models that also the game designer uses to categorize players (e.g., conquerors, seekers, survivors, socializers, daredevil, etc.)[9]
4. *5GPL (Fifth Generation programming language).* The family of programming languages based on solving problems using constraints given to the program, rather than using an algorithm written by a programmer. Most constraint-based and logic programming languages, as well as some declarative languages, are fifth-generation languages. By adding Digital Process Management to 5GL, it is possible to have a comprehensive intelligent viewing capability during the flow of data across the systems, to catch relevant assets and protecting them before cyberattacks occur [43].
5. *Behavioural Security.* It is a new paradigm in security that through biometric and other behavioural measurement techniques tries to consider behavioural aspects of security and privacy in the defence mechanisms. *Behaviour-based intrusion detection techniques assume that an intrusion can be detected by observing a deviation from normal or expected behaviour of either the systems or the users.*[10]

1.5 Cyberterrorism

Current Threats. The most recent researches conducted in the healthcare security field demonstrate that this industry focuses almost exclusively on the protection of patient health records, and rarely addresses threats to or the protection of patient health from a cyber threat perspective.

The likelihood of individuals or organizations conducting activities through the internet will cause physical and/or psychological harm due to specific ideological or religious beliefs is increasing. These activities have become known by the term "cyberterrorism". In the healthcare landscape this can appear in a variety of forms, such as bringing down a hospital computer system or publicly revealing private medical records. Whatever shape it takes, the general effects are the same: patient care is compromised, and trust in the health system is diminished. However, there is evidence to suggest that cyber terrorism related threats are about to happen and that much of the European healthcare system is ill equipped to deal with them. The threat is known since few years and to date still unsolved [44]. Literature reports the following not-anymore hypothetical samples:

- Enemy agents gain access to the immunization records of fighting forces, allowing them to know which biological agents are most likely to decimate troops.
- Patients who underwent abortions at a local clinic receive death threats because an extremist group pilfered their names from the organization's EHR system and posted them online.

---

[9] See DOGANA aDvanced sOcial enGineering And vulNerability Assessment Framework, http://www.dogana-project.eu/.

[10] See for example http://www.sans.org/security-resources/idfaq/behavior_based.php.

- Incorrect dosages of a new medication are administered to patients after a disgruntled employee changes dozens of orders in retaliation for a poor performance review.

Most healthcare systems regularly experience cyberattacks. In many cases, the attacks originate from Eastern Europe and employ automated platforms, but firewalls can thwart the intruders. But for as much attention as organizations pay to automated attacks, the increasing number of targeted attacks pose a larger threat.

Moreover, other types of attacks are possible: "*a disgruntled employee with a list of active passwords and access to a hospital's systems has the potential to inflict far more damage than someone who must first conquer perimeter security appliances and hack into a system. Authorized individuals can download sensitive data, drop nasty viruses into the organisation's network, and even open back doors for others to use*" [44].

The healthcare sector is sensible to trust and an attack could target the trust of people into the system: losing trust in a network's integrity or its data may seem like a secondary concern, but it is really of primary importance in healthcare.

Current Defences. Securing cyberspace is not an easy proposition as the threats are constantly changing, and recognising that cyberterrorism should be part of a broader information technology risk management strategy, there are several "best practices" that can be adopted by healthcare organizations to protect themselves against cyber-attacks. The solutions against cyberterrorism are the same already identified against cybercrime, but in this case, a specific awareness is important: part of good organisational awareness includes examining the different ways a terrorist may be able to access sensitive data.

Moreover, "*with the growth of mobile devices in the healthcare realm, many IT groups no longer have the tight grip on access and storage protocols that they used to. Those other data sources need to be included in ITs overall strategy because it is, unfortunately, a weak link in the chain*" [44].

Future Threats. Large health systems generally have the expertise on staff to ensure that cybersecurity issues are on the organization's agenda and that a fairly robust suite of countermeasures has been put into place. The same thing is not true for smaller hospitals which still forms the backbone of most European national healthcare services. In these cases, the organisations are sometimes stymied by leadership inertia and lack of knowledge. This knowledge gap is fed also by the frantic pace of technology innovation in the healthcare sector. Governments are "*ill-prepared to fight the looming threat of 'online murder' as cyber criminals exploit internet technology to target victims*", warned the European policing agency. In its most alarming assessment of the physical danger posed by online crime, Europol said it expected a rise in "injury and possible deaths" caused by computer attacks on critical safety equipment [32]. The fact that the threat inherent the connection between physical and cyber threat has been seriously recognised by the European community is evidenced by CIP-01-2016 action (*Prevention, detection, response and mitigation of the combination of physical and cyber threats to the critical infrastructure of Europe*).

Future Defences. In 2013, an article in Telemedicine and e-Health [45] was already reporting that "healthcare organisations are at risk for attacks because they increasingly rely on computerised information; share sensitive data across multiple networks; use mobile devices; and are under-protected compared with other, less fragmented industries". Healthcare facilities report more hacking into their clinical data systems, including insertion of malware, denial of service attacks, and computer code attacks to steal or manipulate data, according to the article. A more recent report [15] extends and updates the situation reporting that "after two years of simulating attacks on monitors, health records, surgeries and more, researchers concluded that patients are pretty much sitting ducks". ISE researchers implemented a so-called Patient Health Attack Model which focuses as the primary attack surfaces those that directly affect a patient's health. For example, active medical devices that can be hacked to deliver a lethal dose of medicine, such as an insulin pump, or a heart defibrillator that could be modified or disabled so it can't deliver electrical current to save a patient in distress.

The reported increasing co-operation against cyberterrorism and other large-scale attacks on the Internet [46] is one of the most interesting area of development because "*mutual legal assistance of law-enforcement authorities has to be improved and adapted with regard to technological developments. Security measures for the protection of critical services and infrastructure should be developed. States are internationally responsible for taking all reasonable measures to prevent large-scale cyberattacks from being launched by persons under their jurisdiction or emanating from their national territory*".

Acknowledgements. The research leading to these results has received funding from the European Union Seventh Framework Programme (FP7-SEC-2013) as the CyberROAD project (Development of the Cybercrime and Cyberterrorism Research Roadmap) under grant agreement no 607642.

References

1. Giesecke & Devrient: The Future of Identity Personal information space – The future of identities in a networked world (2013). http://mcaf.ee/l209yu
2. Cooney, M.: Gartner: 10 critical IT trends for the next five years, 22 October 2012. http://www.networkworld.com/news/2012/102212-gartner-trends-263594.html
3. Canina, M.: IndossaMe: il design e le tecnologie indossabili. Milano: FrancoAngeli (in Italian) (2010)
4. Talk to My Shirt Blog (2015). http://www.talk2myshirt.com/blog/
5. Crunchwear (2016). http://www.crunchwear.com/

6. Willemsen, M.: Control Your Mobile Phone or Tablet Directly from Your Brain. NextNature (2013). https://www.nextnature.net/2013/05/control-your-tablet-directly-from-your-brain/
7. Schmidt, A.: Context-Aware Computing: Context-Awareness, Context-Aware User Interfaces, and Implicit Interaction, Interaction Design Foundation (2014). http://www.interaction-design.org/encyclopedia/context-aware_computing.html
8. Frumento, E.: Redefinition of the digital identity through the evolution of modern workforces - Part 1, in Identity, Talk in the Tower. http://goo.gl/AN9043. Accessed 17 Apr 2016
9. Frumento, E.: Redefinition of the digital identity through the evolution of modern workforces - Part 2, in Identity, Talk in the Tower. http://goo.gl/mRf5HV. Accessed 17 Apr 2016
10. World Health Organization: Active Ageing: A Policy Framework. Geneva (2002)
11. Markets and Markets: Wearable Computing Market by Application (Fitness and Wellness, Medical and Healthcare, Enterprise and Industrial, Infotainment, and Others), by Technology (Computing, Display, Networking, and Others), & Geography - Global Forecast to 2020, June 2015. http://www.marketsandmarkets.com/Market-Reports/wearable-computing-market-125877882.html
12. HIPPA Journal: FBI Malware warning issued over CryptoWall Ransomware, in Healthcare Data Security, HIPAA Journal (2015). http://www.hipaajournal.com/fbi-malware-warning-issued-over-cryptowall-ransomware-7095/
13. HelpNet Security: Why cybercriminals target healthcare data. HelpNet Security (2016). https://www.helpnetsecurity.com/2016/01/28/why-cybercriminals-target-healthcare-data/
14. Chesla, A.: Why advanced attack campaigns like security silos, March 2016. http://www.securityweek.com/why-advanced-attack-campaigns-security-silos
15. Vaas, L.: Hospitals vulnerable to cyber attacks on just about everything. Naked Security (2016). https://nakedsecurity.sophos.com/2016/02/26/hospitals-vulnerable-to-cyber-attacks-on-just-about-everything/
16. HelpNet Security: Why cybercriminals target healthcare data, in Don't miss. Help Net Security (2016). https://www.helpnetsecurity.com/2016/01/28/why-cybercriminals-target-healthcare-data/
17. Paganini, P.: CareFirst data breach affects about 1.1M people. Security Affairs (2015). http://securityaffairs.co/wordpress/37005/cyber-crime/carefirst-data-breach.html
18. Richman, J.: Anthem blue cross hack: What you need to know about the health insurers personal information breach. Mercury News (2015). http://www.mercurynews.com/health/ci_27465640/anthem-blue-cross-insurance-hack-what-you-need
19. Weise, E.: Hack at UCLA Health could involve 4.5M people, in USA Today (2015). http://www.usatoday.com/story/tech/2015/07/17/ucla-health-hack-45-million-patients-medical/30304977/
20. Barney, B.: Intrusion detection system: the missing component in healthcare data security. SecurityMetrics (2015). http://blog.securitymetrics.com/2015/12/intrusion-detection-system-missing-security.html
21. Bowman, C.M.: A primer on the GDPR: what you need to know. Privacy Law Blog (2015). http://privacylaw.proskauer.com/2015/12/articles/european-union/a-primer-on-the-gdpr-what-you-need-to-know/
22. HelpNet Security: The unlocked backdoor to healthcare data, in Help Net Security (2016). http://www.net-security.org/secworld.php?id=17062

23. Kemp, C.: Ponemon report shows abysmal state of data security in the healthcare industry - web host industry review, in Cloud Computing, Web Host Industry Review (2015). http://www.thewhir.com/web-hosting-news/ponemon-report-shows-abysmal-state-of-data-security-in-the-healthcare-industry
24. HelpNet Security: Healthcare industry sees 340 % more security incidents than the average industry, in HelpNet Security (2015). http://www.net-security.org/secworld.php?id=18889
25. Kaspersky Lab: Damage Control: The Cost of Security Breaches, in Kaspersky Labs (IT Security Risks Special Report Series) (2015). http://media.kaspersky.com/pdf/it-risks-survey-report-cost-of-security-breaches.pdf
26. Hiltzik, M.: Anthem is warning consumers about its huge data breach. Heres a translation, in Los Angeles Times (2015). http://www.latimes.com/business/hiltzik/la-fi-mh-anthem-is-warning-consumers-20150306-column.html
27. HelpNet Security: Anatomy of a healthcare data breach. Prevention and remediation strategies in ClearDATA (2015). http://net-security.tradepub.com/free/w_clec01/prgm.cgi?a=1
28. Koroneos, G.L.: Enterprise tech spotlight: Wearable security, Phishing targets, healthcare data breaches, in Verizon (2015). http://news.verizonenterprise.com/2015/06/wearable-security-phishing-healthcare-networkfleet/
29. Barney, B.: Healthcare: Recognize social engineering techniques, in Security Metrics Blog (2015). http://blog.securitymetrics.com/2015/08/healthcare-social-engineering.html
30. Cook, C.: The rise of multifaceted social engineering attacks, in Social-Engineer.Com (2015). https://www.social-engineer.com/rise-multifaceted-social-engineering-attacks/
31. Ossola, A.: Hacked medical devices may be the biggest Cyber security threat in 2016, in Popular Science (2015). http://www.popsci.com/hackers-could-soon-hold-your-life-ransom-by-hijacking-your-medical-devices
32. Peachey, P.: Cyber crime: First online murder will happen by end of year, warns US firm, in The Independent (2014). http://www.independent.co.uk/life-style/gadgets-and-tech/news/first-online-murder-will-happen-by-end-of-year-warns-us-firm-9774955.html
33. Frumento, E.: Which could be the consequences of a social engineering attack? Dogana Project (2016). http://www.dogana-project.eu/index.php/social-engineering-blog/11-social-engineering/9-which-could-be-the-consequences-of-a-social-engineering-attack
34. Allen, A.: Billions to install, now billions to protect, Politico (2015). http://www.politico.com/story/2015/06/health-care-spending-billions-to-protect-the-records-it-spent-billions-to-install-118432. Accessed 7 Mar 2016
35. Catalano, A.: Maintaining security during your healthcare merger or acquisition, in Help Net Security (2016). http://www.net-security.org/article.php?id=2356
36. University of Pheonix: More than 75 percent of U.S. Adults express concern about security of health care data, reveals University of Phoenix survey, in University of Phoenix (2015). http://www.phoenix.edu/news/releases/2015/10/us-adults-concerned-about-security-of-health-care-data.html
37. HelpNet Security: The unlocked backdoor to healthcare data, in Help Net Security (2016). http://www.net-security.org/secworld.php?id=17062
38. HelpNet Security: Security risks of networked medical devices, in Help Net Security (2016). http://www.net-security.org/secworld.php?id=18105
39. HelpNet Security: Small healthcare facilities unprepared for a data breach, in Help Net Security (2016). http://www.net-security.org/secworld.php?id=17516

40. Puium, T.: Machine learning used to predict crimes before they happen - minority report style, in ZME Science (2015). http://www.zmescience.com/research/predicting-crimes-before-they-happen-090423423
41. Wang, X., Yang, G., Li, Y., Liu, D.: Review on the application of artificial intelligence in antivirus detection system. In: IEEE Conference on Cybernetics and Intelligent Systems, pp. 506–509
42. ISIGHT Partners: What is Cyber Threat Intelligence and why do I need it? ISIGHT Partners (2014). http://www.isightpartners.com/wp-content/uploads/2014/07/iSIGHT_Partners_What_Is_20-20_Clarity_Brief1.pdf
43. Karisny, L.: Will DPM 5GL save Cybersecurity? Digital Communities (2015). http://www.govtech.com/dc/articles/Will-DPM-5GL-save-cybersecurity.html
44. Knudson, J.: Healthcare information: the new terrorist target. Record **25**(6), 10 (2013). http://www.fortherecordmag.com/archives/0413p10.shtml
45. Harries, D., Yellowlees, P.M.: Cyberterrorism: is the U.S. healthcare system safe? Telemed. e-Health **19**(1), 61–66 (2013)
46. Franken, H.: Increasing co-operation against cyberterrorism and other large-scale attacks on the Internet. Committee on Culture, Science, Education and Media, 8 June 2015. http://www.assembly.coe.int/nw/xml/XRef/X2H-Xref-ViewPDF.asp?FileID=21806&lang=en

European Public-Private Partnerships on Cybersecurity - An Instrument to Support the Fight Against Cybercrime and Cyberterrorism

Nina Olesen(✉)

European Organisation for Security, Brussels, Belgium
nina.olesen@eos-eu.com

Abstract. A European Public-Private Partnership (PPP) is an important instrument for boosting innovation and consolidating the European market and offering in a given sector. When it comes to cybersecurity, the establishment of a PPP is driven by the need to stimulate the competitiveness and innovation capacities of the digital security and privacy industry in Europe, and ensuring a sustained supply of innovative cybersecurity products and services in Europe. Given the growth and severity of cyber-attacks, such an initiative must take into account developments in cybercrime and cyberterrorism, including threats to particularly vulnerable and high impact areas such as critical industrial systems, the issue of trust and privacy, as well as the role of specific threat agents. It is therefore important that all relevant departments of the European institutions and agencies coordinate their efforts and bring in the perspective of Member States so that the full range of cybersecurity issues are considered from the public side, enabling the private sector to focus its efforts on developing a European cybersecurity industry through the adoption of an approach linked to the high and fast growth of technological competence and competitiveness. Only with strong governance and a dynamic approach can a PPP on cybersecurity develop a sustainable Digital Single Market ecosystem in Europe, making it a real and global cybersecurity leader.

Keywords: Public-Private Partnership · Digital Single Market · Cybersecurity · Cybercrime · Cyberterrorism · Industry · Security market · R&D · R&I · Critical industrial systems

1 Introduction

Europe has made important commitments and taken concrete actions towards building a sustainable Digital Single Market (DSM). The European strategy

B. Akhgar and B. Brewster (eds.), *Combatting Cybercrime and Cyberterrorism*,
Advanced Sciences and Technologies for Security Applications,
DOI 10.1007/978-3-319-38930-1_14

developed in this regard [1] comes at the right time as Europe is in danger of falling behind in the international digital economy. The strategy aims at creating the right conditions and a level playing field for advanced digital networks and innovative services while maximising the growth potential of the digital economy. This important objective should, however, be supported by an effort to protect and develop the European Digital Single Market. Against this background, the European Organisation for Security (EOS) [2] has produced, in collaboration with its Members, an extensive in-house study [3] of the European cybersecurity market. In this unique study, EOS gives an overview of the current cybersecurity market and describes the challenges ahead, providing recommendations and concrete actions to be taken in order to raise Europe to its full potential in the global cyber chessboard. The study lays the groundwork for significant collaboration mechanisms between the public and private sector in the years to come.

2 The European Cybersecurity Market

Following the revelations made by Edward Snowden in 2013 [4], the questions of privacy and data protection are of increasing concern to society. Today, thanks to fruitful high level political and societal debates and actions, Europe is seen as a trusted stakeholder in the world when it comes to data security and privacy. This status should be sustained and developed with the support of a strong and competitive European cybersecurity market in line with EU privacy and data protection requirements.

Unfortunately, the European cybersecurity market has inherited some of the problems faced by the general European security market. The cybersecurity market currently suffers from a large fragmentation which is partly due to the fact that security in general and cybersecurity in particular (especially as a component of critical infrastructures and national assets protection) remains a national prerogative. The EUs 28 Member States have different regulations and approaches towards cybersecurity as well as data privacy concerns which inevitably lead to the development of different specific solutions not necessarily competitive on a global scale. At the same time, even though innovation is strong in Europe (coming from ICT labs, SMEs, research centres, and large companies), the necessary funding based on a consistent transnational approach is often lacking. Research and Development (R&D) and Research and Innovation (R&I) in cybersecurity rarely reach market deployment and are exacerbated by weak public procurement policies. There is therefore a strong need for public and private cooperation to focus on advancing the competitiveness, innovation potential, and deployment capacity of the European cybersecurity market [5].

3 The Need for Technological Autonomy

Networks know no boundaries and the continuous interconnection between information systems makes cybersecurity a transnational issue by nature. In addition, the globalisation of trade makes network interconnection and interoperability a necessary requirement between the various economic agents increasing cooperation at regional and international level. Cyber attackers/hackers use this feature to their advantage to bounce from one country to another to cover their tracks.

In this scenario, the weakest link in the supply chain endangers the activity of many stakeholders, especially critical infrastructure managers and operators. Because of the highly fragmented cybersecurity market, European users depend largely on non-European solutions for their cyber protection. The increasing demand for cybersecurity products and services are often met by non-EU originating companies due to a lack of European policies designed to strengthen the European offer. These technologies might potentially include built-in backdoors and with time, increase our vulnerability to the risks posed by cyber threats especially towards vital and critical infrastructures. Some EU Member States like Germany, France, Finland and the UK have started discussions on how to achieve a greater autonomy and authority over ICT services and equipment. Several solutions have been proposed at national level but no convergence has been reached for a common approach based on certified, trusted EU solutions. It is however essential to define a common standardisation procedure for EU products and services among the Member States to avoid further fragmentation and higher costs. It is also of paramount importance that all the players in the ICT value chain, operating or not from a European Member State, adhere to similar requirements concerning data protection and cybersecurity. All market operators of the digital economy should share the responsibility for a secure cyberspace and all players involved must be committed to securing digital products, software and services.

4 Developing Trusted EU Solutions and Securing the Supply Chain

To achieve the aforementioned goals, Europe should find a good balance between the use of certified trusted non-EU technologies and the development of European solutions in vital areas (e.g. ICT infrastructure and public services), and in applications where Europe is a market leader (e.g. aeronautics, car manufacturing, finance services and all sectors falling under the Industry 4.0). In parallel, areas of higher competence in Europe like Identification and Access Management (e.g. smart cards) as well as Data Security (e.g. encryption) should be continuously improved to maintain leadership, while competitiveness should be increased in strategic components for Network Security Systems and Management of Security Services.

The European Organisation for Security (EOS) was created in 2007 by European private sector providers from all domains of security solutions and services. Its 44 members represent all relevant domains of the economy (ICT-Information and Communication Technologies, civil security, energy, transport, finance, services and research) across 13 different European countries. Our work and purpose is to provide a platform of collaborative work, insightful exchange of ideas, and best practices between the European Institutions and European security industry, research centres, universities local clusters and associations.

EOS' main objective is the development of a harmonised European security market in line with political, societal and economic needs through the efficient use of budgets. EOS works towards achieving a better level of technology independence for European strategic autonomy, supporting the development and use of European reference solutions and growth of a genuine European industry.

EOS supports its Members work by providing access to business opportunities by promoting at the highest level the implementation of innovative solutions in priority areas like cybersecurity, border control, civil protection/crisis management, urban security and protection of critical infrastructures.

In this respect, EOS has been actively supporting the creation of a European Public-Private Partnership (PPP) on cybersecurity which will be set up during the course of 2016. This collaborative platform will be a major opportunity to build a stronger technology base and outline a common European industrial strategy to effectively meet the needs and interests of Europe. EOS and its members are confident that the work stemming from this partnership will lay down the basis for a "European Cybersecurity Flagship" harmonising capacity-building in Member States and allowing, by 2025, our industry to become a world leader in key strategic sectors, implementing trusted European cybersecurity solutions and ensuring a greater digital autonomy.

5 EOS' Cybersecurity Flagship Initiative

The Flagship initiative developed and advocated by EOS and its members is built upon two main objectives:

1. The creation of a Flagship initiative for an EU Cybersecurity Investment Programme supported by adequate funding (initial estimate of 13 billion over 10 years), which would be composed of:
   - A Research & Innovation Programme based upon a competitive growth strategy.
   - Capacity deployment across Europe according to an agreed Roadmap, including short term focus on concrete strategic projects on capability and capacity building.

   A PPP in cybersecurity is seen as the initial step of this Flagship.
2. The development of a European Cybersecurity Industrial Policy touching upon several dimensions including: standards, certification and EU labels, innovative funding initiatives, education/training/awareness, support to SMEs and clusters, etc. This Industrial Policy will support the implementation of the DSM Strategy and the EU Cybersecurity Strategy [6] (as well as the Cybersecurity Flagship objectives) at EU and Member State level. Ideally, a PPP on cybersecurity should take into consideration the following elements:

- Market fragmentation. EU Member States have different regulations and approaches towards cybersecurity and data privacy, leading to the development of various specific solutions.
- Pervasiveness of ICT in different products and services with innovation driven by ICT products that are not designed and manufactured in Europe.
- Innovation is strong in Europe but not always properly funded due to a lack of a consistent transnational approach. Results of R&I are hardly reaching the market. Lack of strategy in EU research: several ongoing efforts are identifying technology and societal gaps but the identified R&I priorities are not sufficiently considering the wide economic/industrial perspective to bring the EU industry to a global competitive level.
- Weak entrepreneurial culture, lack of venture capital and seed money.
- EU industrial policies not yet addressing specific cybersecurity issues.
- Sovereignty. Market fragmentation partly due to the fact that security remains, within the EU treaties, a national responsibility.
- Strategic autonomy. The EU is heavily dependent on non-EU technologies in many domains in the ICT and cybersecurity field.

6 A cPPP on Cybersecurity – An Implementation Path for CC/CT Research

Cybersecurity incidents are increasing at an alarming pace with potentially profound effect on daily functioning of society and the economy, both online and offline. These incidents disrupt the supply of essential services such as water, electricity, and healthcare, undermine trust in digital services and products, and lead to financial theft, loss of intellectual property, and data breaches. In addition, as cyberspace knows no borders, the European market for ICT security products and services remains highly fragmented. In order to respond to these challenges, the European Union has called for the establishment of several measures under the Digital Single Market Strategy, including a Contractual Public-Private Partnership for Research and Innovation (cPPP) in Cybersecurity. This is currently being set up by industry under the guidance of the European Commission (DG CONNECT [7]) and is expected to be formally launched by the summer of 2016.

The establishment of the cPPP is driven by the need to stimulate the competitiveness and innovation capacities of the digital security and privacy industry in Europe, and ensuring a sustained supply of innovative cybersecurity products and services in Europe. The intended objectives of the cPPP are to:

- Gather industrial and public resources to deliver innovation against a jointly agreed strategic research and innovation roadmap.
- Maximise available funds through better coordination with Member States.
- Focus on a few technical priorities defined jointly with industry.
- Seek synergies to develop common, sector-neutral technological building blocks with maximum replication potential.

- Obtain economies of scale through engagement with users/demand side industries and bringing together a critical mass of innovation capacities.
- Be a platform to discuss other supporting measures for industry.

Horizon 2020 [8] provides the legal framework for the establishment of the cPPP and could finance activities such as large scale pilots, SME instruments, coordination and support actions (e.g. sectoral clusters in different applications; market knowledge and dissemination), innovation actions, R&D, and standardisation. Other envisaged activities linked to the cPPP, mainly seen as policy support/accompanying actions, include the financing of cybersecurity and of SMEs, investments for deployments (link with other EU and private funds), a strategic and research agenda, regulations in general and in particular for privacy and security by design, standards and certification, EU labels, a catalogue of products and services, awareness-raising, procurement (network of procurers, development of common requirements, etc.), and an EU wide platform for data exchange and better implementation of the Network and Information and Security Directive [9].

While cybercrime and cyberterrorism (CC/CT) will likely not be the prime focus of any of the foreseen working groups of the upcoming cPPP, the topics will certainly be addressed, as horizontal components, given the important threat that CC/CT attacks pose within a wide range of the sectors and application areas (i.e., critical infrastructures) that will be addressed in the cPPP. The instrument will also provide the ideal pathway for the implementation of CC/CT research and accompanying activities, through the link to Horizon 2020 and other EU funding mechanisms. The following sub-chapters outline the main aspects that should be considered in the upcoming cPPP when it comes to CC/CT issues.

6.1 Cybercrime

The growth and severity of cyber-attacks has increased the cost to society significantly. It is estimated that these attacks are costing the global economy billions of dollars each year [10]. A recent study from PwC on CC in the US [11] states that "Most organizations cybersecurity programs do not rival the persistence, tactical skills, and technological prowess of today's cyber adversaries".

CC can be defined as a crime in which computer networks are the target or a substantial tool [12]. A number of different definitions of CC are found in literature, each of them depending mostly on the purpose for which the definition is needed (e.g. focusing on the type of possible offences, or explaining the evolution of the crime, analysing the motivation of the offender). In addition, many international and regional instruments exist regarding this matter, namely the 2001 European Convention against Cybercrime from the Council of Europe (Budapest Convention) [13]; the Commonwealth of Independent States Agreement on Cooperation on Combating Offences related to Computer Information of 2001 (CIS Agreement) [14]; the Arab Convention on Combating Information Technology Offences from 2010 [15]; the Shanghai Cooperation Organization Agreement of Cooperation in the Field of International Information Security of

2010 (Shanghai Agreement) [16]; and the draft African Union Convention on the Establishment of a Legal Framework Conductive to Cybersecurity in Africa of 2012 (draft African Union Convention) [17].

It is very clear from these approaches that a number of general features could be used to describe CC. Focus on the object (material offence), on the individual, thing or value against which the offence is directed. This approach is found in the CIS Agreement (computer information) and in Title One of the Substantive criminal law chapter of the Budapest Convention (computer data and computer system). Another approach considers the computer systems or information systems as an integral part of the modus operandi of the offence [18].

Identifying possible CC offences and their modus operandi do not describe CC acts in their entirety, but it can provide a number of useful general categories into which these acts may be broadly classified [19]. Cybercrime is not a word amenable to a single definition, and is likely best considered as a collection of acts or conducts, rather than one single act [20]. For example, the Council of Europe defines cybercrimes as 'criminal acts committed using electronic communication networks and information systems, or against such networks and systems'. The offences considered cybercrimes under the Budapest Convention are grouped into four categories [21] (Fig. 1):

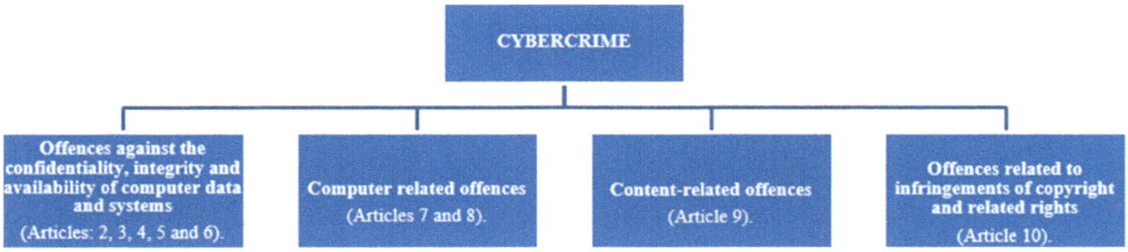

Fig. 1. Budapest Convention categorisation of cybercrime

It is evident that cyber-attacks are not only increasing in numbers and level of sophistication but are also becoming more costly for targeted organisations. It is important to add that the real cost of cyber-attacks is very difficult to estimate due to fragmented and insufficient statistics. Furthermore, many attacks remain unnoticed for years or are simply not reported by the targeted organisations out of the fear of reputational damage. However, many reports give an indication or an estimation of the importance of the problem. Below are some estimations:

- The World Economic Forum warns that over the next six years cyber-attacks could cause losses of up to 3 trillion dollars [22].
- The average annual cost attributed to cybersecurity incidents reached 2.7 million in 2014, an increase of 34 % compared to 2013. Large financial losses were more frequent this year as losses of 20 million dollars or more almost doubled (+92 % compared to 2013) [23].
- McAfee estimates that the cost of CC (and its consequences in restoring services/repairing the system) is between 375 and 575 billion dollars per year [24].

More importantly, cyber-attacks also constitute a direct threat to employment. For example, some companies affected would see a significant number of their jobs threatened: 200,000 in the United States, 150,000 in Europe.
- Lloyd's and the University of Cambridge's Centre for Risk Studies have estimated that a blackout due to a cyber-attack against the US electric grids would cost the US between $243 billion and $1 trillion and would also have a significant impact on the mortality.
- The Center for Strategic and International Studies has estimated the likely annual cost of CC and economic espionage to the world economy at more than $445 billion or almost 1 % of global income [25].

6.2 Cyberterrorism

Various definitions exist for the term Cyberterrorism' (CT), just as different definitions exist for terrorism'. CT is the convergence of Cybercrime and Terrorism [26]. Barry Collin, a senior research fellow at the Institute for Security and Intelligence in California, who is credited with first using the term "Cyberterrorism" in 1997, defined it as the convergence of cybernetics and terrorism. In the same year, Mark Pollitt, special agent for the FBI, offered a working definition: "Cyberterrorism is the premeditated, politically motivated attack against information, computer systems, computer programs, and data which result in violence against non-combatant targets by sub-national groups or clandestine agents".

This term can refer to unlawful attacks and threats of attacks against computers, networks and the information stored therein, with the purpose of intimidating or coercing someone for political or social motives. To qualify as a cyberterrorist attack, it should result in violence against persons or property, or at least cause fear and terror. That includes attacks against critical infrastructures. In instances of CT, technology (most prominently the internet) is used to achieve the same goals as more traditional weapons - i.e., to undermine citizens faith in government by undermining their ability to maintain and provide the critical infrastructure systems that form the foundation of everyday life for regular citizens [27]. Despite a recent rise to prominence, the concept of terrorism being facilitated through the use of technology is not a particularly cutting edge concept and has been anticipated since the 1980's. The US Department of Justice [28] defines CT as the utilisation of network tools to shut down critical national infrastructure or to coerce or intimidate a government or civilian population.

6.3 Critical Industrial Systems (Including Industry 4.0)

Critical infrastructures are vital to the modern society and economy. Most of the critical infrastructures (e.g. water supply, electricity, healthcare, and telecommunication) depend highly on ICS that manage key functions of the infrastructures. As these systems increasingly consist of (interconnected) networks, they have become more vulnerable to threats from outside the infrastructure, such as malware, botnets or denial of service attacks. As ICS manage large-scale physical systems (e.g. nuclear power plants), an attack on an ICS may have serious financial but also societal consequences (e.g. production loss, safety risks, information theft, disruption of key utilities).

Critical industrial systems are systems that are vital for the well-functioning of industrial processes. Most well-known may be the Industrial Control Systems (ICS) that are used to monitor and manage large-scale industrial processes such as manufacturing and product processing (e.g. distribution). ICSs are typically applied to control complex and critical processes such as the production and distribution of electricity, water treatment, oil and gas refining, chemical production and processing, pipeline management and rail electrification. Most ICSs consist of supervisory software installed on (a network of) servers which acquire real-time data from remote devices that control local operations. These supervisory data generally encompass indicators on product, process and environmental conditions (e.g. meter readings, equipment status reports) and are displayed to an operator on (a) central PC(s), often called the control centre. Based on the data retrieved from network devices the control centre sends automated or operator-driven supervisory commands to network devices. These feedback and feed forward loops enable the ICS and the operator to supervise the industrial process and to take action when needed. The types of ICSs which are frequently used (in combination) in industrial production are supervisory control and data acquisition (SCADA) systems, distributed control systems (DCS) and programmable logic controllers (PLC).

From a security point of view, requirements to ICSs have traditionally focused on reliability and process safety. In the early development stages of ICSs, the chance of a potential disruption from outside the company was negligible as ICSs consisted of isolated mainframe computers that were not connected to other systems. Over the years however, the ICSs evolved to interconnected systems sometimes linked to the Internet. In addition, industrial companies started to make use of off-the-shelf IT solutions that are more vulnerable to malware than in-house developed IT, which can affect the availability of ICSs. These developments enlarged the vulnerability of the systems to potential cyber disruptions from outside the company.

In several sectors there is a trend in which industrial devices and machines are increasingly equipped with technologies that acquire real-time and detailed data and distribute these data to other systems over a network (in some cases the Internet). This data is used by enterprises in order to (micro-) manage and control industrial processes, but it is also used increasingly by other actors (e.g. carriers, suppliers, end users) to control and manage their processes or usage. An example may be the energy sector in which smart grid concepts are currently being implemented. Although there are many types of smart grids, central to the concept is the existence of an integrated network infrastructure which enables systems throughout the grid (e.g. energy generators, smart meters, electric vehicles, appliances) to communicate with each other. Control systems constantly measure how electricity is flowing through the grid and enable actors (e.g. ICT systems, operators, but also consumers) to manage the flow. The computing and communication networks that are central to the performance and availability of the smart grid are often considered most vulnerable to security threats and cyber-attacks.

The industrial counterpart trend of the Internet of Things is often referred to as Industry 4.0. This also describes the intelligent - and mostly IP based - networking of machines, devices, smart product tags and other smart things? in industrial scenarios - like in a production or industrial logistics site. These integrated systems are also called I Systems. Again, this increased and more standardised networking holds many advantages and business potentials; however, it also opens attack routes down to core industrial infrastructure e.g. into robots, assembly machines, logistic sorting machines and alike.

A specific characteristic of industrial device communication - as also in some areas of the Internet of Things - is the need for real-time processing - which is supported at the level of the device operating systems but also the networks. This partially hinders the application of standard IT security techniques e.g. SSL encrypted communication channels between devices. As many processes run automatically in an Industry 4.0 scenario, particular attention should also be given to hardware based security - e.g. trusted computing. This can e.g. be used to securely identify devices and validate the integrity of the installed software.

The evolution of ICS technology is largely shaped by the rise of connected systems. Issues with connecting legacy and stand-alone systems to proprietary networks and the internet are not new but will grow quickly as new industrial manufacturing concepts evolve. With developments in embedded systems, cloud and big data infrastructures, and sensor technology as enablers, smart manufacturing concepts will be developed that can be characterised by a rather high level of system autonomy, high level of spatial dispersion and high complexity. For ICS systems, such developments mean that more data streams need to be integrated and processed, with more complex analytics, interpretation and response actions to be performed.

As the number of network nodes increases, the number of potential entry points for attackers also increases. In addition, more interconnections yield more opportunities for DDoS attacks, infection with malicious code and other intrusions. These kind of vulnerabilities might allow attackers to penetrate a network of a smart grid and enable attackers to take over the control and management of (parts of) the grid. Attackers could for instance change load conditions in order to destabilize the grid. Although CC will remain a serious threat in the decades to come, security strategies should not only address deliberate attacks, but also security issues resulting from other causes such as user errors, equipment failures and natural disasters.

6.4 Online Trust and Transparency for Privacy: Trusted Cyber Identities Including Recommendations, Rating, Reputation, and Reasoning for Trust

Electronic or digital identities (eID) uniquely identify individuals or another legal entity, or another type of entity—within a domain, such as a device in a network. Digital identities are based on digital identifiers, which are strings or tokens that are unique within a given scope or context (global to the infrastructure or local within a specific domain, community, application, etc.). Identifiers are the key used by the parties to agree on the entity being represented, and are sometimes a combination of so-called attributes that characterize the entity. Identity management refers to all processes and technologies for the creation, management, and use of digital identities. In practice, it also consists of establishing the identities of the different parties involved in the interaction in order to be able to trust each other's claims. Nowadays, since services and processes cross logical and physical boundaries, and citizens carry out many online interactions requiring their digital identity, Identity Management considerations are especially relevant.

Managing the identification and authentication of users in online environments and the protection of users' privacy are essential functionalities in almost all digital processes. Both in private and in working life, users encounter many situations in which they have to identify themselves to a third party, for example, to obtain a service, to carry out a task or access information. Password-based authentication is the de-facto method of access control in online web services as it is cheap and simple, but also other digital processes (e.g. gaining access to a company's network) are often supported with password-based authentication. Studies show that many users choose passwords that are too simple and expose them to attacks. Even if users select complex passwords, suboptimal security at the side of the party that manages authentication credentials could lead to cyber security incidents.

The key research challenges identified within this area by the FP7 project entitled CAPITAL [29] which has analysed multiple research agendas for trusted cyber IDs are the following:

- The development of rich identification and authentication techniques to ensure privacy, handle identities securely and that have - at the same time - a high level of usability for the end-user.
- As not only humans are digitally connected, but increasingly also all kinds of (new and fast emerging) technological objects, a secure identification and authentication of these objects is ever more important. From the perspective of the internet of things, the development of technologies for identification and authentication that can operate at a global scale is needed. This includes the management of unique identities for physical objects, devices and locations and possible cross-referencing among different identifiers for the same entity and with associated authentication credentials.
- As service coupling (services which are linked to other services, e.g. links between Twitter and Facebook) becomes commonplace, new security issues arise and consequently also research questions on how to solve these security issues (e.g. designing new techniques for interconnecting services in a secure fashion).
- A critical factor in the security of identification and authentication processes is the way users and organisations deal with security issues. Research could

support the development of adequate management techniques and organisational procedures to ensure the correct application of identification and authentication techniques (e.g. guidelines to delegate trust).
- The development of rules and regulation to deal with identity theft, privacy and anonymity rights, as well as private data retention and corresponding access rights.

6.5 Threat Agents

For several years, ENISA [30] has provided an overview of the threat agents currently in the cybersecurity landscape (see Fig. 2 [31]). These threat agents are important elements to consider when assessing and prioritising the future of CC/CT research.

The following is the overview of the threat agents in cybersecurity listed in prioritised order:

- **Cybercriminals** whose objective is to obtain profit from illegal and criminal activities in cyberspace. It is reported that their main motivations are intelligence and monetisation. They are characterised as having large amounts of time and money at their disposal, while being technically highly skilled and well equipped. They have high-performance computing resources and can be part of highly organised criminal groups. Furthermore, it is expected that criminal groups will increasingly engage in this field. They are mostly involved in fraud activities (e-finance, e-commerce, e-payment, ransomware, cybercrime-as-a-service, delivery and development of malicious tools and infrastructures). Cybercriminals are also becoming more and more specialised in their roles such as intermediaries, brokers and solution providers. The possibility of using anonymisation, encryption and virtual currencies makes it possible for the criminals to move in the dark markets?, which in turn makes it very difficult for law enforcement to detect and attribute crimes.
- **Online social hackers** are mostly involved in activities such as phishing and stalking in targeted cyber-attacks. They play a key role in deploying cyber threats. They can be characterised as being skilled in social engineering and understanding the psychology of social targets whilst breaching their privacy. The tools they use include analysis of social engineering information, profiling (logs, social media accounts, breached data etc.). It is reported that the capabilities of this group can be characterised as low to medium regarding the use of technology, however, the social engineering skills are high. It is expected that this type of threat will increase significantly in the future.
- **Hacktivists** are a group of politically motivated threat agents whose motivation derives from political ideology, social justice and sincerity. They use propaganda to influence political decision-making. They are characterised as being dynamic and often lacking a centralised structure. Most cases where their actions are visible are during riots, sports events and other major events that have triggered international attention. The main methods they use are DDoS attacks, leakage, defacement and hacking. It is not easy to pinpoint the

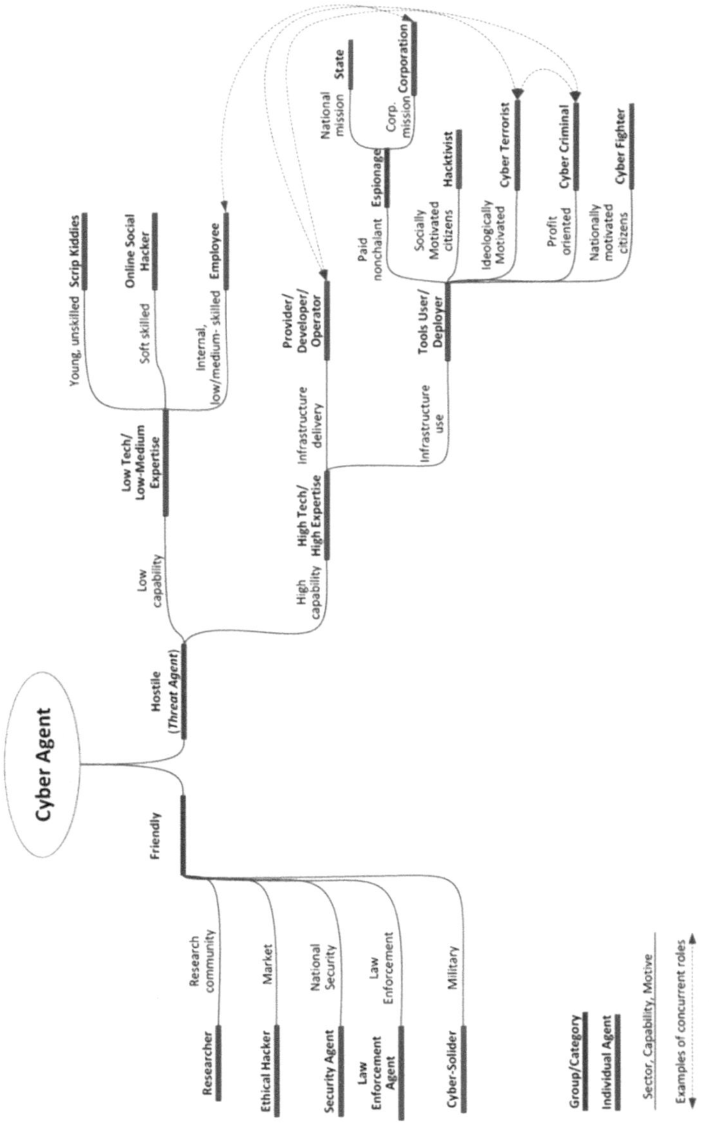

Fig. 2. ENISA cybersecurity landscape - threat agents

profiles of this group since threat agents from other groups might emerge under this group. They try to create as much media attention as possible through the successful attacks on government sites, large companies, media etc.

- **Nation states** are another group of threat agents that have emerged in light of the Snowden revelations in 2013–2014 whose activities are related to national security and intelligence/counter-intelligence. This has ranked on the third position in attribution of cyber incidents in the ENISA report. Various nation states have developed cyber intelligence capabilities but due to its non-transparent nature it can be assumed that the countries with such capabilities are involved in the area of intelligence/counter-intelligence in the cyber domain. The targets of such attacks are information on state secrets, military secrets, data on intelligence, and attacks on critical infrastructures. The degree of success in this case is rated as high and aims at creating intelligence, strategic, psychological and political advantage.
- **Corporations** have been identified as another threat agent group that perform corporate espionage with the aim of collecting business intelligence, competition information, breaching intellectual property rights and causing damage to or sabotaging their competitors. This is a growing trend and due to the availability of budget and information by corporations, this could cause very high costs. This may be performed in close cooperation with nation states and they may use the existing resources of the states to reach their goal.
- **Employees** (current, ex, internal, external) are a group of threat agents that are motivated by extortion, revenge, sabotage or profit. They materialise cyber threats that usually lead to data breaches. This is also called Insider Threat? and it can be both intentional and unintentional. The cost of protecting against such a threat can be quite high which makes it important to identify employee dissatisfaction, knowledge-gaps and setting up alerts when attacks abuse publicly unknown vulnerabilities.
- **Cyber fighters** are another group of threat agents that are nationally motivated citizens who have significant striking power. They are politically motivated and use the technique of sabotage. They may be supporters of totalitarian regimes and act on their behalf. Their activities are reported as being more and more systematic and well organised with increasing maturity and sophistication of attack methods. One example of this group is the Syrian Electronic Army.
- **Cyber Terrorists** is another group that uses large-scale sabotage mechanisms to harm national security and society. Their main target is critical infrastructures and services. They can be characterised as having an indiscriminate use of violence in order to influence decisions and actions of states towards their politically or relationally motivated objectives. National cybersecurity strategies rate this risk as high; however, it seems that risks from CC are much higher at this point. This group uses technology as a means to improve their communication in order to avoid state surveillance; however, by definition this is not seen as a hostile activity. The ability to communicate without any law enforcement surveillance gives them the ability to share information about different tools to launch future attacks.

- **Script kiddies** are usually young individuals who are motivated by the skills of tech savvy individuals who gave lessons to persons, organisations or brands considered outrageous. Due to low levels of knowledge of the use of the hacking tools, low levels of self-control, overestimation of their own skills and the consequences of their activities, they can achieve great impact. Yet, their impact is not considered significant [32].

6.6 Current and Future Threats Patterns

The following list of threats has been identified mainly by the CAPITAL project [29]. It is a heterogeneous list, mixing attack vectors (email viral attachments), payloads (logic-bomb, malware, spyware), attack objectives (DDoS, data breaches, identity theft) and attack patterns (drive-by-download, botnet, man-in-the-middle, social engineering).

- **Advanced Persistent Threats (APT):** Organisations today face what is commonly called "advanced persistent threats" or "APT", programmes particularly pernicious used by an attacker to obtain an illegitimate network access and to remain unnoticed. The objective of the APT is mainly sabotage or recovery of sensitive data and targets organisations with a high informational and financial value, such as R&D centres, financial or defence industries.
- **Botnets:** A network of infected machines instructed to forward harmful material to other computers connected to the internet. Botnets often consist of thousands of hijacked ('zombie') computers, which the user is unaware of.
- **Code Injection:** Code injection is the exploitation of a computer bug that is caused by processing invalid data. Code injection can be used by an attacker to introduce (or "inject") code into a computer programme to change the course of execution. Injection can result in data loss or corruption, lack of accountability, or denial of access. Injection can sometimes lead to complete host takeover.
- **Data Breaches:** A data breach is the intentional or unintentional release of secure information to an untrusted environment. It is a security incident in which sensitive, protected or confidential data is copied, transmitted, viewed, stolen or used by an individual unauthorised to do so.
- **Distributed Denial-of-Service (DDoS) and Denial of Service (DoS):** DDoS aims to flood a target with internet traffic, rendering the service or network unavailable to users. DDoS attacks often rely on human actors to maintain the pressure on the relevant system or network.
- **Email Viral Attachments:** Accessed directly by the user from a received e-mail, viral attachments copy themselves and automatically send themselves throughout the owner's address book. Malware installed by the user themselves is often referred to as a 'back-door' virus.
- **Logic Bomb:** Logic bombs are elements of code inserted into software in order to generate certain results when the code is triggered.
- **Drive-by Downloads:** Unintended download of computer software from the Internet: downloads which a person authorised but without understanding the

consequences or any download that happens without a person's knowledge, often a computer virus, spyware, malware, or crime ware.
- **Exploit Kits: Deliver a malicious payload to a victim's computer.** The kit incorporates tracking mechanisms so that people maintaining the kit know considerable information about the victims arriving at the kits landing page. The information tracked includes the victim country, operating system, browser and which piece of software on the victims computer was exploited.
- **Identity Theft:** Identity theft is a form of stealing someone's identity in which someone pretends to be someone else by assuming that person's identity, usually as a method to gain access to resources or obtain credit and other benefits in that person's name.
- **Information Leakage:** Information leakage is an application weakness where an application reveals sensitive data, such as technical details of the web application, environment, or user-specific data. Sensitive data may be used by an attacker to exploit the target web application, its hosting network, or its users.
- **Malware (worms, trojans):** 'Trojans' are malware which may appear legitimate but can compromise user security by either monitoring user activity, remote control, cyber espionage, or aiding the installation at additional malware. Computer worms are self-replicating malware that spread automatically throughout a computer network; worms may or may not carry payload to further affect the infected computer.
- **Man-in-the-middle attacks:** A man-in-the-middle attack is an attack where the attacker secretly relays and possibly alters the communication between two parties who believe they are directly communicating with each other.
- **Phishing:** Commonly connected to emails, phishing is the process of inducing users to reveal usernames and passwords by pretending to be harmless or official sources yet copying the data. Spear-phishing is an effort to make use at available user data to create 'personalised' bait far the user. Social networking sites are often key sources for this 'personalised' attack.
- **Physical Damage/Loss/Theft:** Millions of mobile phones are lost or stolen every year. A growing amount of lost and stolen phones have their content accessed by someone other than their owners. It also means that attackers may have physical access to the actual device hardware. This is a different threat model than for stationary hardware such as servers and workstations, where physical access is less likely.
- **Poor management (lack of code control mechanisms, lack of security expertise or investment, etc.) and human errors/insider threats:** An unintentional insider threat is a current or former employee, contractor, or business partner who has or had authorised access to an organisations network, system, or data and who, through action or inaction without malicious intent, causes harm or substantially increases the probability of future serious harm to the confidentiality, integrity, or availability of the organisation information or information systems.
- **Rootkit systems:** A rootkit is malware that is designed to conceal operating processes from detection. They are generally used in concert with other

'payload' carrying malware to hide the infection and prevent detection and removal of the primary malware.
- **Social Engineering:** Social Engineering is a human-based information gathering effort designed to obtain confidential information that can be employed for other cyber-attacks. While not relying on information systems directly for the 'attack,' social engineering remains an element within the cyber threat due to its role in gathering information for attacks.
- **Spam:** Electronic spamming is the use of electronic messaging systems to send unsolicited messages (spam), especially advertising, as well as sending messages repeatedly on the same site.
- **Spyware:** Software that collects information on a user's activities without their knowledge. This can often include an attack function designed to disrupt the user's computer activities in addition to information gathering.
- **Targeted Attacks:** Targeted threats are a class of malware destined for one specific organisation or industry. These threats are a type of crime ware of particular concern because they are designed to capture sensitive information. Targeted attacks may include threats delivered via SMTP e-mail, port attacks, zero day attack vulnerability exploits or phishing messages.
- **Zero-day vulnerabilities:** A zero-day attack is a computer threat that exposes undisclosed or unpatched computer application vulnerabilities. Zero-day attacks can be considered extremely dangerous because they take advantage of computer security holes for which no solution is currently available.

7 Inter-institutional Coordination

Within the European Commission, the main actor in the domain of network and information security is DG Connect which is responsible for managing the European Digital Single Market and the NIS Public Private Platform (aiming at implementing the measures set out in the NIS Directive and ensuring a harmonised application across the EU). DG HOME follows CC/CT issues, while other DGs (e.g. DG MOVE, DG MARE etc.) follow the security of cyberspace for their respective application areas. The European Union Agency for Network and Information Security (ENISA) is the primary European cybersecurity agency that was created in 2004 and which is responsible for supporting the European Commission, the Member States and the private sector in addressing, responding to and preventing cybersecurity threats. ENISA advises on legislative proposals, acts as a platform of exchanging information and best practices, and facilitates the Computer Emergency Response Team (CERTs) information exchange both within the EU and across the borders. The inter-institutional alignment of activities is crucial if the cPPP is to cover the full spectrum of cybersecurity issues and priorities at EU level, including specific components such as CC/CT and critical infrastructure protection. If coupled with a strong collaboration with ENISA, Member States, and industry, the inclusion of all needed elements for the successful implementation of the cPPP will be ensured, which will then in turn be reflected in future research work programmes.

8 Conclusion

The cPPP should improve progressively the competence of European industries in critical cybersecurity technologies by 2020, leading them to be among the main competitive global leaders by 2025. For this reason, the cPPP should not be limited to research issues or it will have a negligible impact on the effective and rapid growth of the EU Digital Single Market. Wider objectives, such as linking Horizon 2020 with other funding mechanisms, stimulating the growth of the EU cybersecurity industry and the increase of EU digital autonomy should be pursued. The effective development of a European cybersecurity industry will be attainable through the adoption of an approach linked to the high and fast growth of technological competence and competitiveness. Should Europe not sufficiently master critical technologies and implement validated trusted solutions all along the supply chain, there is a risk that solutions, coming from non-EU trusted providers, purchased on the basis of their economic convenience or for other reasons, could hinder the privacy of European customers and threaten the confidentiality of their data.

While cooperation with the main non-EU industries is needed, some Member States and companies believe that Europe can progressively develop more competence in cyberspace to both, on the one hand, recover market positions using trusted EU solutions, while on the other hand, better controlling the high level of privacy, data management, and the privacy and freedom of decision of EU citizens. This will require a strong political and economic commitment. The cPPP represents the first step towards strengthening the dialogue with the European supply sector for the creation of a major Flagship programme and to supporting a coordinated end-to-end approach. If swiftly implemented with the support of adequate investments to reach ambitious objectives, this would develop a sustainable Digital Single Market ecosystem in Europe, making it a real and global cybersecurity leader.

References

1. Communication from the Commission to the European Parliament, the Council, the European Economic and Social Committee and the Committee of the Regions: A Digital Single Market Strategy for Europe, COM(2015) 192 final. http://eur-lex.europa.eu/legal-content/EN/TXT/PDF/?uri=CELEX:52015DC0192&from=EN
2. European Organisation for Security. www.eos-eu.com
3. EOS Strategic Initiative on "Cybersecurity for a trusted EU Digital Single Market": extended public summary of an EOS Market Study for an EU Cybersecurity Flagship Programme, January 2016. http://www.eos-eu.com/files/Documents/FLAGSHIPS/CYBER/EOS%20study%20on%20a%20EU%20CYBERSECURITY%20FLAGSHIP%20extend%20summ%20Dec2015.pdf
4. Edward Snowden Revelations. https://edwardsnowden.com/revelations/
5. Rebuffi, L.: Towards a competitive European Digital Single Market. Eur. CIIP Newsl. **10**(1), 7–8 (2016)

6. EU Cybersecurity plan to protect open internet, online freedom, opportunity - Cyber Security strategy, Proposal for a Directive. https://ec.europa.eu/digital-single-market/en/news/eu-cybersecurity-plan-protect-open-internet-and-online-freedom-and-opportunity-cyber-security
7. DG CONNECT. https://ec.europa.eu/digital-single-market/en/dg-connect
8. Horizon 2020: The EU Framework Programme for Research and Innovation. https://ec.europa.eu/programmes/horizon2020/
9. Network and Information Security Directive: co-legislators agree on the first EU-wide legislation on cybersecurity. https://ec.europa.eu/digital-single-market/en/news/network-and-information-security-directive-co-legislators-agree-first-eu-wide-legislation
10. Elis, N.: Can Big Data prevent the next Cyber Attack? (2014). http://www.jpost.com/Enviro-Tech/Can-big-data-predict-the-next-cyber-attack-351957
11. US cybercrime: Rising risks, reduced readiness: Key findings from the 2014 US State of Cybercrime Survey. https://www.pwc.com/us/en/increasing-it-effectiveness/publications/assets/2014-us-state-of-cybercrime.pdf
12. Koops, B.J.: The internet and its opportunities for cybercrime. In: Herzog-Evans, M. (ed.) Transnational Criminology Manual, vol. 1, pp. 735–754. WLP (2010)
13. 2001 European Convention on Cybercrime from the Council of Europe (Budapest Convention). http://www.coe.int/en/web/conventions/full-list/-/conventions/rms/0900001680081561
14. Commonwealth of Independent States Agreement on Cooperation on Combating Offences related to Computer Information of 2001 (CIS Agreement). http://www.nti.org/learn/treaties-and-regimes/commonwealth-independent-states-cis/
15. Arab Convention on Combating Information Technology Offences from (2010). https://cms.unov.org/DocumentRepositoryIndexer/GetDocInOriginalFormat.drsx?DocID=3dbe778b-7b3a-4af0-95ce-a8bbd1ecd6dd
16. Shanghai Cooperation Organization Agreement of Cooperation in the Field of International Information Security of 2010 (Shanghai Agreement). http://www.smallarmssurvey.org/?id=977
17. Convention, Draft African Union on the Establishment of a Legal Framework Conductive to Cybersecurity in Africa of 2012 (draft African Union Convention). https://ccdcoe.org/sites/default/files/documents/AU-120901-DraftCSConvention.pdf
18. Podgor, E.S.: International computer fraud: a paradigm for limiting national jurisdiction. UC Davis Law Rev. **35**, 267–317 (2002)
19. UNODC: Comprehensive study on Cybercrime (2013). https://www.unodc.org/documents/organized-crime/UNODC_CCPCJ_EG.4_2013/CYBERCRIME_STUDY_210213.pdf
20. UNODC: Comprehensive study on Cybercrime, p. 41 (2013). https://www.unodc.org/documents/organized-crime/UNODC_CCPCJ_EG.4_2013/CYBERCRIME_STUDY_210213.pdf
21. European Convention on Cybercrime (2001). http://conventions.coe.int/Treaty/en/Treaties/html/185.htm
22. Risk and Responsibility in a Hyperconnected World, World Economic Forum. http://www3.weforum.org/docs/WEF_IT_PathwaysToGlobalCyberResilience_Report_2012.pdf
23. Center for Strategic and International Studies (CSIS). http://csis.org/
24. Net Losses: Estimating the Global Cost of Cybercrime: Economic impact of cybercrime I. http://www.mcafee.com/us/resources/reports/rp-economic-impact-cybercrime2-summary.pdf

25. Report: Cybercrime and espionage costs $445 billion annually. http://www.washingtonpost.com/world/national-security/report-cybercrime-and-espionage-costs-445-billion-annually/2014/06/08/8995291c-ecce-11e3-9f5c-9075d5508f0a_story.html
26. Conway, M.: Cyberterrorism: The Story So Far (2003). http://doras.dcu.ie/496/1/info_warfare_2_2_2003.pdf
27. Brenner, S.: At light speed: attribution and response to cyber-crime/terrorism/warfare. J. Crim. Law Criminol. **97**(2), 379–476 (2007)
28. US Department of Justice, FBI Law Enforcement Bulletin: Cyber Terror. http://leb.fbi.gov/2011/november/leb-november-2011
29. CAPITAL the Cybersecurity Research Agenda for Privacy and Technology Challenges. www.capital-agenda.eu
30. European Union Agency for Network, Information Security. https://www.enisa.europa.eu/
31. ENISA Threat Landscape, p. 59 (2015). https://www.enisa.europa.eu/activities/risk-management/evolving-threat-environment/enisa-threat-landscape/etl2015
32. ENISA Threat Landscape 2014: Overview of current and emerging cyberthreats, December 2014. https://www.enisa.europa.eu/activities/risk-management/evolving-threat-environment/enisa-threat-landscape/enisa-threat-landscape-2014

Are We Doing All the Right Things to Counter Cybercrime?

Michal Choraś[1,2(✉)], Rafal Kozik[1,2], Andrew Churchill[3], and Artsiom Yautsiukhin[4]

[1] ITTI Sp. z o.o., Poznań, Poland
{michal.choras,rafal.kozik}@itti.com.pl
[2] UTP University of Science and Technology, Bydgoszcz, Poland
{chorasm,rkozik}@utp.edu.pl
[3] CBRNE Ltd., London, UK
andrew.churchill@cbrneltd.com
[4] Consiglio Nazionale delle Ricerche, Pisa, Italy
artsiom.yautsiukhin@iit.cnr.it

Abstract. In this paper we present the discussion about the future ideas, needs and trends for cyber security technologies. Our focus is on the future technologies which should be developed in order to further enhance the protection of the cyberspace. Similarly to our work in the FP7 CAMINO project, we follow the comprehensive approach looking at broad range of possible technologies and problems. We termed our approach as THOR since we considered the following dimensions: Technical, Human, Organisational and Regulatory. In this paper we also discuss the idea of the comprehensive approach, since we believe only holistic view on cyber security can improve protection from the cyber threats.

Keywords: Cyber-security · Future technologies · Cyberspace · Technical · Human · Organisational · Regulatory

1 Introduction

FP7 CAMINO (Comprehensive Approach to cyber roadMap coordINation and develOpment) was collaborative project funded under The Seventh Framework Programme of the European Union (in Security theme). The project was coordinated by ITTI Sp. z o.o. and was composed of ten partners from eight countries, supported by further 22 organisations (so-called "Supporting Members"). The major goal of the CAMINO project was to provide a realistic roadmap for improving resilience against cybercrime and cyber terrorism. In other words the project answered the question where should taxpayer money be invested for research purposes. We indicated what research directions could tackle the problems and mitigate the gaps in countering cybercrime and cyber terrorism

B. Akhgar and B. Brewster (eds.), *Combatting Cybercrime and Cyberterrorism*,
Advanced Sciences and Technologies for Security Applications,
DOI 10.1007/978-3-319-38930-1_15

in a timescale up to 2025. The consortium used a holistic approach, analysing functions and capabilities addressing technical and human issues which are interrelated with legal and ethical aspects. On the human front, the project addressed a wide spectrum of players including technicians, end-users and their intermediaries, including administrators, policy makers and regulators. In parallel with looking at the human and technical aspects, the project was focused on strong involvement of various different groups and operators such as LEAs, CERTS, personal users, governments, industry and research and commercial organisations. In this paper we present the idea of the comprehensive approach, since we believe only holistic view on cyber security can improve protection from the cyber threats. Moreover, we discuss the CAMINO Roadmap (research agenda). The full Roadmap (80 pages) is available at: http://www.fp7-camino.eu/assets/files/Book-CAMINO_roadmap_250316.pdf. The paper is structured as follows: In Sect. 2 the idea and practical examples of the comprehensive approach are discussed. In Sect. 3 current cyber security threat landscape is presented. In Sect. 4 the CAMINO THOR approach is shortly over-viewed while the CAMINO roadmap with description of its topics is presented in Sect. 5, before final conclusions are presented in Sect. 6.

2 Current Needs and Challenges in Cyber Security: The Need for the Comprehensive Approach

While discussing, performing research (e.g. national and European projects like CAMINO) and commercial tasks (e.g. consulting, security policies, penetration tests) for cyber security, we postulated the comprehensive approach reflected in the following ideas/points [4,5]:

- Broad view on the subject: do not only focus on critical infrastructures (CI) or CNI (critical national infrastructures). Think about citizens and smaller institutions, especially SMEs upon which European economy is based.
- Broad view on the crime: Do not only focus on the computer crime like hacking and cyber attacks. Do not forget about the computer-related crime (see table below) with IPR violations, cyber stalking, child pornography etc. Beware that the cyber criminals are not all the time the skilled cyber hackers being able to remove traces (fingerprints) in the network. More often, those are cyber amateurs who can be easily tracked and found by specialized forensic officers the bigger problem is the scale and the belief that such criminals can be stopped, arrested and convicted.
- Broad view on the technical aspects: Do not focus on single methods, algorithms and tools for specific tasks. Do implement both online and offline aspects of technical cyber security (see Fig. 1). It happens quite often the researchers focus on the particular solutions they work on and claim it will save the world... however, it will not. One has to assure offline aspects and procedures (risk management, vulnerability management, understanding the context etc.) as well as the online aspects (monitoring, analysis, detection,

Table 1. Cybercrime differentiation

Crimes	
Affecting computers (IT as a target)	Using computers (IT as a tool)
DoS attacks	Posting abusive and untrue content
Developing and distribution of viruses	Cyberstalking
Unauthorized access from a remote machine, e.g. guessing password	Crimes affecting copyrights (piracy)
Unauthorized access to local superuser (root) privileges e.g., various "buffer overflow" attacks	Sexual child abuse
Probing: surveillance and other probing, e.g., port scanning	Illegal trades
	Scam and financial frauds

reaction and remediation). With the aspects of analysis and detection, one should also implement both signature-based and anomaly-based approaches to detect intrusions in the computer networks and systems.

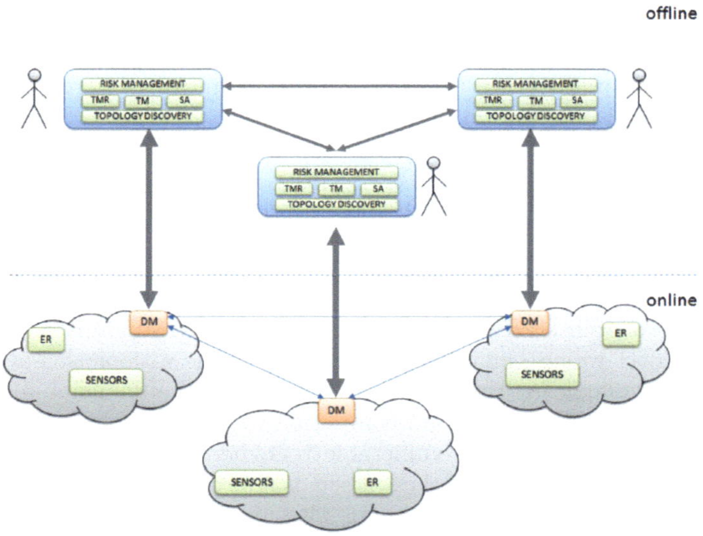

Fig. 1. Online and offline security convergence [4].

– Broad view on the needed investments: do not plan budgets and spend money only on technical solutions and tools, technical consulting etc. There is a need for other efforts such as increasing awareness of the users (internal and external) and training (should start in schools). It is claimed very often that users (humans) are the weakest link in security and currently behavioral attacks are a major threat (e.g. phishing or even Stuxnet are good examples). Therefore, investments in training of societies and users should increase.

While preparing the research agenda (CAMINO) roadmap offered to the EC as the result of project CAMINO, we tried to take into account the above ideas as well as other suggestions by cyber security experts (e.g. raised at our workshops). The roadmap items (topics) are presented in Sect. 5 (Table 1).

3 Cybercrime and Cyber Terrorism - Threat Landscape

Current road mapping initiatives with the identification of main research gaps and challenges were the subject of analyses performed in the first months of the CAMINO project. The current section is focused on the presentation of main conclusions formulated in as results of the CAMINO analyses to summarise key areas, technologies and threats impacting cybercrime and cyber terrorism nowadays.

Firstly, in we analysed a number of cyber security roadmaps (also sector-specific ones), current and completed R&D projects and international strategies [1–3, 7–15]. The common aspects discussed in these documents and analysed in various projects are:

- Evaluation of system security,
- Improvements of analytical tools for security monitoring,
- Security-related information sharing mechanisms,
- Increasing of the security awareness,
- Standardisation in the field of cyber security,
- Application of Security/Privacy-by-design principles,
- Identity management,
- Critical Infrastructure Protection.

These topics were our starting point for defining the CAMINO Roadmap scope.

In the project we also analysed risks related to the various classes of assets. As a result, we diagnosed that payment systems (in the financial and banking domain), embedded systems, cloud computing services and systems processing personal data are particularly vulnerable to cybercrime and cyber terrorism threats. Therefore, protection of these assets is addressed within particular parts (topics and objectives) of this Roadmap. Also, means to reduce risks connected to these assets are reflected by the milestones defined in the proposed research agenda timeline.

The study of state of the art cyber security technologies allowed us to identify several key areas that due to their emerging status and maturity level should be specifically addressed by the Roadmap. These are:

- Cyber fraud prevention technologies,
- Denial of Service (DoS)/Distributed Denial of Service (DDoS) Protection,

- Internet of Things (IoT) Security,
- Intrusion Detection Systems,
- Advanced Persistent Threat (APT) Detection,
- Cloud Forensics,
- Cryptography,
- Technical Security Standards,
- Big Data Security Analytics,
- Countering ransomware,
- Cloud Security.

Finally, we performed a number of surveys and face-to-face interviews with experts from different sectors related to cyber security and the fight against cybercrime and cyber terrorism.

4 Comprehensive CAMINO Approach and Roadmap

In this section we present the Roadmap topics divided into four THOR dimensions.

The THOR approach comes from the comprehensive view mentioned in Sect. 2. We are sure that only the comprehensive approach to cyber security can improve European cyber space and its cyber resilience.

Our approach for the CAMINO roadmap development is based on the THOR concept. THOR dimensions are the foundation of the CAMINO roadmap scope and structure. THOR dimensions address the following aspects:

- (T)echnical related to technology, concrete technological approaches and solutions that can be used to fight against cybercrime and cyber terrorism,
- (H)uman related to human factors, behavioral aspects, privacy issues, as well as raising awareness and knowledge of society with regards to cybercrime and terrorism threats,
- (O)rganisational related to processes, procedures and policies within organisations, as well as cooperation (public-private, public-public) between organisations,
- (R)egulatory related to law provisioning, standardisation and forensics.

Each topic addressed in our roadmap corresponds to the particular subsections in Sect. 5. In the original roadmap, particular topics are presented in a unified way, including:

- Summary of key research objectives related to a given topic.
- Summary of stakeholders with their roles and who should participate in the specific research subject.
- Detailed timeline for concrete milestones and specified for three different timespans (2017, 2020 and 2025). Such timelines briefly explain the current situation in a given topic and the expected (desired) end-vision at 2025, after the roadmap milestones achievement.

– Summary of research activities that should be performed leading to the defined milestones achieved.

Within the technical dimension, some of the proposed topics for future development and promoting are focused on big data and forensic aspects, improvement of authentication/authorisation mechanisms, security engineering and testing capabilities, as well as on means to effective fight against malwares, botnets and APTs (Advanced Persistent Threats). The human dimension emphasises need for mechanisms regulating use and reuse of personal data and for training and raising cyber security awareness. Topics from the organisational part are focused on societal and cultural aspects of cyber security, on adaptation of the organisations in the light of international nature of cybercrime and cyber terrorism, as well as on cooperation between organisations (e.g. SMEs) and supporting EU institutions (Fig. 2).

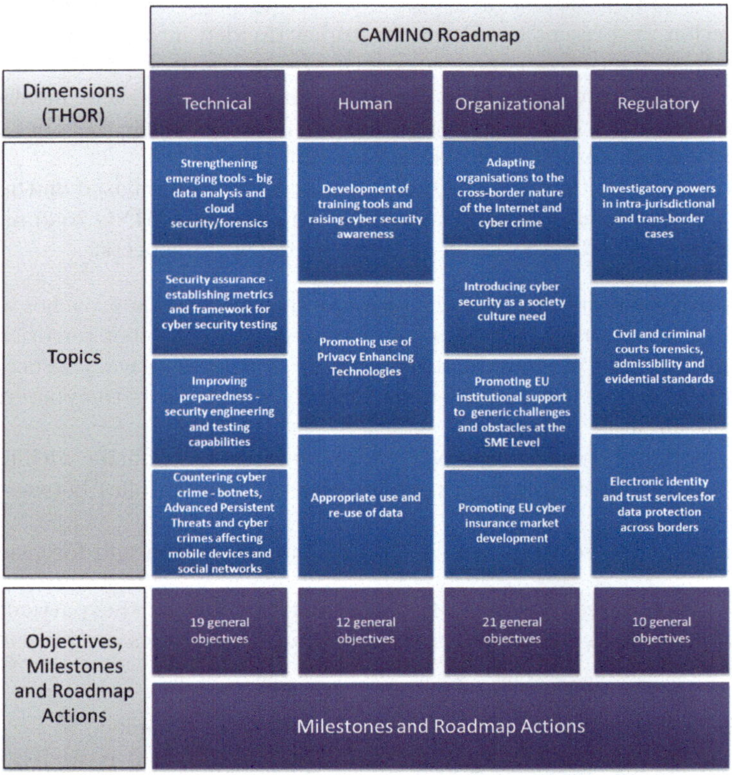

Fig. 2. CAMINO roadmap structure.

Finally, the regulatory dimension is composed of the following topics: investigatory powers aspects, interoperability of Common and Roman code law, forensics and evidential standards, as well as standards for data protection across borders.

5 Roadmap Topics Description

5.1 Technical Activities Overview

Strengthening Emerging Tools - Big Data Analysis and Cloud Security/Forensics. Cyber attacks may not be visible on a small scale due to their nature or intensity (e.g. amount of traffic they introduce). Therefore, recently the techniques for using big data tools are being adapted. The recent research shows that deep analysis of large volumes of data (received from different segments of IT networks) has a unique capability of revealing interesting patterns. This concept is recently adapted to many cyber security areas, namely: spam detection, botnets detection, malwares analysis, web-based infection, network intrusion detection systems.

Security Assurance - Establishing Metrics and Framework for Cyber Security Testing. The IT world is becoming more dynamic, distributed and heterogeneous. This evolution implies novel security challenges, especially for security assurance. New methods for authentication, authorisation and trust management must deal with lack of pre-defined trust assignments and be ready to establish new relations with immediate effect. Moreover, establishing such relations requires reliable knowledge about previously unknown parties. This observation is also applied to security, in order to ensure the clients that outsourced business will not be compromised, even when it is under control of partners. In order to achieve this, information about incidents should be shared. The shared information can be used to get the correct assessment of security within an organisation, issue an insurance policy and strengthen the security of the Internet as a whole.

Improving Preparedness - Security Engineering and Testing Capabilities. One of the most important and demanding aspects in every product, system or organisation is quality; guaranteeing fundamental characteristics such as reliability or availability in any system, moreover if it is a security one, it is an essential part of revealing the confidence of the development team in their system and/or product. Therefore, activities focused on maintaining and improving this quality are needed and the most effective ones are testing and simulation processes. Concepts such as automated tools or cyber exercises between companies will help to raise the awareness of not only people responsible for cyber security but also of the rest of the staff. And finally, in order to promote and encourage the realisation of these necessary actions, proper regulations and standards should be written and discussed, thereby achieving a desirable and prepared environment to benefit all these good practices.

Countering Cybercrime - Botnets, Advanced Persistent Threats and Cybercrimes Affecting Mobile Devices and Social Networks. Nowadays, one of the main challenges affecting the fight against cybercrime is considerable with an increasing amount of evolving malware samples. Evolution and changeability of malwares and botnets (e.g. new, fast-evolving botnet architectures) are also factors that should be addressed by the research communities to more effectively fight against cybercrime. This is particularly important in the context of limitations of existing signature-based scanners and malware detectors. On the other hand, cybercrime also affects mobile devices and in the near future will affect micro devices (now not often connected to the Internet), that will be exposed to cyber attacks in conjunction with growing popularity of the IoT (Internet of Things) concept.

5.2 Human Activities Overview

Development of Training Tools and Raising Cyber Security Awareness. One of the most fundamental aspects of improving society's defences against cybercrime, as with protecting against any other new and evolving threat, is to ensure that users and those involved are properly kept abreast of the nature of the threat and the underlying rationale of the defensive steps being taken to mitigate it.

Whilst almost all new legislative changes are accompanied by training and situational awareness as part of their lifecycle, few technological changes sufficiently incorporate this vital feature into their own roadmaps. This is true both of the new possibilities opened up through greater online access to data, but also to the tools being rolled out to support the intended security behind them.

Promoting Use of Privacy Enhancing Technologies. With surveillance powers and techniques a very current topic, both from perceived excessive use in some quarters and inadequate interpretation of available evidence in others, the roadmap towards more effective implementation of Privacy Enhancing Technologies is inexorably entwined with the development of forthcoming legislation and its regulatory interpretation [6].

In particular, DPR, eIDAS, and Payment Services Directive 2's early adoption through SecuRe Pay, introduces requirements for the adoption of PETs, albeit through the adoption of undetermined techniques or technologies and in advance of their formal ratification into EU or Member State legislation. These advance regulatory roadmaps provide an interesting and often unexpected set of requirements to the organisations handling sensitive personal data. Regulatory requirements to assist consumers in remaining anonymous, for example with merchants online must also be seen in the light of requirements passed under the 4th Anti-Money Laundering Directive, which entered into force on the 26th June 2015, and which Member States have two years to enact.

Appropriate Use and Re-use of Data. Under a range of current regulations and industry standards, across a wide and varied range of industries, the use of data is frequently, but not universally, restricted to the use originally intended when data was collected. Users also face a range of opt-ins or opt-outs for the use, or subsequent re-use, of this data. The advent of big data has made the search for new uses of data held on existing systems a growth industry (see under "Technical" above), but there are strong Human and Ethical concerns raised through this re-use. The application of these existing data sets for LEA purposes has caused some debate, and our Roadmap will provide pointers to those issues that need to be addressed and to what timescale.

5.3 Organisational Activities Overview

Adapting Organisations to the Cross-Border Nature of the Internet and Cybercrime. Nowadays, competitiveness is global, so any company may receive an attack from anywhere on the planet. Therefore, most importantly, regulatory differences between countries should be understood and organisations should be aware of this fact and accordingly protect their assets and intellectual property. Therefore, organisations need to think "cross-border" regarding cybercrime and protect their networks globally.

Introducing Cyber Security as a Society Culture Need. The use of new technologies is now not only present in the office, at home and at professional level but also during free time for children and adults and also to interact with the public sector, with banks, supermarkets and online stores. Moreover, these different functions overlap and initiatives such as BYOD are becoming more popular every year, mixing personal with professional activities. Therefore, cyber security is now crucial in terms of securing all aspects of day-to-day functions and should be introduced as a new culture capability.

Promoting EU Institutional Support to Generic Challenges and Obstacles at the SME Level. A common and unified institutional support is needed to promote changes at enterprise, company and SME level. The creation of an expert committee at the request of interested countries could contribute to overcoming these obstacles and challenges at a European level. In addition, an information sharing platform would support the approach and collaboration of interested parties prompting easier sharing of efficient ideas and problems. This support will assure the minimum protection needed in these matters.

Promoting EU Cyber Insurance Market Development. It is widely accepted that achieving perfect security is impossible. Security incidents and data breaches will occur regardless of the security controls and practices applied (though with much lower frequency). Thus, organisations have to deal with the residual risk. Recently, insurance, a common approach for residual risk, was applied to the cyber world. The developing cyber insurance market faces a number of unique challenges such as "heavy information" asymmetry, lack of statistical data, interconnected security and correlated risks, rapid change of risk landscape and un-clear underwriting language etc.

The market in the USA is becoming increasingly mature with \$2,75 billion in premiums for 2015 [16,18] whereas the EU market is considerably less at \$150 million for 2014 albeit increasing at the rate of 50 % to 100 % per annum [17,18]. There are a number of steps which can be taken in order to help the EU market to flourish. The enforcement of a data breach notification law (which has currently passed the first reading in the European Parliament) will boost the EU cyber insurance market as the 2003 California bill did in the USA. Furthermore, information sharing on incidents, their consequences and prerequisites will help insurers get reliable statistical evidence. More advanced economical and regulatory models, together with technological advancements, will help reduce the effect of risk correlation. Last, but not least, scientific studies are required to assess possible behavior within the market place and identify incentives for individual organisations to increase their security level as well as the overall social benefit.

5.4 Regulatory Activities Overview

Investigatory Powers in Intra-jurisdictional and Trans-border Cases. Steps must be taken to instigate adequate investigatory powers as well as their use by LEA's members regarding cyber-enquiries. The pace of regulatory reforms, the balance between abstraction and establishment of investigatory powers and the need for a training policy need to be taken into consideration. The effectiveness of international cooperation in transborder cases, paramount to successfully prosecuting cybercrime, may be augmented in years to come if the EU takes advantage of the shift in the views on reciprocity issues by key players such as China. Then again, improved data exchange between EU and National LEA's comes not without risk to Fundamental Rights, one of the keystones of European culture. Efforts must be made in order to find a regulatory and technical framework allowing the juggling of augmented data exchange capabilities and respect of Fundamental Rights.

The regulatory driver towards greater levels of security in the face of cybercrime, such as promotion of more secure end-to-end encryption services and more advanced malware analytics, is being actively promoted by emerging EU Regulations.

One notable example, both of the regulatory move towards more coherent policy, but equally of some of the pitfalls faced, is the general Data Protection Regulation (DPR). The headline grabbing threats of penalties of up to 4 % of global turnover combined with a first attempt at global enforcement have caused a great deal of concentration of minds on the need for organisations to protect their customer's Personally Identifiable Information (PII), the contents of which are of interest to cyber-criminals such as log in details and payment card information.

The DPR replaces the wide range of somewhat divergent nationally transposed Data Protection Act implementations stemming from the pre-existing

Data Protection Directive. These national interpretations have led to a sometimes confusing array of data standards being applied across the 28 member states, with occasional attempts by the Commission to bring individual interpretations into line.

In principle these differences should be removed through harmonising Regulations rather than Directives. However, the permission of 'exemptions' from the DPR in cases of national strategic interest is a potential source for ongoing differences in treatment across the European Union. One exemption that has the potential to cause ongoing confusion, and indeed friction, between Member States is in the lengths to which Member States are permitted to 'infringe' on the Data Protection rights of their citizens/subjects.

Interception powers and prohibitions vary from Member State to Member State, ranging across the spectrum of the privacy versus national security debate. The DPR's drafting clearly sits more at ease with the view that privacy is paramount. However, the Member State with the highest adoption of e-commerce, and hence key area of attraction from cybercriminal perspective, is the United Kingdom, where the Investigatory Powers Bill is in advanced stages of development. Whilst not commenting on the technical nature of such requirements, the parliamentary committee has agreed with the government's intention to seek access to protected communications and data when required if supported by a warrant, but not in requiring encryption keys to be compromised or system backdoors to be implemented. These 'lawful intercept' requirements, common in many nations, yet anathema to others, could become the source of tensions in the managing of 'exemptions' under DPR where cybercrime is noted as being required as an exemption in the national interest.

Whilst clearly not within the European Union, current news from the US involving the three major global operating systems, most notably Apple in the San Bernadino' terrorism case, suggest this issue could have wider global ramifications. Tim Cooks recent announcement that Apple will refuse to comply with the DoJ ruling, and indeed choose to modify subsequent iterations of iOS so that Apple could not comply (with similar such rulings, and interestingly some EU Member State laws) has highlighted the ongoing public policy debate over whether the rights of the State to Lawful intercept or the individuals' right to privacy should have primacy.

As noted in more detail below under data sharing across borders, the EU-US Privacy Shield seeks to address these concerns. Under the Privacy Shield U.S authorities affirm absence of indiscriminate or mass surveillance', though this clearly does not equate to a requirement to restrict targeted interception or surveillance, as envisaged in the San Bernadino case in the US, or within the UK's Investigatory Powers Bill, which proceeded to further reading on 15th March 2016 and is expected to enter force in summer 2016.

Civil and Criminal Courts Forensics, Admissibility and Evidential Standards. At present, there exists a wide variety of standards and best practices for information security and digital evidence gathering, amongst which the following ones can be emphasised:

- "Cobit, Framework for IT Governance and Control", Information Systems Audit and Control Association, ISACA
- "ISO/IEC 27002:2005. Information technology. Security techniques. Code of practice for information security management", International Organization for Standardization (ISO) International Electro Technical Commission (IEC)
- "Forensics sound techniques in the collection and analysis of digital and multimedia evidence", Scientific Working Group Electronic Evidence
- "NIST Special Publication 800-61. Computer Security Incident Handling Guide", "NIST Special Publication 800-86. Guide to Integrating Forensic Techniques into Incident Response" and others, United States National Institute of Standards and Technology (NIST)
- "Searching and Seizing Computers and Obtaining Electronic Evidence in Criminal Investigations", Computer Crime and Intellectual Property Section, Criminal Division, United States Department of Justice
- "Forensic Examination of Digital Evidence: A Guide for Law Enforcement", Office of Justice Programs, National Institute of Justice United States Department of Justice
- "BS 10008:2008. Evidential weight legal admissibility electronic information", British Standard Institution (BSI).

This variety hinders the adoption of common standards and procedures for a strong foundation of cooperation and an effective fight against cybercrime and cyber terrorism at Pan-European level. This type of crime is particularly decentralised and not restricted to any frontier. The admissibility of digital evidence in Courts is still sometimes dependent on case-by-case analysis by experts who lack a common reference framework. Thus, the challenge is to achieve a common understanding by adapting current Member States criminal procedures. The achievement of a European Forensic Science Area has become a priority for the European Union. Last but not least, the respect for fundamental rights and freedoms of citizens must always be maintained as a basic and key principle.

Electronic Identity and Trust Services for Data Protection Across Borders. A majority of classes and applications of cybercrime and cyber terrorism contain a misrepresentation of identity or attempt to authenticate for access to goods or services to which the attacker has no legitimate use. There currently exist a plethora of standards to identify and authenticate a genuine user as to who he or she claims to be and their access rights in the given circumstances. At present there is no interoperability with poor controls over the degree as to what constitutes 'strong authentication' sufficient for each application. However, within the European Union, the eIDentity, Authentication & Signatures Regulation, launched in October 2014 seeks to address these issues. Our CAMINO Roadmap will take account the timetable for its implementation and the external steps necessary to ensure international promotion.

Equally, with the payments industry now being required to look at early adoption of the Second Payment Services Directive (PSD2), the Identity/Authentication roadmap has moved forward dramatically for one of the key cybercrime asset classes and one of the most likely candidates for higher level eIDAS requirements.

The European Central Bank and European Banking Association's announcement on 19th December 2014 that Secure Retail Payment (SecuRe Pay) Strong Authentication requirements would be put in place from 1st August 2015, several years in advance of PSD2's expected ratification, let alone mandated implementation, was thought to show how quickly cybercrime and the standards to address it move. Yet by the final ratification in October 2015, just two months later, the SecuRePay minimum requirements for multi-factor authentication had been augmented with an additional requirement for dynamic linkages between the payer, payee, and transaction, a major additional security step to further secure against man-in-the browser attacks.

Standards development is underway in both levels of assurance for eIDAS classification, and, whilst member states start transposition of strong authentication into national legislation the European Banking Authority has carried out (to February 2016) a Request for Information on cyber security standards, expected to lead to a formal consultation during the summer.

Meanwhile the striking down by the European Court of Justice in October 2015 of the Safe Harbour arrangements with the United States, where the storage of EU citizen's data in the US, or access of such data by the US, was deemed compatible with EU requirements has led to a re-examination of the standards of trust in data sharing across borders. Following the Schrems case a re-evaluation of transatlantic data sharing was initiated, with the potential threat (and in some jurisdictions probably still threatened) that companies using US based servers or services were in strict breach of the ECJ ruling on minimum data protection requirements.

6 Conclusions

In this paper we presented the cyber security research agenda (the CAMINO roadmap) specifying our suggestions related to the future efforts in fighting against cybercrime and cyber terrorism. The roadmap is focused on four key pillars of cyber security research, presenting the main objectives, problems, challenges and associated stakeholders from each dimension: Technical, Human, Organisational and Regulatory. These four dimensions constitute the CAMINO THOR approach that is basis for this roadmap, as well as for other research activities performed during the whole project. The ideas beyond this comprehensive approach are also presented and discussed.

We have presented the cyber security research agenda (the CAMINO roadmap) specifying our suggestions relating to future efforts in fighting against cybercrime and cyber terrorism.

Each of the four THOR dimensions described in the roadmap follow the same structure. Firstly, the top priority areas (topics) in the THOR dimensions have been defined. In summary, there are 14 key topics in the CAMINO roadmap. Topics from the Technical Dimension are focused on big data and forensic aspects, improvement for authentication and authorisation mechanisms, security engineering and testing capabilities, as well as means for an effective fight against malware, botnets and APTs (Advanced Persistent Threats). The Human Dimension emphasises the need for mechanisms regulating the use and reuse of personal data and training and raising cyber security awareness.

Topics from the Organisational Dimension part of the roadmap are focused on societal and cultural aspects of cyber security, on adaptation of the organisations in light of the international nature of cybercrime and cyber terrorism, as well as on co-operation between organisations (e.g. SMEs) and supporting EU institutions. The development of the cyber insurance market is also one of topics in the Organisational Dimension.

Finally, the Regulatory Dimension is composed of aspects of investigatory powers, forensics and standards of evidence and data protection across borders.

For each topic, the roadmap specifies a number of objectives with assigned milestones and actions to achieve those milestones. In total, the Project CAMINO has identified over 60 objectives and over 250 milestones considered as micro-steps in our research agenda, leading to a more effective fight against cybercrime and cyber terrorism up to 2025.

The policy of Project CAMINO was to ensure wide consensus and agreement on the CAMINO roadmap encapsulating suggestions from relevant experts and stake-holder groups. The CAMINO Roadmap has been validated with feedback from experts as part of the evolution of the research agenda. In addition, CAMINO topics incorporated into the CAMINO-COURAGE-CyberROAD joint roadmap were assessed as the top priority research agenda points. In particular challenges relating to big data analysis, cloud forensics and to the cross-border nature of the use of the internet by cybercrime and cyber terrorism were evaluated as the most important and urgent problems to be solved. In addition, we spent effort to avoid the situation where various means to counter cyber threats might be seen as individual silos or islands rather than a coordinated and joined up approach where all parties talk with each other. Therefore, some top level ideas include:

- Effective solutions, procedures and regulations for LEAs (e.g. what types of cybercrime should be investigated and by whom, what means/techniques are allowed and could be used as evidence in the courts etc.).
- Effective solutions, procedures and regulations for prosecution we need well trained prosecutors and well defined procedures for collecting and evaluating evidence.
- Effective solutions, procedures and regulations for courts and judges to clearly state what types of evidence can be admitted by courts to avoid the situation where the courts do not understand the cases.
- Effective solutions, procedures and regulations for transborder cooperation and information sharing.

These are just some examples of the required improvements and actions to form an effective and comprehensive system. In particular, such a system should address the current needs and challenges that facilitate requirements for improvements to legal systems and related processes that impact upon all phases of cybercrime cases. One of the main efforts to be done is the improvement of digital forensic products, services and procedures. It is important to ensure an adequate flow of information at different stages of any investigation - from disclosure of crime, securing and preserving evidence and its processing, up to the judicial decision. In this context, it is also important to ensure and develop appropriate levels of knowledge and expertise across all the actors involved in the judicial process. Major improvement in information sharing and cooperation between victims, LEAs (the Police), the prosecution and forensic experts and finally the judges and courts is needed.

The CAMINO comprehensive roadmap can be now used by national funding agencies, by the EC to structure future calls by ENISA, EDA, etc. It can and is also used by national bodies working on national doctrines and strategies (such as Ministry of Digitalization and National Security Bureau in Poland). The suggested research items are also targeted at the cyber PPP board in order to help structure the future cPPP initiatives.

The important feature of our approach is the comprehensiveness of the roadmap, since we believe that only holistic solutions can really help counter cybercrime and cyber terrorism.

Acknowledgement. The research leading to these results has received funding from the European Union Seventh Framework Programme (FP7-SEC-2013) as the CAMINO project under grant agreement no 607406.

References

1. Batz, D., et al.: Roadmap to achieve energy delivery systems cybersecurity, Technical report, Department of Homeland Security, Cyber Security R&D Center (2011)
2. Berenson, J.: The Roadmap to Secure Control System in the Transportation Sector, The Roadmap to Secure Control Systems in the Transportation Sector Working Group (2012)
3. Bhuyan, M.H., Bhattacharyya, D.K., Kalita, J.K.: Network anomaly detection: methods, systems and tools. IEEE Commun. Surv. Tutorials **16**(1), 303–336 (2014)
4. Choraś, M.: Comprehensive approach to information sharing for increased network security and survivability. Cybern. Syst. **44**(6–7), 550–568 (2013)
5. Choraś, M., et al.: Comprehensive approach to increase cyber security and resilience. In: Proceedings of ARES (International Conference on Availability, Reliability and Security), Touluse, pp. 686–692 (2015)
6. Choraś, M., Kozik, R., Renk, R., Holubowicz, W.: A practical framework and guidelines to enhance cyber security and privacy. In: Herrero, A., Baruque, B., Sedano, J., Quintan, H., Corchado, E. (eds.) International Joint Conference CISIS 2015 and ICEUTE 2015. AISC, vol. 369, pp. 485–496. Springer, Heidelberg (2015)

7. Eisenhauer, J., Donnelly, P., Ellis, M., OBrien, M.: Roadmap to Secure Control Systems in the Energy Sector, U.S. Department of Energy, U.S. Department of Homeland Security (2006)
8. EU NIS Platform Working Group 3, Secure ICT Research Landscape Deliverable. https://resilience.enisa.europa.eu/nis-platform/shared-documents/wg3-documents/state-of-the-art-of-the-secure-ict-landscape/at_download/file
9. European Union Agency for Network and Information Security (ENISA), National Cyber Security Strategies in the World. http://www.enisa.europa.eu/activities/Resilience-and-CIIP/national-cyber-security-strategies-ncsss/national-cyber-security-strategies-in-the-world
10. Johnson, S., Larson, B., Edwards, D., Morley, K.: Roadmap to Secure Control Systems in the Water Sector, Water Sector Coordinating Council Cyber Security Working Group (WSCCCWG) (2008)
11. Markatos, E., Balzarotti, D.: A Roadmap for Systems Security Research, SysSec, FP7 NoE Project
12. National Institute of Standards and Technology (NIST) 2014, NIST Roadmap for Improving Critical Infrastructure Cybersecurity
13. Pederson, P., Roxey, T., Gray, J.: Cross-sector Roadmap for Cybersecurity of Control Systems. Industrial Control Systems Joint Working Group (ICSJWG) (2011)
14. U.S. Department of Homeland Security 2009, A Roadmap for Cybersecurity Research
15. U.S. Department of Homeland Security 2010, Dams Sector Roadmap to Secure Control Systems
16. Betterley, R.S.: The Betterley Report: Cyber/privacy insurance market survey (2014). http://betterley.com/samples/cpims14_nt.pdf. Accessed 22 Apr 2016
17. Jones, S.: Lloyds CEO Sees Cyber Insurance to Surge After Attacks, Bloomberg Business (2014). http://www.bloomberg.com/news/articles/2014-10-08/lloyd-s-ceo-sees-cyber-insurance-to-surge-after-attacks. Accessed 22 Apr 2016
18. Marotta, A., Martinelli, F., Nanni, S., Yautsiukhin, A.: A Survey on Cyber-Insurance, Consiglio Nazionale delle Ricerche, IIT TR-17/2015. Technical report (2015)

Consolidated Taxonomy and Research Roadmap for Cybercrime and Cyberterrorism

Babak Akhgar[1(✉)], Michał Choraś[2,3], Ben Brewster[1], Francesca Bosco[4], Elise Vermeersch[4], Vittoria Luda[4], Damian Puchalski[3], and Douglas Wells[1]

[1] CENTRIC (Centre of Excellence in Terrorism, Resilience, Intelligence and Organised Crime Research), Sheffield Hallam University, Sheffield, UK
B.Akhgar@shu.ac.uk
[2] University of Science and Technology, UTP Bydgoszcz, Bydgoszcz, Poland
[3] ITTI Sp. z o.o., Poznań, Poland
[4] UNICRI, United Nations Interregional Crime and Justice Research Institute, Turin, Italy

Abstract. In this concluding chapter, we consolidate the broad spectrum of challenges discussed throughout this book towards the formulation of a number key priority topics to be addressed by future research related to cybercrime and cyberterrorism. During this process many of the specific areas that need to be addressed are defined across four interlinked dimensions; technological, regulatory, organisational and human. In the process of identifying the nature of the challenges posed, the scope of the research and initiatives needed in order to progress measures targeting them, as well as the required impacts needed in order to ensure the significance of those initiatives. Initial sections of the chapter recapture, from a definitional perspective, the definitions of cybercrime and its constituent elements towards establishing a harmonised taxonomy of terms that we can use to inform the future work being proposed.

Keywords: Cybercrime · Cyberterrorism · Cybersecurity · Roadmap · Research · Taxonomy · Classification

1 Introduction

In this concluding chapter, we consolidate and take influence from the themes and ideas being presented throughout the volume, alongside previous work and existing research, in order to build and present what have been identified as the highest priority challenges for future research and practice to address. Throughout this volume we have focused on a number of key themes, with some widening our appreciation of approaches focused on increasing our knowledge of the Cybercrime (CC) and Cyberterrorism (CT) field holistically, whilst others have in depth, discussed specific areas such as issues associated with public-private partnerships, attack attribution and malware detection, to mention but a few.

B. Akhgar and B. Brewster (eds.), *Combatting Cybercrime and Cyberterrorism*, Advanced Sciences and Technologies for Security Applications,
DOI 10.1007/978-3-319-38930-1_16

The respective impacts related to issues of discrimination, victims' rights, data protection, illegal content and national security are all unpacked in considerable detail in section 2 of the book, providing fresh insights towards contemporary cyber challenges for society, in addition to its ability to respond, prevent and maintain the security of its citizens. While later sections have covered topics highlighting technologies, scenarios and effective practices towards the development of a research and policy roadmap, discussing the potential ways in which these priority areas can be proactively addressed through future initiatives.

This chapter first provides an extended discussion of the domain of CC/CT, and the many facets that underpin our understanding of the field, towards establishing a conceptual taxonomy of terms that can be used as the basis of future work, and drawing upon established typology's and definitions. In later sections we present a discussion of the various approaches used to identify pertinent challenges and priorities, and the measures taken to distil the various inputs from these approaches. This process led to the eventual definition, and subsequent validation, of the specific subject areas themselves alongside an appreciation of the challenges that contributed to their identification as a research priority before highlighting the potential areas which the undertaking research and other initiatives in this domain could provide significant societal benefit.

2 Understanding Cybercrime and Cyberterrorism: Towards a Taxonomy

The central theme of this book lies in discussing the very nature and impact of CC, alongside the emerging threat of CT. Presently, there are many factors that contribute to the ways in which, CC and CT takes place, and, how we as societal stakeholders are addressing it. This varies, from prevention through to response and recovery. In the processes that border the roles and involvement of technology, policy, legislation and other factors. To effectively evaluate this array, it is necessary to unpack and clarify the terminology we use to discuss these issues, as our understanding of the techniques, typesets and characteristics of crime are pivotal to our efforts to educate, prevent and prosecute. Therefore, in this chapter we build upon the baseline definitions presented earlier in the book to present the elements of a conceptual taxonomy framework as a means to define and contextualise the field of CC and CT as a basis to aid our understanding of these issues in more detail, whilst making some initial steps towards harmonising this terminology so that it can be understood and applied holistically.

2.1 Existing Classifications

In this initial section we attempt to broadly categorise CC/CT based on specific crime characteristics. These include; content related offences, offences against the confidentiality, integrity and availability of computer systems and data, (and other types of characteristic) taxonomies. These categorisations draw upon existing definitions from the literature, both from academic, practitioner and legal contexts, to establish a baseline definition and theoretical understanding of CC and CT. Almost every introduction of new technical concepts (such as the switch to IPv6) and new services (such as social networks, Mobile/Smart devices, Mobile Phone Application such as NFC, Cloud Computing and more recently the notion of Big Data) holds an impact over the way crimes are committed and/or the ability to investigate/prevent incidents. CC and CT use of the Internet encompass broad areas of human-machine interaction within complex socio-technical systems.

2.2 Cybercrime

CC has been defined by the ITU[1] and the Budapest Convention on Cybercrime[2] as; crimes committed using information and infrastructure networks as a means of transmission, targeting real-world facilities (control rooms, critical infrastructures, etc.) as well as IT facilities (database, intellectual property, computers, SCADA systems etc.). Across a variety of targets, cyber criminals are deploying a comprehensive set of means, often also involving human behaviour as an additional, enabling facilitator (through for instance phishing, spam and other solicitation that in turn entices individuals to provide privileged data or entry points in networks). Criminality generally aims to exploit breaches in any system in order to make profit, or to use innovation technologies to increase their impact, like any business. Terrorists, although different in their nature than criminals, dispose of the same tools to threaten countries, organisations, infrastructures, and citizens. Therefore, fighting CC and CT poses an extraordinary challenge to public authorities, industries and research organisations, which must remain ahead of cyber criminals across a whole range of topics, including the interaction between different processes.

The growth and severity of cyber-attacks has increased the cost to society significantly. It is estimated that these attacks are costing the global economy billions of dollars each year[3]. As terminology has evolved, academic efforts have been undertaken to define the term "cybercrime". CC can be defined as a crime in which computer networks are the target or a substantial tool.[4] A modern approach is to recognise that CC is not necessarily a legal term of art, but rather an aggregate term for a collection of acts committed against or through the use of computer data or systems. Other approaches focus on offences against computer information, or the use of information resources for illegal purposes.

A variety of definitions for CC are defined throughout the literature, with differences mostly dependant on the purpose for which the definition is needed (e.g. focusing on the type of possible offences, or explaining the evolution of the crime, analysing the motivation of the offender).

---

1 http://www.itu.int/ITU-D/cyb/cybersecurity/docs/Cybercrime%20legislation%20EV6.pdf.

2 http://www.europarl.europa.eu/meetdocs/2014_2019/documents/libe/dv/7_conv_budapest_/7_conv_budapest_en.pdf.

3 N. ELIS (2014), *Can Big Data prevent the next Cyber Attack?*

4 B.-J. KOOPS (2010), *The Internet and its Opportunities for CC*, p. 737.

Referring to international and regional instruments in this matter, we have the European Convention against Cybercrime of 2001 (Budapest Convention); the Commonwealth of Independent States Agreement on Cooperation on Combating Offences related to Computer Information of 2001 (CIS Agreement); the Arab Convention on Combating Information Technology Offences of 2010; the Shanghai Cooperation Organization Agreement of Cooperation in the Field of International Information Security of 2010 (Shanghai Agreement); and the draft African Union Convention on the Establishment of a Legal Framework Conductive to Cybersecurity in Africa of 2012 (draft African Union Convention).

The CIS Agreement has the objective to regulate any 'offence relating to computer information' and 'CC' that could be described as a 'criminal act of which the target is computer information'[5]. The Shanghai Cooperation Organization Agreement describes the 'information offences' as 'the use of information resources and (or) the impact on them in the informational sphere for illegal purposes'.[6] The draft African Union Convention similarly to the Budapest Convention makes a distinction between 'offences specific to information and communication technologies', and 'adapting certain offences to information and communication technologies'[7].

It is very clear regarding these approaches that a number of general features could be used to describe 'CC'. Focusing on the object (material offence), on the person, thing or value against which the offence is directed could all be included for instance. This approach is found in the CIS Agreement (computer information) and in Title One of the substantive criminal law chapter in the Budapest Convention (computer data and computer system). Another approach is through considering the computer systems or information systems as an integral part of the *modus operandi* of the offence [14, p. 267–317]. This approach is taken in Titles two, three and four of the substantive criminal law chapter of the Budapest Convention, as well as in the Shanghai Agreement and in the draft African Union Convention. Identifying possible CC offences and their *modus operandi* do not describe CC acts in their entirety, but it can provide a number of useful general categories into which these acts may be broadly classified.[8]

The majority of the definitions analysed as part of the process conducted in preparing this work reflect the content of the Budapest Convention. The Budapest Convention was drawn up by the Council of Europe (CoE) in 2001 (including the participation of observers Canada and Japan) was the first international treaty seeking to address these types of crimes and harmonise national laws. Within it, are the offences that are considered CCs. Listing the offences constituting this crime also allows for defining the scope of specialised investigative

[5] *Commonwealth of Independent States Agreement* (2001), article 1(a).

[6] *Shanghai Cooperation Organization Agreement* (2010), articles 1, 2 and Annex 1.

[7] *Draft African Union Convention* (2012), Part III, Chapter V, Sections II, Chapters 1 and 2.

[8] UNODC (2013), *Comprehensive study on CC.*

and international cooperation powers, which are better focused on electronic evidence for any crime[9].

Because of the wide variety of acts that constitute CC, it is not a word particularly amenable to a single definition, and is likely best considered as a collection of acts or conducts, rather than as a single act[10]. For example, the CoE defines CCs as 'criminal acts committed *using* electronic communication networks and information systems, or *against* such networks and systems'. The offences considered CCs under the Budapest Convention are grouped into four categories, these are shown in Fig. 1:[11]

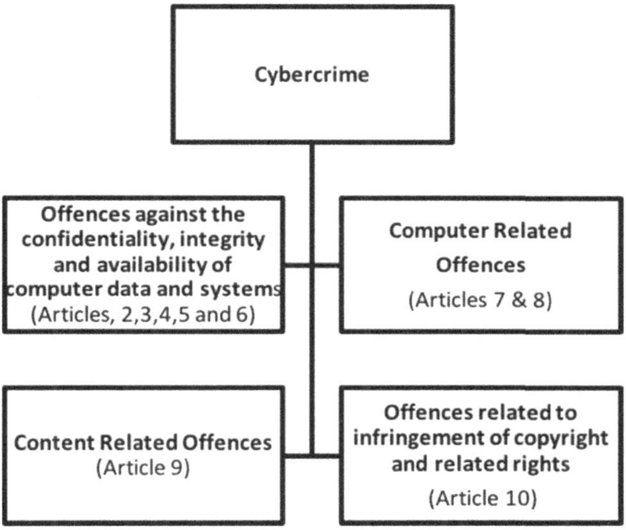

Fig. 1. Categorisation of Cybercrime Offences according to the Budapest Convention.

These categories, and the Budapest Convention from which they are taken, serve as guidelines for any country developing national legislation related to CC whilst also forming a framework for international cooperation between state actors. As a result, the language used, the categorization that is reflected within these categories, and the articles from which they are taken is evident throughout international studies, academic research and other works that are focused around CC, and thus can be used as a key component of any attempt to categorize CC related offences moving forward.

The United Nations Office on Drugs and Crime (UNODC) groups 14 offences considered under the banner of CC into three broad categories. These categories are shown in Fig. 2.

[9] *Ibid*, p. xvii.
[10] *Ibid*, p. 41.
[11] *European Convention on CC* (2001). Available at: http://conventions.coe.int/Treaty/en/Treaties/html/185.htm.

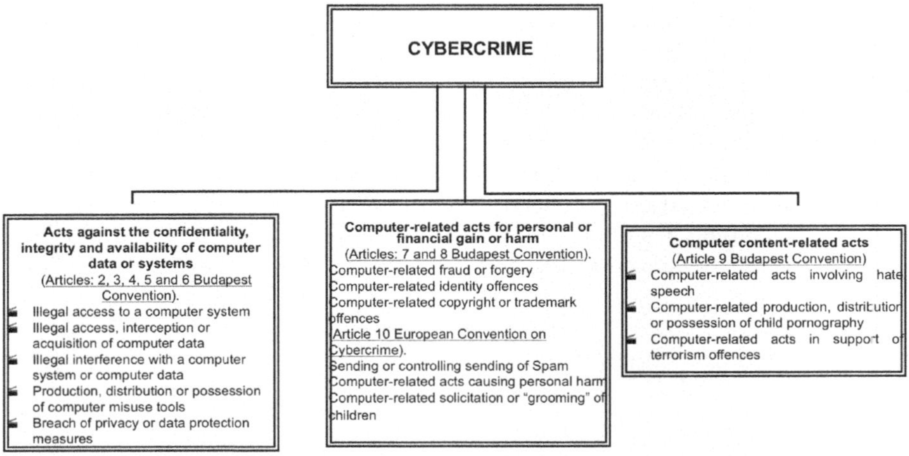

Fig. 2. UNODC categorisation.

The UNODC categorisation is based upon the Budapest Convention. However it differs slightly by combining two of the categories (computer related offences and offences related to copyright infringement) under the banner of 'Computer related acts for personal or financial gain or harm'. Although different in this sense, the terminology and grouping used in the categorisation is taken from the CoE Convention. It is notable that many countries did not identify a large range of offences outside of the 14 listed by the UNODC's original categorisation. To a degree, a consensus exists; one of the reasons for this is that as of December 2015, 58 States had ratified the Budapest Convention, while a further six states had signed but not ratified it[12]. The comprehensive study made by UNODC took that into account for the development of its categorisation. The categorisation of the Budapest Convention is also included in the convention developed by the International Telecommunication Union (ITU).

The prominent literature includes several categorisations regarding 'types' of CC. Chawki [4] divides them into two very broad categories depending on whether or not there is potential for the presence of violence during the criminal act. This perspective brings added interest when considering the categorisations in that seek to group crimes according to the role of technology in facilitating them. Wall [16] describes an approach defining three generations of CC, crimes in the machine (computer content), crimes using machines (computer related) and crimes against the machine (computer integrity). The most prominent differentiation in this context comes from the idea of cyber enabled crime where computers are used to increase the scope and impact of existing forms of criminality, and cyber dependant crime where computers are considered a dependant

[12] Council of Europe (2016), 2001 Budapest Convention on CC - Chart of Signatures and Ratifications of Treaty 185. Available at: http://www.coe.int/en/web/conventions/full-list/-/conventions/treaty/185/signatures?p_auth=CHRqyFpJ.

factor. It is also useful to distinguish the different criminal activities based on the motivations and *modus operandi* of the perpetrators. The most common distinction by scholars regards the use of Internet as a tool or as a target. In addition to computer networks as an instrument or object of a crime, Parker [13] adds a third grouping where computers are the *environment* of crime. Again, this typology provides scope for comparison with those previously discussed as it seeks to define the role of the internet in a crime, as either the place where the crime is committed (Environment), the tool being used to commit the crime (Instrument), or the target of the crime (Object). Although this particular model focuses on the role of the internet, it could be applied to the categorisation of the general role of technology in CC.

Finally, Wall offers a chronological typology, focusing on the evolution of opportunities offered by CC [17] quoted by Koops [11, p. 739]. This typology approaches the same topic of Hargreaves and Prince, but through a different perspective, choosing to define four categories that move from left to right depending on the significance of technology in the facilitation of the crime. The left hand category focuses on traditional 'cyber enabled crimes', and the right focuses on crimes that are wholly 'dependent' on technology/the internet (completely virtual spaces), a categorisation that again shows significant overlap but which is a potentially more granular approach than those defined previously.

2.3 Cyberterrorism

The term 'cyber terror' appeared for the first time in the mid-eighties. Since then the notion has been misused a number of times. As there is no internationally agreed definition of terrorism this categorisation proved to be more challenging and includes an extremely diversified typology. Perhaps one of the more widely accepted definitions is that of Denning's testimony [8] before the Special Oversight Panel on Terrorism which describes CT as the 'convergence of terrorism and cyberspace', generally understood to refer to unlawful attacks and threats of attack against computers, networks, and the information stored therein when done to intimidate or coerce a government or its people in furtherance of political or social objectives. Further, and drawing parallels to the 'violent or potential violent crimes' defined under Chawki's typology [4] under this definition, to qualify as CT, an attack should result in violence against persons or property, or at least cause enough harm to generate fear. Serious attacks against critical infrastructures could also constitute acts of CT under this definition, depending on their intended impact. Attacks that disrupt nonessential services or that are mainly a costly nuisance would not.

Of course, various definitions exist for the term 'CT', just as different definitions exist for 'terrorism'. CT is we established previously is the convergence of CC and terrorism [5].

Barry Collin, a senior research fellow at the Institute for Security and Intelligence in California, who first used the term "CT" in 1997 was attributed, defined it as the convergence of cybernetics and terrorism. In the same year, Mark Pollitt, special agent for the FBI, offered a working definition, describing

CT as the premeditated, politically motivated attack against information, computer systems, computer programs, and data which result in violence against non-combatant targets by sub-national groups or clandestine agents.

According to the literature analysed, this term refers to unlawful attacks and threats of attacks against computer, networks and the information stored therein, with the purpose to intimidate or coerce someone for political or social motives. To qualify as a CT attack, it should result in violence against persons or property, or at least cause fear and terror. That includes attacks against critical infrastructure. In instances of CT, technology (most prominently the internet) is used to achieve the same goals as more traditional weapons - i.e. to undermine citizens' faith in government by undermining their ability to maintain and provide the critical infrastructure systems that form the foundation of everyday life for regular citizens[13]. Despite a recent rise to prominence, the concept of terrorism being facilitated through the use of technology is not a particularly cutting edge concept and has been anticipated since the 1980's. The US Department of Justice[14] defines CT as the utilisation of network tools to shut down critical national infrastructure or to coerce or intimidate a government or civilian population. The ENISA (2013) typology of 'cyber agents' identifies a number of components that help to define Terrorism within the context of a Cyber-attack (CT attack); however, it does not necessarily consider one important facet of Terror-related attacks, the motivations that underpin them[15]. These motivations are commonly embedded within the extreme Ideological standpoints of the individuals and groups that commit them, and the radicalised political and religious perspectives by which they are moulded. Some scholars focus on the use of computer technology' for terrorist purposes and identify three categories:

1. Weapon of mass destruction;
2. Weapon of mass disruption; and
3. Weapon of mass distraction.

In addition, two facets of terrorist use of technology are identified:

1. Terrorist use of computers as a **facilitator** of their activities, and
2. Terrorism involving computer technology as a **weapon** or **target.**

According to the Centre for the Study of Terrorism and Irregular Warfare at the Naval Postgraduate School in Monterey, California,[16] CT capabilities can be grouped into three main categories and five types of attacks[17]:

[13] S. BRENNER (2007), "At Light Speed". Attribution and Response to CC/Terrorism/Warfare.

[14] US DEPARTMENT OF JUSTICE, FBI Law Enforcement Bulletin: Cyber Terror.

[15] N. VEERASAMY (2010). Motivation for CT.

[16] S.A. JALIL (2003), Countering Cyber Terrorism Effectively: Are We Ready To Rumble?, GIAC Security Essentials Certification (GSEC) Practical Assignment, Version 1.4b, Option 1.

[17] https://www.giac.org/paper/gsec/3108/countering-cyber-terrorism-effectively-ready-rumble/105154.

Categories:

- **Simple-Unstructured**; that is the capability to conduct basic hacks against individual systems using software and tools that has been created by others. These types of attacks are commonly facilitated by groups, individuals or organisations that have limited learning and command and control capabilities.
- **Advanced-Structured**; that is the ability to conduct more sophisticated attacks against multiple systems or networks and possibly to create basic software and tools, or modify existing tools in order to conduct more sophisticated attacks.
- **Complex-Coordinated**; this is the capability to commit coordinated attacks capable of causing disruption and or damage against integrated and heterogeneous defences. This is synonymous with the capability to create sophisticated software and tools and the capability to conduct target analysis and advanced organisational learning.

Types of Attack:

- **Incursion**; attacks carried out with the purpose of gaining access or penetrating into computer systems and networks to get or modify information.
- **Destruction**; attack used to intrude into computer systems and networks with the main purpose of inflicting severe damage or destroying them.
- **Disinformation**; attack used to spread information (true or false) that can have a severe impact to a particular target, such as creating chaos.
- **Denial of Service (DOS)**; attacks with the purpose of disable or disrupt the online operations by flooding the targeted servers with a huge number of packets (request) that would lead to the servers to being unable to handle normal service from legitimate users.
- **Defacement of websites**; attacks in which a website can be changed totally for propaganda purposes or redirect to other one.

As with CC, the use of the internet to commit acts in support of terrorism extends beyond its application as a means to conduct cyber-dependant attacks. Those involved with terrorism also continue to exploit the online environment as a means to promote and support acts of terrorism (UNODC[18]). This approach has resulted in the identification of six sometimes overlapping categories: propaganda (including recruitment, radicalisation and incitement to terrorism); financing; training; planning (including through secret communication and open-source information); execution; and cyberattacks.

Under these definitions, it's possible to argue that we are yet see a tangible, admissible instance of actual 'CT', although with the proliferation of IoT, and the increased reliance on interconnected devices (for instance, we have seen recent articles on how pacemakers, airplanes and car control systems can be hacked

[18] https://www.unodc.org/documents/frontpage/Use_of_Internet_for_Terrorist_Purposes.pdf.

from the outside), that the threat of actual CT attack is becoming all the more inevitable. Meanwhile, we continue to see the use of the internet as a means to finance, plan, recruit and execute terrorist plots as the boundaries between our physical and online existences continues to blur – both within social and business contexts.

Veerasamy, Grobler and Von Solsm [15] provide a framework to conceptualise many of the different elements of CT This is perhaps the most high level, and holistic of the typology's and categorisations that have been identified in this chapter and as a result has been used as part of the foundation of the CC/CT taxonomy, a snapshot of which is presented below, incorporating many of the elements identified throughout previous chapters. As the complete taxonomy itself is far too large to represent meaningfully in the body of this document, we draw upon two specific example, one related to CT, and one to CC, in order to demonstrate at a high level which factors are included in the taxonomy, whilst also providing some specific examples.

Although we identify the the harmonisation and simplification of the terminology used in reference to the subject area of CC and CT, as specific research priorities in the next section of this chapter, the discussion around the various facets that build our understanding of the field (motivations, legal categorisations, actors, etc.) serve a broader purpose, setting the scene for the full spectrum of disciplines that are touched upon by the following research items. Although our categorisation appears as four distinct silos, in reality the topics within them are fundamentally interdisciplinary in nature, touching issues related to technology and regulation as well as the organisational and human elements as a result of their role in modern business, and, as a fundamental fabric of contemporary society; enabling work, leisure and interaction.

3 Research Item Formulation

One of the major characteristics of the consolidated roadmap is the user-centred methodology of the approach. In order to be considered truly meaningful it was imperative that the outputs represent a broad view of the subject matter, through engaging with an extensive and disparate group of stakeholders and experts representing a diverse range of professional fields. In other words it is essential to ensure both the quantity of perspectives on the problem, but also the diversity of those sharing their views.

The resulting items have been subsequently distilled, qualitatively analysed and where appropriate aggregated in order to form a number of topics underpinning the research roadmap. The exact process for defining research items is dependent on the project from which they are originally derived. The methods utilised here further the development of the themes described earlier in Chap. 3. Building on the outputs of the data collected during the COURAGE project, the data from the CAMINO project was also utilised. This include a mix of questionnaire data, face to face meetings, phone interview transcripts, in addition to accounts of key points and trends identified in the current literature, scientific

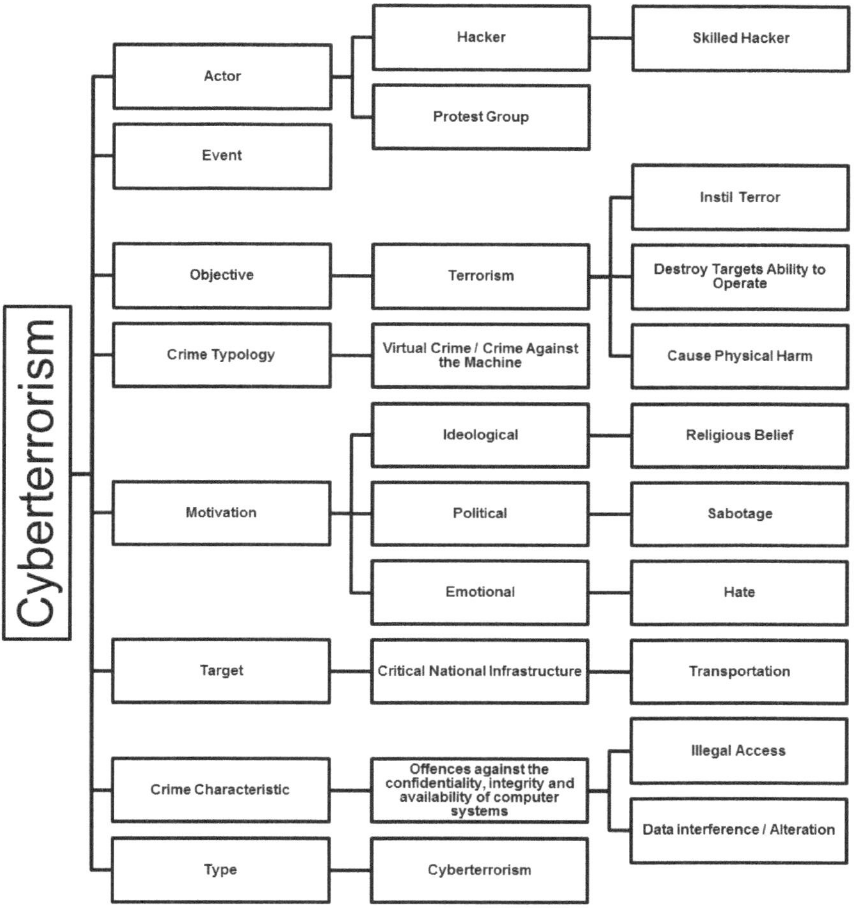

Fig. 3. Cyberterrorism example.

papers and policy reports. Post analysis, the items presented in later sections of these chapters were discussed, reviewed and revised at length after further consultation at a number of specialised organised conference workshops, focus groups and knowledge exchange events. While many of these events covered the full spectrum of 'THOR' dimensions, others were targeted at specific aspects such as critical infrastructure (Bern, DE) and regulatory aspects (London, UK). Intermediate versions of these outputs were also presented at a number of public conference events across Europe, including Den Haag, Montpellier and Toulouse (ARES 2015) resulting in the production of other associated publications [7].

Another key aspect of the consolidated roadmap centres on the 'THOR' dimensions. These dimensions demonstrate the holistic view presented by the roadmap identifying interdisciplinary challenges that are not solely focused on

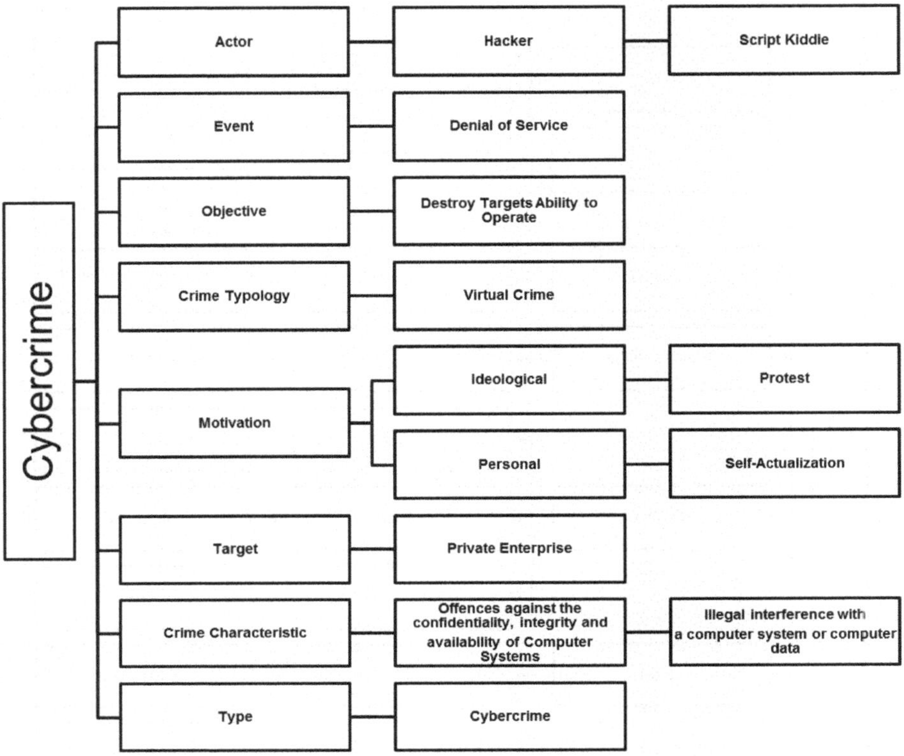

Fig. 4. Cybercrime example.

technology as the most crucial obstacle. The core topics discussed later in this chapter are a direct result of the consolidation of much of the empirical work conducted in service of the COURAGE and CAMINO FP7 projects. These areas form the basis of a larger piece of work informing future policy and research initiatives at an EU level. To ease this process, all the inputs were presented across the four aforementioned dimensions; Technological, Human, Organisational and Regulatory (THOR) using a format which draws many parallels for those familiar with existing security research funding calls under the Societal Challenges pillar of Horizon 2020[19]. In this format the following characteristics are defined for each research area:

1. **Specific Challenge -** Provide background information and insights into the problem domain, the specific challenges and issues being faced as a result, and

[19] https://ec.europa.eu/research/participants/portal/desktop/en/opportunities/h2020/calls/h2020-sec-2016-2017.html#c,topics=callIdentifier/t/H2020-SEC-2016-2017/1/1/1/default-group&callStatus/t/Forthcoming/1/1/0/default-group&callStatus/t/Open/1/1/0/default-group&callStatus/t/Closed/1/1/0/default-group&+identifier/desc

an overview of what the proposed research should address. In the research items presented this is extended to include the findings of prominent pieces of existing work - from both research and practitioner perspectives.
2. **Scope -** Set the boundary for what the research should aim to achieve and the specific outcomes which are expected/needed of the research in order to sufficiently address the specific challenge previously outlined.
3. **Expected Impact(s)** - Outline explicitly the expected beneficiaries of the research, how it will provide value and how this could be achieved.

In order to distil and consolidate these outputs a basic thematic analysis was conducted, initially across the COURAGE and CAMINO outputs, however this will later be adapted to incorporate the output of CyberRoad upon publication of its research items in May 2016. The basic premise of the thematic analysis is to identify thematic similarity across different data, and is widely used as a qualitative data analysis approach, considered to be effective means of identifying the subtle intricacies of meaning within a dataset [2,3].

4 Research Items

As a result of the consolidation process, twelve research items have been identified. Summaries of the items are included below, broadly categorised across the four 'THOR' dimensions.

4.1 Technical

Strengthening Emerging Tools for Big Data Analysis, Cloud Forensics and Security

Cyber-attacks are not always immediately visible due to their nature or intensity (e.g., amount of traffic they introduce). Therefore, recent techniques using big data tools, pattern recognition and machine learning have been adopted. In depth analysis of large volumes of data (received from different segments of IT networks, distributed heterogeneous sensors, etc.) has a unique capability of revealing interesting patterns, trends and relationships. This concept can potentially be adapted and applied to many cyber-security areas, namely: spam detection, botnet detection, malware analysis, web-based infection, network intrusion detection systems, etc.

This topic is focused particularly on the correlation of capabilities for big data analysis and scalability of big data tools and methods. This allows the cyber security system to have a deeper insight into different layers of the monitored system and as a result to provide better situational awareness. Potentially, such advanced techniques could detect distributed malicious activities in cyber space (e.g. propaganda on social media networks, cyber-attacks on critical infrastructures, etc.) related to hybrid conflicts and are now considered essential for homeland security.

The topic includes also consideration on the challenges related to the realistic workload conditions of the currently used test-beds that have to operate in real-time or near real-time. However, this performance requirement is not always possible to be met and depends heavily on the type of attack to be detected. For instance, analysing and tracking e-mail spam will require a lot of evidence patiently collected over a longer period of time. Moreover, it is important to provide an efficient security mechanism for communication channels and data storage within big data infrastructures, since the collected and analysed information is considered of great value. Therefore, it is postulated to have a European view on mechanisms and procedures for retention and processing of that kind of data.

All these objectives will not be possible to achieve without the close cooperation between cyber security solution vendors and big data tools providers, without mutual involvement in respective communities and without agreement on sharing datasets and tools for benchmarking purposes. Within this topic, we also expected the important role of national and European legislators and regulators to participate in, coordinate and promote development of the large-scale realistic test beds.

As a result of the recommendations given in this topic, we expect that typical network monitoring solutions and early warning systems should evolve to context aware systems which allow the user to identify current cyber security problems and what is more important – their roots. The second important expectation is the test beds community using wide variety of data samples (data sets) containing different malware, real and synthetic network traffic characteristics (or other challenging problems) that should be widely available to researchers.

Establishing Metrics and Frameworks for Cyber-Security Testing

One of the most important and demanded aspects in each product, system or even organisation is the quality; guaranteeing fundamental characteristics such as reliability or availability in any system, moreover if it is a security one, is an essential part of revealing the development team's confidence in their system or product. Therefore, activities focused on maintaining and improving such quality are needed, and the most effective ones are testing and simulation processes. Concepts such as automated tools or cyber exercises between companies will help to raise the awareness of not only cyber security responsible people, but also of the rest of the staff. And finally, in order to promote and encourage the realisation of all these necessary actions, proper regulations and standards should be made and discussed, and thus achieve a desirable and prepared environment to benefit all these good practices.

Therefore, the key points of this topic include; Security-by-Design frameworks, development of representative security metrics, the sharing of information about vulnerabilities, in addition to building open test beds for testing cyber security. Furthermore, issues of access control and trust management in distributed environments are also addressed. Finally, the ultimate goal of development and implementation of the specified topic milestones is objectiveness and measurability of cyber security not only for assurance purposes, but also using security metrics in security related contracts between product consumers and providers.

After successful development and implementation of the objectives specified in this topic, the approach to cyber security should become objective and quantifiable. Security should no longer be an ad-hoc practice where controls are installed only because recently there had been a data breach or incident. The proposed models and methods should help to indicate the existing security problems, helping to select the most appropriate (from security point of view) products and considering the influence of a specific security component within a complex system (e.g. on a corporate network or complex business process, also within the critical infrastructures). Investments in security should become more rational and supported by hard evidence. Moreover, the sharing of information about security should raise the 'security level' of products, whilst sharing the common data for cyber security testing which should raise effectiveness, reliability and objectivity of tested results. Additionally, the benchmark data (including realistic traces) should have to be provided and often updated to reflect traffic characteristics, behaviour of the users and new services.

Countering Cybercrime Affecting Mobile and IoT Devices

One of the primary technical challenges society faces in countering CC is the vast and continually expanding number of malware samples. The diversity and evolution of malware and botnets (e.g. new and quickly developing botnet architectures) are also key factors that should be addressed by research communities towards enhancing our capability to prevent and counter CC. This is particularly important in the context of the limitations for existing signature-based scanners and malware detection platforms. There is continued evidence of the expanding impact on the wider spectrum of internet-connected devices. Mobile and IoT devices, which traditionally were not networked, are now exposed to cyberattacks in a parallel manner to traditional computer systems. Research taking place to address this topic needs to focus primarily on; the development of novel and improved methods to detect and prevent malware and other malicious software, particularly taking into account that targeted at mobile and small/micro devices. The expanded scope of this topic must also take into account the specific needs of stakeholders, particularly looking at areas where there is a perception that issues related to the rate of change of technology have traditionally posed a particular problem. For instance, developments in the use of technology to facilitate crime poses the requirement for investigators to expand their capabilities and skill sets in areas such as digital forensics and other specialist areas. Although focused on technology, this challenging area also requires an interdisciplinary approach, with education and awareness raising a baseline method of improving societal resilience to attacks. Alongside this, public/private sector cooperation must be considered a vitally important mechanism in ensuring that private sector internet security organisations play a key role in assisting and cooperating with law enforcement.

Returning to the technical challenges themselves, the growing resilience and stealth of modern botnets which benefit from the use of P2P architectures, and techniques such as fast-flux, DNS, Domain Generational Algorithms (DGAs), encryption of command and control challenges and others means that significant investment needs towards enhancing the existing approaches. Methods that do not rely so heavily on signature based detection techniques should be further developed; meanwhile security-by-design principles require implementation at a more practical level.

4.2 Human

Collective Awareness and Education for Increased Societal Resilience to CC/CT Threats

This section focuses on the identification and facilitation of new approaches to enable the improved resilience of society to cybersecurity threats, through increasing the awareness and education levels of stakeholders across society; ranging from non-technical citizens, right up to security professionals, policy makers and the full spectrum of private sector industry and critical infrastructure providers. Prevention strategies, and in this context, particularly those associated with increasing awareness and standards related to online safety and information security play an important role in improving societal resilience to CC, whilst 'human security' specifically is an important factor as popular attack vectors such as social engineering and phishing continue to exploit human security vulnerabilities[20]. Under this topic research should focus on the identification of new approaches to increasing societal awareness, and subsequent readiness, to deal with cybersecurity threats. Where necessary, the impact of new and emerging technologies and behavioural changes that occur because of them should be identified and considered. The research proposed should identify and address awareness and education requirements across all identified sectors, for example; national teaching curricula, law enforcement and other public and private sector institutions, etc.

The ubiquity of the internet continues to drive the requirement for an increase in awareness of CC/CT, and poses legitimate challenges across society resulting in a specific necessity to develop and improve collective understandings of existing approaches. Prevention strategies, and in this context, particularly those associated with increasing awareness and standards related to online safety and information security play an important role in improving societal resilience to CC [9]. Although many schemes, from those aimed at raising basic levels of 'grass roots' awareness (European Cyber Security awareness Month [ECSM[21] right through to those aimed at organisations (cyber security essentials[22]), including those as a result of research (i.e. FP7, H2020, DG Home, DG Connect etc.), are

[20] http://www.mcafee.com/uk/resources/reports/rp-quarterly-threats-aug-2015.pdf.
[21] https://cybersecuritymonth.eu/about-ecsm.
[22] https://www.gov.uk/government/uploads/system/uploads/attachment_data/file/317480/Cyber_Essentials_Summary.pdf.

taking positive steps to develop awareness and education levels we lack metrics to assess the widespread impact of them. Therefore, it is imperative that relative successes of previous initiatives are evaluated so that effective practices can be taken forward and widely propagated throughout society, including in national education curricula.

New Standards for Private Data Minimisation, Appropriate Use and Re-use of Data and Privacy Enhancing Technologies

With surveillance powers and techniques, a very current topic, both from the perceived excessive use in some quarters and the inadequate interpretation of available evidence in others, the roadmap towards more effective implementation of Privacy Enhancing Technologies is inexorably entwined with the development of forthcoming legislation, and the regulatory interpretation of these. In particular, DPR, eIDAS, and the Second Payment Services Directive early adoption through SecuRe Pay, introduce requirements for the adoption of PETs (Privacy Enhancing Technologies), albeit through the adoption of undetermined techniques or technologies, even in advance of their formal ratification into EU or Member State legislation. These advance regulatory roadmaps provide an interesting, and often unexpected, set of requirements to organisations handling sensitive personal data.

A further contemporary issue raised in this topic, is the fact that under a range of current regulations and industry standards, across a dynamic range of industries, the use of data is frequently, but not universally, restricted to the use originally intended when data was collected. Users should face a comprehensive range of opt-ins or opt-outs to the use, or subsequent re-use, of this data. The advent of big data has enabled the search for new uses of data held on existing systems a growth industry, but there are considerable Human and Ethical concerns raised regarding this re-use.

In this context we propose to focus on three main aspects: data minimisation tools and techniques, anonymisation/pseudonymisation techniques and encryption management. As for data minimisation, we propose identification and assessment of such tools, and assessment of the appropriate use and limitations of PETs in relation to recently developed emerging technologies. Those objectives should lead to the incorporation of privacy enhancing features ensuring minimum sharing of data as a default/standard of applications. Similarly, regarding data minimisation techniques, anonymisation capabilities ought to be incorporated as a standard privacy enhancing functionality of applications. Ideally, these techniques must be preceded by identification and evaluation of techniques and tools currently in use, and on the horizon. The ultimate goal of the research in the field of encryption; is the adoption of the guidance on the appropriate use of techniques for the encryption of the private communication and data and ensuring the support in development of new encryption protocols.

Moreover, we believe the relevant organisations (such as national DPAs) and the EU should publish and promote practical guidelines and recommendations for privacy and data minimisation in the IT systems used within various sectors such including public administration etc. [7].

In result, we expect that the adoption of PETs should not only reach such a point as to better protect users' privacy but also to have matured to the extent that they have regained a degree of control over previously exposed data, and the ability to exploit it to their own personal expectations.

Definition, Characteristics and Behaviours of the Offenders and Victims in CC

The scale and proliferation of Internet use as a means to facilitate crime has also introduced new challenges for the social and behavioural sciences, in addition to the technological and criminological disciplines we normally associate with studies in the domain. Due to the potential overlaps and absences of clarity in distinguishing between CC, CT, cyber warfare, and often the inability to immediately identify the origin of an attack means that there is a significant benefit in assessing its impact towards discerning the potential motivations behind it. The enormous widespread impact exerted by modern CC means that individuals and groups involved in committing, responding to, and preventing events, is equally expansive. The sheer quantity and diversity of the number of criminals and victims of CC means that despite the importance of analysing the various different actors, there is still significant scope to further progress our understanding.

In order to develop and deliver improved intervention and prevention measures, this topic furthers the requirement for future research to help build our understanding of the diverse range of actors involved. With the proliferation of crime-as-service models, CC is no longer the preserve of the technically skilled. As these skills and knowledge are no longer prerequisites, the market now affords the would-be criminal access to tools, services and even individuals for hire in order to facilitate attacks on their behalf in exchange for payment [12].

Furthermore, as the concept of CC penetrates into the social and psychological spheres it's no longer suitable to assess its impact solely based on its economic impact more work is needed to establish the underlying factors that contribute to the profiles of victims and offenders alike, in addition to establishing human, environmental and other PESTLE factors that drive CC. One of the reasons for the current gap in research in this area is the lack of understanding of the physical and virtual areas where CC and CT take place and, increasingly, the areas where the two converge. As we rapidly approach a time and place where all crime has some associated element of 'cyber', a greater understanding of the specific social impact challenges is imperative as although this ubiquity may be well defined from a criminological perspective the social and psychological aspects as not as clear [1]. Thus, of particular importance is learning about victim and offender profiles in respect of criminal adaptation and exploitation of technologies. Whether these need to differ from traditional methods in order to be effective needs to be evaluated.

Unfortunately, rates of prosecution and low levels of reporting create a paradoxical challenge in our understanding. Indeed, the low level of awareness acts as both a contributor to, and, an impact of this challenge. As has been discussed in earlier sections of this chapter, emphasis must be placed on increasing education and awareness levels.

Subsequently, from the culmination of these factors, it is incredibly difficult to create realistic and representative offender profiles. The emerging field of 'Cyberpsychology' adopts a multidisciplinary approach to understanding this intersection between technology and human behaviour. This less technologically focused approach is especially vital when contemplating issues such as radicalisation, where work is needed to further our understanding of how technology in this context affects the social and psychological constructs of the process. Future research will take into account ongoing case studies, such as the Hacker Profiling Project (HPP)[23]) and further refined with cooperation alongside multiple, multi-national and multi-sector expert focus groups. The further development and progression of this methodology may better inform collective understandings and responses to CC, through focusing on its manifestation on human action rather than something that is just a technological phenomenon.

4.3 Organisational

Adapting Organisations to the Cross-Border Nature of the Internet and Cybercrime and Cyberterrorism

Nowadays, competitiveness is global, so any company or system can receive an attack from anywhere on the planet. Therefore, it is vitally important that regulatory differences between countries are known and understood, and in consequence, organisations should be aware of this fact and protect their assets and intellectual property appropriately. Organisations need to adapt, protect their systems and networks, and to cooperate effectively cross-borders in fighting CC within the framework of the existing law at the time of the event.

Therefore, key research points of this topic concern homogenisation of law (national and EU), cooperation between Law Enforcement Agencies (LEAs) from different countries and continents, CERTs, governmental cooperation in terms of cross-border monitoring and information sharing within proposed frameworks. Current diversity of national laws is often an obstacle for cross-border CC investigation and prosecution. There is a need to unify different legislation in order to remove those obstacles, or to understand and overcome the differences that cannot be quickly removed.

Tools and techniques allowing for the collection of evidence of crimes, not only from the victims perspective, but also from other entities, such as various ISPs, compromised web servers in different countries are needed. Therefore, top priority objectives in this topic include the interoperability of forensic tools and best practices at the cross-border level, including automatic services responding to the cyber security incidents.

[23] http://www.unicri.it/special_topics/securing_cyberspace/current_activities/hackers_profiling/.

Research agendas in this topic should also address; incentive-based cooperation for information sharing and development of appropriate balance between such incentives-driven good practices and mandatory information sharing procedures. The information exchange that is currently mandatory and enforced by law and regulations is not the only way to foster the cooperation on cyber security/terrorism detection and prevention. There is a need to encourage interested parties to acknowledge such exchanges are universally beneficial. In other words, there is a need to identify the incentives for information sharing in order to make this exchange mutually beneficial rather than a burden to organisations.

Potential impact of work to be done in this topic includes benefits for law enforcement agencies, CERTs, ISPs, and IT-based organisations. Each of these stakeholder groups aims at effective detection, monitoring, prevention and reaction to cyber-attacks. Such collaboration should be supported by the mutually recognised tools and shall be included in the framework of national and international laws. The result of successful national and international collaboration should reduce possibilities for cyber criminals to hide behind borders and feel invincible, extending the arsenal of IT-based organisations to protect themselves from CC.

Creating User-Friendly Terminology, Language and Features to Assure a Better Understanding of Cyber Security Challenges

The definitions and understanding of the terminology used in reference to CC and CT are, in some instances, inconsistent across EU Member States, potentially causing confusion and in extreme cases hinder law enforcement, prosecution and international cooperation efforts due to the ambiguity surrounding the subject area in general. Harmonising terminology in both areas of CC and CT is crucially important in defining how law enforcement and the public and private sectors should cooperate in an EU and broader international context. Without a clear understanding of the characteristics that distinguish them, these areas will likely remain difficult to addresses properly across at a holistic level. The absence of equal representation and understanding of terms from both areas of CC and CT, the lack of definition of terms and the different taxonomies in current use in the field has been repeatedly identified as a problem by academia, LEAs, and by entities representing legal and ethical organisations as well as from the critical infrastructure stakeholders.

In this topic, it is proposed that efforts should be made to increase levels of knowledge/information exchange among stakeholders, leading to the provision of harmonised and standardised terms through the development of a new taxonomy framework that involves all aspects of CC and CT, specifying their differences and commonalities. Crimes such as online fraud, hacktivism, terrorist activity preparation, DDOS attacks and the dissemination of online illegal content may all be broadly considered as cyber-attacks but each requires significantly different mitigation and prevention strategies. The nature and extent of each type of crime needs to be universally understood so that they can be prioritised and dealt with appropriately.

Clear, unambiguous and universally understood terms and definitions are necessary to enable those measuring, predicting, combating, investigating and prosecuting crime, to do so effectively. Information sharing is crucial and general understanding of requirements and purposes is required, across all jurisdictions and entities. The scope of this category of research should focus on the development of a new taxonomy framework and EU-wide harmonised terminology that involves all aspects of CC and CT and specifies their differences and commonalities. A key issue in this area is to identify how to realise such harmonisation. In particular research should evaluate the possibility of a top-down approach, propagated from the EU into national legislation and policy.

Promoting EU Institutional Support to Generic Challenges and Obstacles at the Enterprise/Company/SME Level Including Incentives for Cyber Insurance

Common and unified institutional support is needed to promote changes at the enterprise, company and SME levels. The creation of an expert committee at the request of the main involved countries can contribute to overcoming these obstacles and challenges upon the European level. Additionally, an information sharing platform may assist in the approach and collaboration between interested parties, making quick and efficient ideas/problems sharing possible. This support platform should assure the minimum cybersecurity protections required by the involved parties.

It is worth considering that no security strategy will ever be flawless, it is widely accepted that achieving perfect security is impossible. Security accidents and data breaches will occur regardless the amount of security controls and practices applied (though with much lower frequency). Thus, organisations have to deal with the residual risk, and the emphasis be placed on maximum damage resilience and mitigation. Recently, we have seen that insurance, a usual treatment approach for residual risk, has been applied to the cyber world. The developing cyber insurance market faces a number of unique as well as usual (for insurance) challenges. In particular, heavy information asymmetry, lack of statistical data, interconnected security and correlated risks, rapid change of risk landscape, unclear underwriting language, etc. Currently enterprises, companies and SMEs find it difficult to execute appropriate strategies to fight against CC. Finding support from EU, governments and regional entities in order to establish an adequate level of cyber security at enterprise, company and SME level should be the answer to this challenge.

One of the major research objectives to be achieved in the upcoming years should be towards developing and establishing effective, bi-directional communication between organisations and EU institutions. The other challenge that enterprises, companies and SMEs have to face is the lack of the qualified human resources. It is difficult to find IT staff with expertise in cyber security. EU certification programme in the cybersecurity domain would solve this problem. However, there is a need to create new cyber security curricula for kids in schools, for young people in high-schools, and advanced programmes for students at universities, as well as for postgraduate studies.

Moreover, collaboration between both sides (enterprise and EU institutions) would be good practice to implement communication, certification programmes and compliance agreements.

Resultantly, the achievement of objectives and milestones defined for this topic should allow the European enterprises, companies and SMEs to obtain valuable support from the EU to integrate new cyber security initiatives, and, to raise the overall level of their security and security/trust of their customers. Additionally, growing cyber insurance market and development of cyber insurance as a reliable tool for the management of cyber risks will be beneficial for both cyber insurers (increased confidence in the procedures to follow) and insured (who will be protected from unexpected threats). The use of insurance policies should help the insureds to manage risks in a more predictable way and governments should benefit from increased productivity within the economy and from its law abiding market participants resulting in a more secure society.

4.4 Regulatory

Dealing with Different Levels of Legal Frameworks for Illegal Content: Questions of Geolocation and Jurisdiction

CC is an inherently cross border issue, with a given incident potentially involving a number of different countries and territories each with their own legal frameworks and jurisdictions. This 'internationalisation' of crime creates new challenges for law enforcement. This includes issues such as the reporting and deletion of illegal content, the collection of court evidence, cross-border accessibility of data and other issues. In this research topic, the identification and development of new methods that enable LEAs to gather and share information across geographic borders resulting in improved cooperation among international and public/private authorities and to support the development of new standards for harmonising collaboration between the private sector and law enforcement.

In addition, the absence of physical proof and the frequent anonymity of perpetrators complicate the task of LEAs in collecting admissible evidence against cybercriminals. Geolocation technologies are limited in tracking down cybercriminals. Indeed, even if each computer on the Internet has a unique Internet Protocol (IP) number revealing their geographic location, cybercriminals could either be physically mobile or have the necessary skills and tools to avoid being tracked and act anonymously online (e.g. through the Darkweb). The potential impact of the introduction of the IPv6 Internet addressing scheme is not yet fully understood.

The ways in which illegal content and illegal activity are perceived and dealt with in different countries means that there are important questions about jurisdiction and the application of national laws to online content and activity. Furthermore, there are issues about the legal basis of prosecuting illegal content hosted in one jurisdiction and accessible in another, as well as procedural questions relating to which country enforces laws where content-related or activity-related offences cross more than one border. In incidents such as the Belgian Yahoo! Case, these questions have been well researched [10], but jurisdictional issues need periodic review in light of changes in the Internet landscape. Research should investigate how to deal with such lack of physical proof and cross-jurisdictional cases (e.g. examine the opportunity of introducing a dispute settlement procedure for illegal content cases).

Electronic Identity and Trust Services for Data Protection Across Borders

The research community will need to address the technical standards agreed for the degrees of identity and authentication, and the circumstances under which each of those is appropriate. The research community will play a vital role in this area as what is perceived to be 'uncrackable' in some Member States (or nations outside the European Union) could have relatively trivial flaws when looked at from outside. A majority of classes and applications of CC and CT contain a misrepresentation of identity or attempt to authenticate access to goods or services that the attacker has no legitimate use for. Currently a plethora of standards exists that enable the identification and authentication of genuine users. At present there is no interoperability of these, and poor controls over the degree to what constitutes 'strong authentication' sufficient for each respective application. The main challenges include agreement of various standardisation bodies on levels of interoperation according to adopted security model and ensuring similar levels of certainty to be adopted in each EU Member State. Other challenge to be addressed through the research in this topic is the alignment of credential management practices within EU with wider global standards and agreement on data protection equivalency.

The proposed research identified in this topic includes the timetable for the implementation of eIDentity, Authentication & Signature regulations, and the steps necessary to ensure its impact internationally. Equally, with the payments industry now being required to look at early adoption of the Second Payment Services Directive (PSD2), the Identity/Authentication roadmap has moved forward dramatically as one of the key CC asset classes, and one of the most likely candidates for higher level eIDAS requirements. Objectives of fostering international management of e-identity related interoperability should also require US adoption of EU industry standards.

It is also necessary, to include biometric human identification techniques (both traditional such as fingerprints, faces, as well as new emerging modalities and approaches) within the framework of the future effective electronic identity and trust.

Research in these topics will result in internationally recognised and mutually collaborative sets of private and publicly issued identity credentials. These may subsequently provide degrees of certainty according to underlying enrolment and security standards at subsequent authentication. In general this approach may offer wider international opportunities for harmonisation of data protection law across borders (EU-EU and EU-US).

Comprehensive Legal System to Fight Against Cybercrime and Cyberterrorism

This topic reflects the current needs and challenges that facilitate requirements for improvements to the legal systems and related processes that impact upon all phases of CC cases. One of the main efforts to be done in this area is the improvement of digital forensic products, services and procedures. In particular, it is important to ensure an adequate flow of information at the different stages of the investigation - from disclosure of crime, securing and preserving evidence and its processing, up to the judicial decision.

Unfortunately, the current situation in many countries is characterised by the low or even lack of the cooperation at the consequent stages of the legal system. The challenge is to better organise the following chain:

$$\text{victim} \longrightarrow \text{the police} \longrightarrow \text{prosecution} \longrightarrow \text{the court.}$$

In many countries, there are police officers dedicated to specifically countering CC and CT. However, there is still a need for in-depth training of the police staff so more of them can better understand and handle CC and the victim. Indeed, there are few prosecutors working specifically on CC – usually the prosecutors handle a very broad range of crimes, and some of them do not understand the specifics and technical complexities of CC and CT, particularly the differences in the chain of evidence. Similarly, also due to the human-age factor, many judges do not understand the specifics of the CC/CT domain, nor are fluent with the relevant vocabulary used by cyber security experts. Therefore, better organisation of the prosecution and court responsibilities to handle CC is postulated. Moreover, further training is needed, to improve and increase the admissibility of the forensic evidence in courts.

The ultimate goal is to assure the society and citizens, that legal system can understand and protect the victims of the CC, and that cyber criminals can be effectively sentenced.

Therefore, in this context it is also important to ensure and develop appropriate levels of knowledge and expertise across all the actors involved in the judicial process. The major improvement in information sharing and cooperation between victims, LEAs (the Police), the prosecution and forensic experts and finally the judges/courts is needed.

5 Validation

The outputs of the final planned refinement and validation workshop serve as a significant milestone in the production of the consolidated research roadmap presented here. As part of a wider knowledge exchange event, the workshop session presented the current draft of the research agenda items to ~70 conference delegates, consisting primarily of 'cyber' stakeholders connected (be this informally as part of extended networks, or as part of the projects direct consortia and advisory boards) to each of the projects. Stakeholders were asked to rank each research item in terms of the extent to which it should be considered a priority (i.e. its significance as a topic to be addressed), and where it should be placed in terms of urgency, (relative to the other items), to be conducted along a three point scale of 'low', 'med' and 'high'.

In total, 40 complete responses were received. These are visualised below in Figs. 5 and 6 using divergent stacked bar charts. For both graphics, the goal midpoint is set as the central point of the 'med' ranking in order to more clearly demonstrate the differentiation between 'low' and 'high', with the medium value contributing equally to both sides. From the urgency plot in Fig. 5 we can note some interesting characteristics about a number of the research items. First, we can see that resolving cross border issues and increase awareness and education levels across the board were seen as the most urgent, with more than 60 % of respondents marking each of these research items as 'high' in terms of urgency.

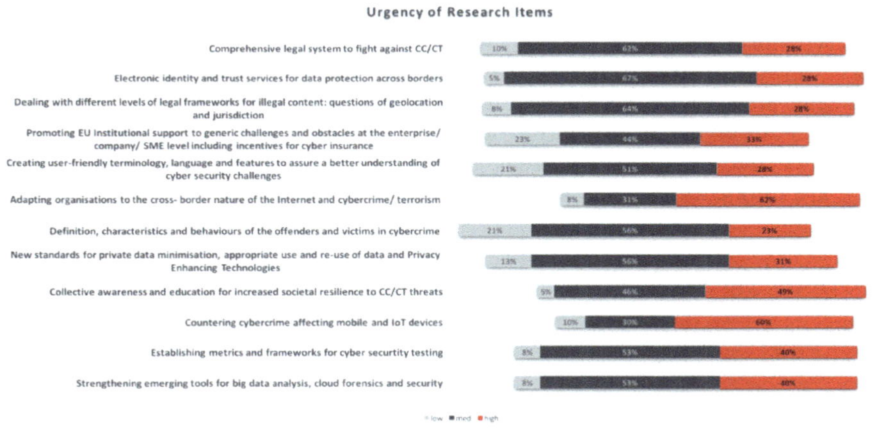

Fig. 5. Urgency of research items.

In Fig. 5, the same method is used to plot the extent to which each of the research items should be considered as a priority. In this plot, we can again see that cross-border issues ranked highly along with strengthening tools for big data analysis, cloud forensics and security, and issues concerning mobile and IoT devices with each of these topics having been marked as 'high' priority by the respondents. It also worth observing that when considering the medium and high prioritization collectively; awareness and education scores highly.

Although we can draw some basic insights as to which items featured more prominently with the participants, it is important to acknowledge that the results

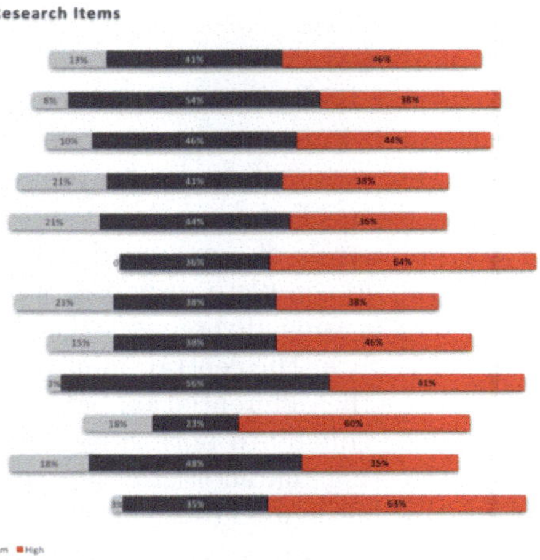

Fig. 6. Prioritisation of research items.

themselves are somewhat superficial as given the option, under most circumstances human inclination is to ask for things as soon as possible as opposed to waiting for it. This is reflected in the data as all items featured prominently as either high or medium priority of urgency. However, we can take away some confidence that the identified areas are recognised by the participants as areas which are important and pose as a significant societal challenge.

6 Concluding Remarks

In this chapter we have defined twelve specific topics that pose a prominent and significant challenge to modern society. These challenges can in some capacity, be addressed and assisted through further research. Each individual challenge has been presented across four interdisciplinary dimensions, highlighting challenges such as education, data protection and privacy, technical prevention and detection measures and more across a range of academic disciplines, from criminology and information security, to law and the social sciences as an integrated roadmap for research. In defining these topics, we categorise the aspects of CC and CT towards the development of a harmonised taxonomy – setting the scene for identification of the many facets of the C/CT domain.

Acknowledgement. The research leading to these results has received funding from the European Union Seventh Framework Programme (FP7-SEC-2013) under grant agreement numbers 607949 (COURAGE) and 607406 (CAMINO).

References

1. Aiken, M., et al.: A consideration of the social impact of CC: examples from hacking, piracy, and child abuse material online. Contemporary Social Science, pp. 1–19. http://www.tandfonline.com/doi/full/10.1080/21582041.2015.1117648. Accessed 21 Apr 2016
2. Boyatzis, R.E.: Transforming qualitative information: thematic analysis and code development, Sage (1998)
3. Braun, V., Clarke, V.: Using thematic analysis in psychology. Qual. Res. Psychol. **3**(2), 77–101 (2006)
4. Chawki, M.: A critical look at the regulation of cybercrime. ICFAI J. Cyberlaw. **IV** (4) (2005)
5. Conway, M.: What is Cyberterrorism? The story so far. J. Inf. Warfare **2**(2), 33–42 (2003)
6. Choras, M., Kozik, R., Torres Bruna, M.P., Yautsiukhin, A., Churchill, A., Maciejewska, I., Eguinoa, I., Jomni, A.: Comprehensive approach to increase cyber security and resilience. In: ARES 2015, pp. 686–692 (2015)
7. Choraś, M., Kozik, R., Renk, R., Holubowicz, W.: A practical framework and guidelines to enhance cyber security and privacy. In: Herrero, Á., Baruque, B., Sedano, J., Quintián, H., Corchado, E. (eds.) International Joint Conference CISIS 2015 and ICEUTE 2015. AISC, pp. 485–495. Springer, Switzerland (2015). ISBN: 978-3-319-19712-8
8. Denning, D.: Testimony before the Special Oversight Panel on Terrorism. US House of Representatives, Committee on Armed Services (2000)
9. Europol, The Internet Organised Crime Threat Assessment (2015). https://www.europol.europa.eu/iocta/2015/. Accessed 21 Apr 2016
10. Koops, B.J., Brenner, S.W.: Approaches to cybercrime jurisdiction. J. High Technol. Law **4**(1), 189–202 (2004)
11. Koops, B.J.: The Internet and its opportunities for Cybercrime. In: Herzog-Evans, M. (ed.) Transnational Criminology Manual, vol. 1, pp. 735–754. WLP, Nijmegen (2010)
12. Manky, D.: Cybercrime as a service: a very modern business. Computer Fraud & Security (2013). http://www.sciencedirect.com/science/article/pii/S1361372313700538. Accessed 21 Apr 2016
13. Parker, D.B.: Threats to computer systems (No. UCRL-13574). California Univ Berkeley Lawrence Livermore Lab (1973)
14. Podgor, E.S.: International computer fraud: A paradigm for limiting national jurisdiction (2002)
15. Veerasamy, N., Grobler, M., Von Solsm, B.: Building an Ontology for CT (2012)
16. Wall, D.S.: The rise of the Internet as a crime problem. In: Handbook of Internet Crime, Vancouver, pp. 88–102 (2010)
17. Wall, D.S.: The Transformation of Crime in the Information Age. Polity, Cambridge (2007)

Author Index

Akhgar, Babak, 39, 295
Ariu, Davide, 53
Armin, Jart, 135, 175

Bosco, Francesca, 97, 295
Brewster, Ben, 39, 295
Brynielsson, Joel, 209

Choraś, Michał, 193, 279, 295
Churchill, Andrew, 279

Didaci, Luca, 53
Drobniak, Szymon, 17

Franke, Ulrik, 209
Freschi, Federica, 53, 237
Frumento, Enrico, 53, 237
Fumera, Giorgio, 53

Gasper, Ulrich, 81, 97, 117
Giacinto, Giorgio, 53

Jaroszewski, Przemyslaw, 175
Jerman-Blažič, Borka, 157

Kemp, Benn, 117
Kert, Mari, 81
Kijewski, Piotr, 135, 175
Klobučar, Tomaž, 157
Koops, Bert-Jaap, 3
Kozik, Rafał, 193
Kozik, Rafal, 279

Luda, Vittoria, 97, 295
Lyle, Alison, 81, 97, 117

Maciejewska, Iwona, 193
Mazurczyk, Wojciech, 17
Moore, Sean, 17

Olesen, Nina, 259

Puchalski, Damian, 295
Roli, Fabio, 53

Roosendaal, Arnold, 81

Spasova, Albena, 117

Thompson, Bryn, 135

Urbanowicz, Janusz A., 175

Vaciago, Giuseppe, 97
Varga, Stefan, 209
Vermeersch, Elise, 97, 295

Wells, Douglas, 39, 295

Yautsiukhin, Artsiom, 279

The manufacturer's authorised representative in the EU is Springer Nature Customer Service Centre GmbH, Europaplatz 3, 69115 Heidelberg, Germany. If you have any concerns regarding our products, please contact ProductSafety@springernature.com

Printed and bound by CPI Group (UK) Ltd, Croydon, CR0 4YY
30/09/2026
02225572-0012